8/14

Science, Skeptics, and UFOs

A Reluctant Scientist Explores the World of UFOs

B. Timothy Pennington, Ph.D.

Illustrations by B. Timothy Pennington and Jason Behr

First published by Dog Ear Publishing
4010 W. 86th Street, Ste H
Indianapolis, IN 46268
www.dogearpublishing.net

ISBN: 978-1-4575-2312-0

This book is printed on acid-free paper.

Printed in the United States of America

Acknowledgements

Thanks to my wife, Esther, for her support and for enduring my days of obsession on the computer, and my unruly collection of books, articles and webpages.

A special thank you to my daughters, Laurie, Katie and Ashley, for their unconditional support for the project, and to friends and family that encouraged me in my endeavor, especially those who shared their personal experiences with me.

Thank you, Jason Behr, for assisting me with the images and illustrations.

TABLE OF CONTENTS

Science, Skeptics, and UFOs

A Reluctant Scientist Explores the World of UFOs

B. Timothy Pennington, Ph.D.

INTRODUCTION

(How I Came to Write This Book)

"Science is a wonderful thing if one does not have to earn one's living at it."

—*Albert Einstein (1879-1955) German Physicist, General Relativity, The Photoelectric Effect*

Science is a wonderful thing indeed, and today there are many opportunities for rewarding careers in a broad spectrum of scientific fields. I earned my living as a chemist for some 39 years. Thirty of those years were in the chemical industry, and 31 of those years were in basic or applied chemical research. Early in my career, I taught college-level chemistry, wrote scripts, and helped produce chemistry videotapes under a grant from the National Science Foundation. Over the years I have worked in academic, government, and industrial laboratories with all types of substances from the extremely dangerous to the benign. I earned my living doing science, mainly in the chemical industry, and over the years I saw many ups and downs in the fortunes of the companies I worked for.

Most of my career was spent with Olin Chemical in Lake Charles, Louisiana, but the Lake Charles plant was split up and sold off to other companies before the end of the 1990s, first to Arco Chemical and then to Lyondell Chemical. Projects and priorities changed, and the higher cost of natural gas at that time meant that the Lake Charles plant was struggling financially. The last straw came in 2005 in the form of Hurricane Rita (the other hurricane to hit Louisiana in 2005), which damaged the plant site to such an extent that the decision was made to shut it down. Fortunately, I was offered and accepted a position in the Houston area at the Lyondell Channelview North Petrochemical Plant.

Things were going well until the financial meltdown near the end of 2008. Earlier in 2008, Lyondell and Basell Polymers merged to become LyondellBasell. The debt involved in financing the deal could not be serviced, and LyondellBasell declared bankruptcy in early 2009. Voluntary early retirement packages were offered soon afterwards. I accepted an early retirement package in September of 2009, and for two years I worked on home improvement projects (honey-dos) and pursued my interests in jogging, yoga and gardening. However, something began nagging at me in the back of my mind. I felt a bit like a coward, somewhat intellectually dishonest. I had not felt this way during my career because I had thrown all of my intellectual energy into the work at hand. That is what enabled me to survive all of the chaos that is the modern workplace.

What were these nagging thoughts that continued to invade my consciousness? These thoughts were about UFOs of all things. During the 1970s, I had personally witnessed a number of sightings of the unexplained phenomena known as UFOs. Every day these sightings continued to consume more of my thoughts. There had been a number of sightings by family members as well. I had wondered about these various sightings from time to time over the years. In 1977, a book caught my eye entitled *UFOs Explained* by Philip J. Klass. I eagerly read the book, but I came to realize that in none of the explanations did the book seem to be able to explain what we had seen.

I had, on several occasions over the years, read science journal editorials bemoaning the fact that such a high percentage of the American public believed in the nonsense of unexplained phenomenon such as UFOs. Those editorials always left me conflicted, because if anyone had seen what we had seen, they would know that there was something unexplained out there. The topic arose at work on a few occasions over the years, and I would respond that I had observed things that seemed to defy scientific explanation, or at least any scientific explanation that I was aware of.

In August of 2011, I made a decision to write about the family experiences in order to formally document what we had observed. I arranged to interview the family members involved, and record all of their observations. During this time period, I also realized that I should undertake a literature search just as I would have done for any topic I was researching during my career. Since our UFO sightings occurred mainly along a one-mile stretch of a lonely, country road near the family farm in Smith County, Mississippi, I began a search on the internet for Mississippi UFOs. Six widely reported cases were found, and four of them had occurred in the 1970s. After reading about each case, I realized that these four cases had aspects that clearly showed them to be related to our family sightings. Before the literature search, I had been aware of only two of the well-known Mississippi cases, and had been unaware of any relationship to my family's sightings. I was immediately intrigued by the information that further searching might possibly reveal. In the process of formally interviewing family members, something that I had not done previously, I soon found out that I only knew a fraction of the story, the story that will be revealed in this book. One family member's story was so disturbing to me that it took me a month or so to recover sufficiently to get a good night's sleep and to decide on a path forward for the book. Thus, I embarked upon a deeper quest into UFOs and related phenomena than I had ever intended. This book is the result of my family's experiences, and my research into the controversy of UFOs and all of its implications.

Chapter 1

Mr. Philip J. Klass – Arch UFO Skeptic

"......intellectual capacity is no guarantee against being dead wrong."

—*Carl Sagan (1934-1996) American Astronomer in Cosmos, P. 20*

As I indicated earlier, I read the paperback book *UFOs Explained* by Philip J. Klass in 1977. [1] I was not familiar with Mr. Klass previously, but Philip J. Klass was to become the principal UFO debunker in the world for some 30+ years. His day job was as the senior editor of the respected defense publication *Aviation Week and Space Technology* in Washington D.C. His passion was debunking UFO reports, apparently at his own expense, although some have suggested that he was secretly working for the CIA. However, absolutely no evidence has been found at this writing of such a connection. While the man and the book were brilliant, I realized that Mr. Klass's explanations could not explain the majority of our sightings. In a case covered in the book that I had become somewhat familiar with, Mr. Klass left out some very pertinent details. I found this very troubling. I also found very troubling that he had developed a set of UFO-logical principles, as he called them, which seemed to be a way of prejudging cases instead of letting them stand or fall based upon their individual merits. On balance though, I believe Klass had a beneficial influence upon ufology by making everyone sharpen their thinking.

Mr. Klass died in 2005, and was in good standing in the skeptical community who overlooked his fanatical tendencies. It might seem surprising to the reader after seeing what I have to say later, but if I were leading a team investigating UFO cases, I would have considered having him on the investigation team up to a certain point in time. After that time, I probably would have parted company with Mr. Klass, but maybe not. He would have pointed out any inconsistencies and offered dissenting viewpoints that are valuable in a balanced investigation. I always thought vigorous discussion of opposing viewpoints was necessary, but sometimes messy and often a heated aspect of working through a problem or issue. Klass would certainly provide an interesting viewpoint and plenty of heated discussion. He would not, however, accept any evidence that contradicted his prior beliefs, which were set in stone that all UFO reports could be explained as man-made or natural phenomena. Early in his debunking career, Klass maintained this position even if he had to evoke unexplained or ill-defined natural phenomena. These proposed phenomena included atmospheric plasmas, such as ball-lightning and balls of energy emanating from the earth, or atmosphere called earth-lights by later investigators. Klass actually was open minded enough early in his career to realize that some UFO cases involved some type of unexplained phenomena, but holding onto that view, and to the title of chief UFO debunker, were mutually incompatible.

Every investigator must start from a skeptical position on a UFO case and determine if there is a way to explain the case using known phenomena or man-made technology. During the research for this book, I was surprised to find that most skeptics ignore difficult-to-explain witness observations or other data, and will most times dismiss credible witnesses' statements altogether in order to make the case into something that can be explained. Most skeptics and debunkers readily admit to dismissing eyewitness testimony in this way. Mr. Klass was to become the master practitioner of this technique along with Robert Sheaffer, another prominent UFO skeptic and debunker, and a number of others. Witness after witness complained of Klass ignoring parts of their statements. The skeptic justifies this by saying that the witness must be mistaken because what they are describing is not known to modern science or technology. This skeptical argument actually involves circular logic, which is a type of logical fallacy. Most of us are guilty of logical fallacies at various times. Ignoring inconvenient facts is commonly done by most of humanity. It is how our society operates. In TV commercials, advertisers push their strong points and never mention their weaknesses. Most of modern life involves selling your strong points to someone else and hoping that they do not see your weaknesses—at least not until after signing on the bottom line.

In the UFO debate, most books or articles simply ignore the points made by those expressing an opposing point of view. The pro-UFO group plays to their audience, and the skeptics play to their audience. Both may make brief references to the other, but seldom dwell on the opposing view. Few discussions encompass both arguments, and try to make sense of them or try to test the hypotheses involved, but that is what a completely scientific approach would attempt. In all facets of life, most people try to convince others of something by picking and choosing the points they want to make. A more balanced story, or the opposing view, is usually omitted or presented in a distorted manner.

In a truly scientific debate, the listener must also be objective. In order to be well informed, one must seek out the entire story, not just the part one wants to hear. When it comes to UFOs, the science establishment is not listening to the entire story. In fact, the science establishment does not seem to want to listen to any of the story. A number of times, well regarded individual scientists (Hynek, McDonald, Vallee, Sturrock, Friedman, Maccabee and others) have attempted to gain their attention. It is difficult, because mainstream science's position on UFOs is that the witness is either mistaken, or deranged, or perpetrating a hoax. In this view, unexplained phenomena of any type are not possible, so the preference is to ignore such things. Mainstream science seems to prefer the deranged explanation whenever possible, even to the point of listing the sighting of a UFO as a possible symptom of a mental disorder.

Klass was embraced by mainstream science as an actual ufologist who opposed the extraterrestrial hypothesis (ETH) as an explanation for UFOs, something he did with vigor, but Klass's views from early on were more than a little off message from the standard scientific position on the subject. In Klass's opinion, psychology had little to do with UFO reports, as he stated in an interview with Nova Online in the early 2000s. [2] He said 97 to 98 percent of witnesses were normal and well balanced. In fact, the science establishment believes most people that report UFOs are either deranged or are hoaxers, and a relative few are sincere, normal and well balanced who are simply mistaken. This is just the opposite of what Klass said. Klass had dealt directly with the witnesses, although most of the time he did a telephone interview. In certain cases he would talk to the witnesses directly, and he would be quick to call a case a hoax when there was no other way to explain it.

The psychological explanation, which was rejected by Klass and most American ufologists, became ingrained in European ufology in the 1990s and was called the "new ufology," where it was maintained that all UFO reports were imagined in the mind of the observer. [3] This view came to be known as the psychosocial, PSH, or psychocultural hypothesis, PCH, and makes for a very easy investigation. All one has to do is assign it to the "nut case" basket. This leads me to believe that there were some "nut cases" in Europe, and that they were not primarily the UFO witnesses. Anyone that actually believes the psychosocial hypothesis needs to seriously re-examine that position. The early history of the UFO era completely disproves the psychosocial hypothesis, as will be shown later. The psychosocial hypothesis is simply not logical, although it may be psychological or what I like to call psycho-illogical. UFO researchers of any stripe can recognize most of the "nut cases" and screen them out pretty quickly, but maybe that is not so easy in Europe.

I actually trust Klass's opinion on this more than I trust any of the psychosocial crowd as they point to the true "nut cases" and indicate that anyone who sees a UFO belongs in that group. Klass talked to the witnesses and knew that they were sincere and not fantasizing. He would even tell you that the witnesses saw what they said they saw up to a point, but then he would use as much or as little of the witnesses' information as he needed to turn the sighting into ordinary phenomena (a so-called prosaic explanation).

Mr. Klass left few stones unturned when searching for a conventional explanation, but he turned a blind eye to evidence or aspects of a case that pointed to a non-conventional explanation, and omitted it from his reports. Most cases that he could not explain with his preexisting concepts, he would call a hoax. He did give himself a psychological out in his early career in that he realized some of the sightings were of unknown phenomena, so he hypothesized that ball-lightning or some as yet ill-defined natural phenomena later called earthlights/atmospheric plasmas could explain most of those. In the 1980s and later, he seemed to drop the ball-lightning/atmospheric plasma explanation. This was apparently due to receiving flak from all sides about his atmospheric plasma hypothesis. Ufologists supporting an unknown or extraterrestrial hypothesis, ETH, could easily show that ball-lightning from power lines could not explain the sightings, and mainstream scientist did not accept it either, as it is very doubtful that true ball-lightning can be produced from power lines. Thus, Klass's "ball-lightning" explanation actually suggested that some UFO sightings were caused by unexplained phenomena, which was in conflict with the position taken by mainstream science and the skeptical establishment.

Klass's first UFO case was the Exeter, New Hampshire case of 1965 [4], which occurred nearly two decades after UFOs had burst on the scene and grabbed public attention shortly after WWII. In his first case, Klass realized the witnesses were seeing something real, and he postulated that the witnesses were seeing ball-lightning from nearby power lines. Looking at what Klass said himself, he came up with the idea of ball-lightning from power lines before he had spent any time at all investigating the case. The ball-lightning explanation was weak at best, because the phenomena were seen over the better part of an hour, apparently moving purposely, and was not always near the power lines. To top it off, ball-lightning is a short-lived phenomenon, and even today has no scientific consensus explanation. In fact, there are many proposed explanations for ball-lightning from qualified scientists, and none involve electric power lines. Klass came to realize that he himself was saying that some UFO reports seemed to be related to unexplained phenomena. Later in his career, realizing the implications of what he was contending, Klass let the ball-lightning/atmospheric plasma suggestion quietly drop, presumably to be more readily embraceable by skeptics and mainstream science.

Klass's investigating and debunking techniques were his own, and were sometimes similar and sometimes quite different from some of the earlier scientific debunkers. The first debunker was the astronomer Dr. J. Allen Hynek who worked for Project Blue Book for some 18 years explaining UFO cases as known phenomena until he finally became a believer that some UFOs were real and were unexplained phenomena. [5] Hynek, while mild mannered and reserved, and almost the complete opposite personality type as Klass, actually approached the cases in a very similar way. He distorted what witnesses told him until he could fit the case into a known phenomenon category. Klass considered Hynek to be a scientific traitor when Hynek reversed his position on UFOs.

Another prominent scientist of the 1950s into the 1970s that was one of the primary debunkers was theoretical astronomer and astrophysicist Donald H. Menzel, who preceded Klass by a decade or more. [6] Menzel was famous for putting forward the idea that all UFOs were mirages or reflected or refracted images, even for his own UFO sighting from an airplane. Menzel's mirage explanation requires what is known as a temperature inversion in the atmosphere, and while it certainly can explain some cases, there are many cases it cannot explain, as we will get into later. [7] Some cases that involve a light that seems to be following a car at night can be due to refraction of a light source some distance behind the car. If one stops the car and the light flies on by and maneuvers around in some way, the probability of refraction drops to nearly zero.

Another skeptic was nuclear physicist, Edward U. Condon, who became known in the UFO world because of the so-called Condon report. The Condon investigation was an effort headed up by the University of Colorado under the guidance of Condon supposedly to objectively investigate UFO reports for the Air Force. [8] The problem was that Condon had no intention of doing an objective investigation, as the so-called "Trick" memo showed very clearly.[9] The memo came to light due to the efforts of early ufologist Donald Keyhoe (originally a skeptic) and an article in *Look* magazine by Raymond Fowler. [10] By the late 1960s, the subject of UFOs had come to have a very bad stigma in scientific circles, and all Condon wanted to do was dismiss the subject. The entire effort was driven by the desire of the Air Force to get out from under the burden of having to investigate UFO reports from the public where some 90 to 95% of the reports were explainable as natural phenomena or due to man-made objects. [11]

In 1968, both Menzel and Condon dismissed the necessity for science to formally study UFOs, something I doubt that Klass actually agreed with, as he studied them for some 35+ years apparently using his own time and money to do so. As pointed out above, Klass, early on, maintained that some UFOs were examples of the poorly understood phenomena known as atmospheric plasmas, which he believed were generated from power lines. (Klass was looking for a cause of the phenomena that was nearly everywhere where people would be. Electric utility lines fit the bill most of the time.) Klass's hypothesis was in sharp contrast to saying that there was nothing in the UFO phenomena for science to study. Early in his career, Klass recognized that there were ill-defined or unexplained phenomena present in some UFO reports, even though he shied away from that position as his career went along. Mainstream science, by the late 1960s, just did not want to recognize that there was anything of substance present in such reports, and Klass eventually realized that to be the standard bearer of the skeptical/debunker community he would have to adopt that position as well.

In his 1974 book, Klass presented a set of UFOlogical principles. Klass's so-called UFOlogical principals were logical constructs that allowed him to prejudge any UFO case. In effect though, these UFOlogical principals are very much like prejudging a court preceding or a lab experiment before the evidence had been presented or the data collected. Even the introduction in Klass's book talks about using the principles 'derived' by Mr. Klass so that they can be intelligently applied to any UFO case with the implication, unstated, that all cases can be debunked as conventional phenomena, or as the catch-all ball-lightening, or ill-defined earth-generated balls of energy for the unexplainable.

Scientific American magazine gave Klass's book a good review in its May 1975 issue, but with some reservations thoughtfully included—thankfully. [12] These included his tone in the book, which was more indignant "than seems wise" in their words. *Scientific American* made the point that each case and each witness must be listened to with patience, and the evidence must be examined carefully case by case. The reviewer unceremoniously dismissed Klass's evidence for his hypothesis of ball-lightning from electrical power lines. *Scientific American* called Klass's ball-lightning from power lines "….a case too strongly based on a few erroneous reports and doctored photographs, some of which he mentions here." Klass had published an earlier book espousing the ball-lightning hypothesis. [13] Even *Scientific American* saw some of the problems with Mr. Klass's tone and behavior. The real problems would become evident a year or so later to any who would look objectively at his actions.

However, Klass's proposal of ball-lightening and balls of plasma or energy from the earth is absolutely a valid hypothesis that must be considered seriously. The significance of his atmospheric plasma hypothesis was that Klass, at that time, firmly believed some UFO reports were real and had been generated by some sort of unexplained phenomena, although he tried to play down the unexplained part later. This is a huge, huge admission from a skeptic/debunker, and this admission seems to have gone mostly unrecognized for its significance. It was mostly opposed by ufologists who called it explaining the unexplained by evoking something else unexplained, and by mainstream scientists who called it an unlikely phenomenon.

Nevertheless, Devereux and Brookesmith, using the work of Persinger and Derr as a basis, greatly expounded on the so-called earth-lights hypothesis in their extremely interesting and entertaining 1997 book *UFOs and Ufology: The First Fifty Years* without giving Klass that much credit for his support of the idea. [14] Although Klass was clearly espousing essentially a very similar idea nearly 25 years earlier, he had backed away from the hypothesis by that time. Klass was not the first to suggest this idea as it arose sometime in the mid-1950s. It is not clear to me who first espoused the idea. The bottom line is that earth-lights appear to exist based upon the work of Derr and Persinger and upon Devereux's and some field researchers' direct observations related in *UFOs and Ufology*. Ball-lightning certainly exists, as a number of credible witnesses over several centuries have seen it, including some scientists who have observed the phenomenon. While still very ill-defined and without a consensus scientific explanation, these balls of plasma energy can explain some UFO reports comprising strange nocturnal lights, so Klass was right in a limited way.

Using his UFOlogical principles, Klass said that it was not necessary to study every case. Instead, he implied that one need only to classify a case according to type and stamp 'debunked' across the file. Klass usually investigated only the big headline cases and in fairness, from a practical standpoint, he could not investigate every case. How Klass used these UFOlogical filters was on somewhat shaky ground scientifically from any objective standpoint. *Scientific American* was subtly pointing this out in their review. The UFOlogical filters imply that nothing outside the scope of what was known or accepted at the time could ever be found or could ever occur. One must go into these things with an open mind, and not with dogmatic views that cannot be swayed by any evidence. The investigator must not function as either a true believer or as an unmovable skeptic, but rather must be able to let the data do the talking. This was the position that was taken in the *Scientific American* review, much to the credit of the reviewer, Philip Morrison. However, I do disagree completely with some of the excerpts from Klass's book that Morrison put into his review, as he assumed Klass was completely objective in his investigations when that was not exactly the case. Morrison praised Klass a little too much, and Klass's book obviously reinforced Mr. Morrison's own preconceived negative ideas about UFOs. But who can really fault him? It is only human nature to praise that which reinforces your own ideas or feelings about something. I will get back to that point over and over again in this book.

For the reader to fully understand Mr. Klass, they would have to read Jerome Clark's 1981article in *Fate* magazine entitled "Phil Klass and the UFO Promoters" now available on the ufoevidence.org website. [15] Particularly telling in Jerome Clark's article are the comments of Allen Hendry, longtime investigator for the Center for UFO Studies, CUFOS, founded by Hynek. Hendry is said to be one of the most objective investigators ever in the UFO field. He condemns Klass's modus operandi most convincingly. The reader would also need to read the last 85 pages of the expanded edition of Travis Walton's book *Fire in the Sky* that came out in 1997 to see the interaction between Klass and those that he gave what J. Clark called "The Treatment." [16]

Another insight into Klass came in 2006 when his FBI file was released through the Freedom of Information Act and put on-line by CUFON, The Computer UFO Network, www.cufon.org. [17] The FBI file contained some very interesting information. Klass was incensed that the FBI had published in their "Law Enforcement Bulletin" of February 1975, an article by Dr. Hynek detailing how a law enforcement agency can call in a UFO organization to investigate UFO reports, thus freeing the law enforcement agency to get on to more pressing matters. To Klass, the FBI was endorsing the reality of UFOs when really all they were doing was advising law enforcement on how to deal with UFO reports in the post Project Blue Book era. The FBI conclusion after dealing with Klass: "In view of Klass's intemperate criticism and often irrational statements he made to support it, we should be most circumspect in any future contacts with him." The FBI file also noted that twice Klass published classified information in *Aviation Week* magazine without receiving permission to do so. He was not prosecuted, because the agencies involved did not want to have to release even more classified information in order to pursue the case.

Yes, Klass's behavior was quite over the top when it came to UFOs. He would find a way to fit a UFO case into a known phenomenon category, including ignoring key pieces of witnesses' testimony. I have come to find that most skeptics/debunkers of any unexplained phenomena almost immediately dismiss the witnesses' statements by saying that the witnesses were mistaken, delusional, or were some sort of fanatic. Usually with no further investigation, they then proceed to give their version of what they believe the witnesses actually saw, leaving out whatever was in the witnesses' accounts or other details of the story that do not jive with their explanation. At times, Klass operated along these same lines. In one case that I have become very familiar with, the Pascagoula abduction case, a rookie newspaper reporter and a veteran magazine reporter conducted a vastly superior, much more objective and thorough investigation than that of Klass. Klass simply labeled the case a hoax and dismissed it, even though there was sufficient supporting evidence to show that something unusual had happened. This case displayed Klass's tendency to disparage or to try to discredit witnesses. Very few people have lived perfect lives, so he could usually find something. In high visibility cases, he would give some witnesses "The Treatment." He went over the line in harassing witnesses, and was accused of trying to bribe a witness in the Travis Walton abduction case. He would make the most highly trained Air Force personnel out to be as bumbling as the Three Stooges, as he did in the famous 1957 RB-47 case, to bring the case around to his point of view. His belief was that there could be no explanation other than known phenomena or a possibly logical extension of known phenomena, and he would twist every case to fit into these categories. Favorite explanations were the planets Venus or Jupiter, and fiery meteors known as bolides.

In fairness, Klass had debunked some puzzling UFO cases that turned out to be known phenomena, and another case that was the work of a hoaxer (tiny match stems in the middle of an area supposedly scorched by a UFO). On balance though, other serious UFO researchers have disagreed with Klass over the years and pointed out where he was wrong in a case here and there, but most mainstream scientists did not look at these critiques. The science establishment's views on the subject of UFOs were, and are, quite closed minded. It is fortunate that science's consensus views on many other subjects are correct most of the time, because they are vigorously and thoroughly researched from every conceivable angle, but such is not the case for UFOs. There has been little sustained, systematic research concerning UFOs backed by mainstream science. Thousands upon thousands of investigations of sightings and incidents have been performed over some 65 years, but that is similar to investigating a nighttime hit and run auto accident without the tag number or a good description of the offending vehicle.

Klass was the scientific establishment's knight in shining armor ready to slay these delusions of the masses, and mainstream science was not interested in his methods. Klass studied many cases, and he was brilliant in his fanatic efforts to classify them as misunderstood ordinary phenomena, either natural or manmade. Klass knew a lot about radar, and after reading his work you would wonder how we could ever get any useful information about UFOs from radar, or get any useful information from radar at all for that matter. According to Klass, radar was always wrong when it showed an unidentified target. Every UFO hit on radar was, according to Klass, a malfunction or due to a temperature inversion or an operator error. He seemed to think that a radar operator could find an unknown target anytime someone suggested they had seen something out of the ordinary in the sky. One of his UFOlogical principles even states this. If you think about it, this makes no logical sense and it amounts to prejudging every case involving radar. How could any scientist of the time read Klass's book of 1974 and not see Klass's blatant biases? They could because his biases reinforced their own beliefs either consciously or subconsciously, and reinforced their comfort zone.

When I read Mr. Klass's book in 1977, it seemed that the explanations just did not fit our sightings, and not even the ball-lightning or earth-lights could explain the range of sightings. However, I was a young scientist and felt overmatched by this tenacious engineer who had received mixed but overall good reviews for his book from the mainstream scientific establishment. The praise was for bashing ufology. The criticism was for his overbearing fanatical tone and suggesting some sightings contained something real and unexplained (ball-lightning). I was reluctant to publically come forward to engage Mr. Klass and other skeptics at that time, and the fact of the matter was that I probably did not have time to do justice to the subject of UFOs during my career. I would totally immerse myself into my current projects, and there was always the thought that I might damage my career if I came forward on the subject of UFOs. Yes, I admit it. I was very reluctant to say very much about UFOs. I was reluctant to take the risk, and that was the way things stood until August of 2011.

CHAPTER 2

Why the Science Establishment Abandoned UFOs: Projects Sign, Grudge, Blue Book, and the Robertson Panel

"If an elderly but distinguished scientist says that something is possible, he is almost certainly right; but if he says that it is impossible, he is very probably wrong."

—*Arthur C. Clarke, (1917-2008) Science Fiction Writer, Inventor*

During the course of my research, I have run across several skeptical blogs that take great pleasure in saying that ufology is not scientific. They apparently do not realize that they are using the fallacy of circular logic, in this case, to put down ufology. Their logic is based upon the fact that mainstream science ignores ufology just as it does most unexplained phenomena. This is a very unscientific attitude on the part of the science establishment, but because UFOs have been ignored or, more correctly, have been made into a taboo subject in the view of mainstream science, the skeptics say ufology is unscientific. The skeptics do not seem to understand the situation or the meaning of the word 'scientific.' One practices the scientific method by investigating something logically and objectively without reliance upon prior beliefs, except to form a hypothesis, and then think of a way to test the hypothesis. Anyone can use the scientific method—it just requires logic and reasoning to be objectively applied to the subject at hand. It does not matter what the subject to be investigated is. What matters is the manner in which the investigation is carried out. Dismissal based upon prior beliefs without investigation or explaining without investigation is irrational or emotional, but certainly not scientific. It would seem that the scientific establishment had little concern for UFOs from just about the very beginning, but that is not quite an accurate assessment. However, things did go awry just a few years after the phenomena gained widespread attention in 1947. How did this present situation come to be?

From a broader view one must realize that modern science is annoyed by any unexplained phenomena embraced by the public before the phenomena has been properly vetted by the science establishment. Since UFOs were quickly embraced by the public at large before the science establishment could hardly get involved, the science establishment reacted skeptically and negatively from the beginning. Mainstream science's view is that they are the gatekeepers of knowledge, and they will say what is real or not real. The actual gatekeepers are the relatively few individuals at the top of the scientific establishment, the editors of scientific journals, along with the most highly regarded scientists and reviewers of those journals.

If these individuals cannot be convinced, then the subject matter does not show up in scientific journals. Most young scientists follow their lead and the reason is simple: money and status. Scientists have to be funded by someone or some institution in order to carry out their studies. Those controlling the purse strings are greatly influenced by the high priests of the science establishment as to who gets funded and who does not. Very few taboo subjects get funded. The enterprise of science is a highly controlled endeavor, and yet ufology has been driven in part by what can only be called maverick scientists convinced of an underlying validity to UFOs, such as Hynek, Harder, Vallee, Hendry, McDonald, Friedman, Sturrock, Salisbury, Maccabee, and others, along with a host of talented investigators and historians, including Keyhoe, the Lorenzens, Bloecher, Pflock, Philips, Randle, Jacobs, J. Clark, Bullard, Kean, and many others. [18]

The most formidable opponents to making the topic of UFOs acceptable for scientific study are the editors of scientific journals. In the last 35 years only one ufologist/scientist has had a degree of success with these guardians of scientific knowledge, and that scientist is Dr. Bruce Maccabee, optical physicist USN retired, who pursued his interest in UFOs with his own time and money. I will talk more about those articles later in the book. It seems that even the skeptics should concede that at least some of ufology is scientific.

I encountered the resistance of the science establishment to the subject of ufology during my career, but never understood exactly how it came about. During my research into UFOs, I finally came to better understand how the current situation came into being. I would dare say that there are many people who are interested in UFOs who are not aware of this fascinating history. In order to understand it, we must go back to 1947 and the Air Force's attempt to investigate UFOs. It can be fairly stated that the present-day tone was being set by the end of 1953, only six years after the world's attention was first drawn to those strange objects reported to be flying through the skies.

The Kenneth Arnold sighting, and an incredible 800 more in the summer of 1947, brought our nation's attention to the flying saucers (a term invented by someone in the news business from Arnold's comments that they were flying like saucers skipping across water) and flying discs said to be filling our skies. [19] In the famous Twining memo, Lt. Gen. Nathan F. Twining stated that the phenomenon was real and that a detailed study of the matter should be undertaken. [20] The Air Force responded by creating the classified Project Sign in early 1948, just after separating itself from the Army. [21] Project Sign, which became known to the general public as Project Saucer, was staffed with top-notch personnel in aeronautics and other specialties. J. Allen Hynek was one of those specialists in astronomy. While Project Sign was created to focus upon the flying saucer phenomenon, there was some indication that the Army Air Corp had been investigating strange things in the sky before 1947 (the Air Force was part of the Army until late 1947). The ghost rockets in Europe in1946, and particularly in Sweden, would have been of considerable military interest, as were the Foo Fighters of

WWII. [22] The idea that the ghost rockets were a result of the Soviets experimenting with leftover German V-2 rockets did not pan out, and they remain mostly unexplained to this day.

The idea circulating around early on was that these unknown flying craft, these flying saucers, seen over the US were controlled by the Soviets. Since this idea was shown to be incorrect, as were a number of other explanations, certain Project Sign investigators were some of the first to come to the extraterrestrial hypothesis (ETH) by the process of elimination. The flight behavior and flight capabilities reported by credible witnesses strongly pointed in that direction. A number of people working for Project Sign were open minded, and realized that there were few other options that could explain what appeared to be technology far beyond our earthly capabilities based upon the sightings and observations of highly credible military personnel, scientists, airline pilots, engineers, and technicians. However, they may have been a little premature is sending their conclusions up the chain of command. In hindsight, perhaps Project Sign personnel reported their conclusions a little too quickly in an "Estimate of the Situation" report issued in September 1948. [23] "Estimate of the Situation" reports were the standard way of reporting the results of a military investigation. Although a copy of the report apparently does not exist today, the first public disclosure of the report's existence was in Edward Ruppelt's 1956 book *The Report on Unidentified Flying Objects,* which had to be cleared through military censors to be published. [24] Ruppelt, the first director of Project Blue Book, plays a prominent role in the UFO drama and will be discussed in detail shortly. The estimate of the situation report stated that the best explanation was the phenomena were spacecraft from another world. The report went up through channels and created a divided opinion in the upper brass. The report made it all the way to Chief of Staff Gen Hoyt S. Vanderburg, who rejected it due to the lack of physical evidence (absolute proof beyond any doubt) and called for a scaled-down report. The Project Sign personnel pushed back (before that term was in existence) and held to their positions in a passively resistant manner. Ultimately, the competing hypotheses of ETH and the explainable earthly phenomena hypothesis were hotly debated behind the scenes. The ETH lost out, although it had a minority in the intelligence community that supported it. The main Project Sign personnel that stubbornly supported the ETH explanation got reassigned to various locations, which were usually less desirable than the Air Intelligence headquarters at Wright-Patterson field near Dayton, Ohio. In other words, they were punished for their resistance.

The name of the project was changed to Project Grudge in late 1948, and the next few years were called the "Dark Ages" of UFO investigations by Ruppelt. The remaining investigators realized that they should not say much, and that to survive they would just leave puzzling cases as unexplained without much comment. Otherwise, they would be the next to be reassigned. Many times they were just going through the motions and were not doing thorough investigations. This continued until several inexplicable UFO incidents were witnessed by a number of high-level officers. These officers came to realize that the Project Grudge reports were not thoroughly investigated and that many reports were simply rubber stamped as explained, any explanation would do, without serious investigation.

Project Grudge was renamed Project Blue Book in late 1951 with the mandate to conduct as an objective investigation as possible into all credible reports, and this was how the project operated for two years or so until sometime in 1953 or 1954. Project Blue Book under Ruppelt reinvestigated all the credible earlier cases from 1947 onward through 1952. These are Project Blue Book's most thoroughly investigated cases with access to secret mogul balloon and skyhook balloon launches, rocket launches, and other secret activities. All manner of technical experts, analysts, and resources were available to work on the cases. After January of 1953, the UFO world changed again, and by the mid-1950s emphasis was once again to stamp a case as explained if at all possible, as resources for investigations began to dwindle away. The case load was too great to thoroughly investigate all the cases. I would guess that only the most difficult to explain cases with the most credible of witnesses were listed as unexplained in the Project Blue Book files from that time forward. How could there be such reversals in the mode of operation of the project in such a relatively short period of time? It was due to several things: the science establishment, the CIA, Cold War paranoia, and the belief that an answer to the puzzle could be readily found in short order.

Ruppelt was clearly the person for the job of doing objective investigations. He almost immediately came up with the term "Unidentified Flying Object" (UFO) to be used as the general term for the phenomena. He intended it to be pronounced U-foe, but that did not stick. U-F-O did. With Ruppelt in charge of Project Blue Book, more thorough investigations were done, the witnesses thoroughly investigated, and only cases with high credibility witnesses were considered. One case, the scoutmaster case from Florida, was called the "The Best UFO Hoax in History" by Ruppelt, even though there was physical evidence to support it. The witness was known for tall tales and stretching the truth on occasion. The physical evidence, scorched grass roots, only came to light because of thorough investigation methods by Project Blue Book personnel. Even though this evidence was such that it could not be hoaxed, Ruppelt would not include the case in the official Blue Book files because of the character of the witness. Karl Pflock discussed the case in 1997, and strongly suggested that it was not a hoax, but was rather a ufologist's worst nightmare of a real case with an unreliable witness. [25]

Ruppelt points out that radar-visual cases only became known to the public in 1952. There had been a number of radar cases up to that time and some were radar-visual cases involving classified military personnel encounters. The radar-visual cases in Ruppelt's book should be required reading for anyone getting into the UFO controversy. Radar-visual cases appeared in a big way in1952, and in a very public way. The famous Washington, DC cases occurred in July of 1952, and were nowhere near being the best radar-visual cases to occur that year according to Ruppelt, but were the best publicized, and the public was becoming alarmed. The CIA took notice and had come to fear what might happen to national security if the public panicked over UFOs. [26] The CIA also believed that the Soviet Union could use UFOs to cover up some of their activities and to create anti-US propaganda. The CIA feared anarchy in the US, if the public's concern were to turn into a true panic over UFOs. My question is this: If there was nothing to UFO reports, why was the CIA worried? If there was nothing at all to them, then the craze would rather quickly die down. One might conclude that the only reason to be worried was if the CIA knew that there was something real behind the reports as Lt. General Twining had stated.

Most of the UFO cases in Ruppelt's book involved government employees, scientists, technicians, pilots, and military personnel, because of their credibility. The military was very concerned about the apparent national security risks associated with strange aircraft filling our skies, and military intelligence recognized that these reports did not arise from our own secret advanced technology.

Ruppelt died in 1960 of a heart attack at about 37 years of age, but three additional chapters were published in the second edition of his book in that same year. [27] He gave some more unexplainable cases, but in the last chapter he tried to debunk everything he had written up to that point. The entire tone of the book changed in that last chapter as if someone else was writing it. Ruppelt was still under the control of Air Force security censors, which had the policy at the time (1960) of trying to discredit UFO reports and prevent any real information from reaching the public. In this last chapter, he gave some of the most absurd explanations ever put forth for some of the famous cases, and he tried to dismiss UFOs with sweeping statements, which are nearly always fallacies, such as people are terrible observers, and radar operators are so fallible that you cannot rely on their statements. Earlier in his book he had pointed out that radar operators are responsible for the lives of everyone who flies and he said that there were no explanations for the radar-visual cases involving highly trained radar operators, pilots, ground crews, and flight crews. If he wanted to debunk UFOs, he needed to put forth reasonable explanations point by point for the radar-visual cases he discussed earlier in the first edition of the book. Instead, in that last chapter he gave skeptics and debunkers sweeping dismissal statements, which are known as the class of fallacies called sweeping generalizations that were later used by Klass and others, and are used to this very day by nearly all skeptics to cast doubt upon a case.

The CIA was behind the scenes pulling strings after the alarming UFO year of 1952. The CIA organized the Robertson Panel, named after Dr. Howard P. Robertson, to analyze the UFO situation. [28] Robertson was a prominent physicist and mathematician, and was in the employment of the CIA, who charged him with bringing a panel together that would produce a report that would reduce public concern and show that UFO reports could be explained as conventional phenomena. It was an attempt by the CIA to calm the public by having prominent scientists debunk the subject of UFOs. Otherwise, why in the world would the CIA be involved in a purely scientific matter? The panel was composed of Robertson and:

Luis Alvarez, physicist, radar expert, and later Nobel Prize winner;

Lloyd Berkner, physicist;

Frederick C. Durant, CIA officer, secretary to the panel, missile expert;

Samuel Abraham Goudsmit, Brookhaven National Labs nuclear physicist;

Thornton Page, Deputy Director, Johns Hopkins Operations Research Office, astrophysicist, radar expert; and

J. Allen Hynek, partial member, presenter to panel, astronomer, Project Blue Book. (Only Hynek had any previous involvement with UFOs. Most of the panel was chosen because they had been disinterested in and had little knowledge of UFOs up to that time).

The scientific review was really a public relations operation. The entire exercise was a type of fallacy known as the appeal to authority, scientific authority in this case. Months and years of detailed technical work that was labored over by highly qualified specialists were dismissed in minutes or seconds. In January of 1953, the panel met for 12 hours in all on four separate occasions and reviewed 23 cases. The panel rejected some of the most rock solid, radar-visual cases on record, an analysis by Major Dewey J. Fournet that took months of work dealing with UFO flight paths that concluded there was intelligent control involved, and 1,000 hours of technicians' efforts in verifying two UFO films, the Great Falls, Montana film of August, 1950, and the Tremonton, Utah film of July, 1952, to be authentic showing unknown flying objects. The panel also rejected the connection between UFOs and nuclear radiation spikes that is, in my view, some of the best evidence ever collected to unequivocally show the presence of unexplained phenomena. Officially, the panel seemed to reject everything that was brought before them in a very cavalier manner. They were charged with calming the fears of the populace not in doing a valid scientific review. Behind the scenes, things were not so clear cut, as I will get to later.

The Robertson panel was hastily assembled and met in early1953, some two years before the report by the Battelle Memorial Institute, *Project Blue Book Special Report Number 14*, was published covering the years 1948 through 1953. [29] It took almost two years for a highly trained team to study 3,201 cases, but the Robertson panel could perform their study and draw conclusions in 12 hours while looking at only two or three percent of the unexplained cases up to that time. Ruppelt did put forward what he felt were the most thoroughly examined cases and the best of the unexplained cases on hand. The average time the Robertson panel spent considering each of these cases was about 31 minutes. The Battelle Memorial Institute staff spent far longer in individual staff time investigating each of the unexplained cases. It takes no time at all to dismiss a case in a cavalier manner rather than to investigate a case; only a few minutes are needed after the completion of the presentation.

In *PBB Special Report Number 14*, about 20 % of the cases were classified as unknown, and 9% had insufficient data for a grand total of nearly 30% unexplained cases. The Secretary of the Air Force at the time, Donald A. Quarles, misled the press and public, and said that only 3% of the cases were unknowns, and that those could be explained if more data were available. Quarles was talking about what he believed to be true of the current PBB investigations at the time, and not about *Special Report Number 14*. Actually, *PBB Special Report Number 14* showed that the more thoroughly a report was investigated and the more information that there was available, the more likely it was to be classified as unexplained. This has been pointed out by a number of UFO researchers over the years. Both parts of Quarles' statement were false if applied to *Special Report Number 14*, but he knew that very few people would actually read the report. Why would you confuse the issue unless you were trying to cover up something?

January 1953 was a very monumental turning point in the UFO story. The impact upon the UFO controversy cannot be overstated. The reverberations are still being felt today. The net result of the Robertson panel report was to help kill Ruppelt's plan to scientifically study UFOs instead of just investigate reports. Northern New Mexico, with all of its secret government installations, had been a hotbed of UFO activity from 1947 into 1952. Many of the scientists and other personnel working at these facilities had seen UFOs and/or "Green Fireballs."[30] Some had even measured nuclear radiation spikes on several occasions as UFOs passed over their facilities. In terms of demonstrating that UFOs are an unknown phenomenon, this evidence was rock solid, and it was duplicated at another secret government laboratory that was also a hot bed of UFO activity, Oak Ridge National Labs in Oak Ridge, Tennessee. [31] Again, this was irrefutable proof that UFOs represented an unknown phenomena, and it was all collected before 1953. They searched for other possible causes for the radiation spikes, such as instrument malfunction, sunspots, balloons, and aircraft with specialized equipment, and could find no other explanation. The odds turn out to be literally astronomical against these radiation spikes being random events. As many as 15 separate spikes occurred at two different government laboratories at the same time that a UFO was observed to be passing nearby. At least one of the cases involved radar detection of the UFO so that a new type of case came into being, a radar-visual-radiation spike case. The odds against all of this just being random, separate, unconnected events is literally about a quadrillion to one. Ruppelt knew about all of this, and had planned to put in place a series of sighting towers with specialized equipment for measuring wavelengths and intensity of light, detecting nuclear radiation, magnetic effects, and heat effects, along with special cameras on the radar network across the area. The justification for the plan was greatly weakened with the Robertson Panel's official report and that report, coupled with some early technical failures of the equipment in the field, doomed Ruppelt's plan.

Yes, the science establishment on the face of it appeared to have struck a blow against open, objective study of a widely reported phenomenon. That was not quite the case. According to Ruppelt, the panel remarked that the data presented to them was very interesting, but was not strong enough for them to go out on a limb. Scientists are almost always afraid to go out on a limb when considering someone else's data for the first time. They usually want absolute proof in several unrelated ways to do that. It does not matter if logical deduction indicates that there probably is something real behind the data. In that case they will encourage further study, but will not wholly embrace the current data. Ruppelt reported in his book that the panel actually told him he should not only undertake his study, but he should request additional resources to expand the study. However, the public recommendations did not reflect this, as they were controlled by the CIA and meant to calm the public. The almost unbelievable statement was made that UFOs were not a threat to national security (a statement at odds with the real view of the security agencies, if only because they clogged up reporting channels), and that statement probably killed support for Ruppelt's planned study more than anything else. The public impression was that the Air Force took the recommendations as if they were the final word on the matter. After all, the science establishment is never wrong is it, especially when directed by the CIA!?!? If Ruppelt was correct in his book, the science establishment actually said that there was some intriguing information already on record concerning UFOs and further studies should not only be done, but should be expanded. It did not happen.

The CIA had molded a false impression of the general scientific consensus resulting from the Robertson panel, and that impression took hold, and holds sway, to this very day. Can the scientific establishment settle upon consensus opinions that are wrong? It certainly can and has been wrong many times over the course of history. The so-called absolute truth of science is constantly being tweaked and is under revision all of the time. As later studies are done, the errors and misconceptions of the past are corrected. Stanton Friedman and Kathleen Marden have written an entire book on the subject.[32] Some examples of scientific truths that were proven false are: rocks are not able to fall from the sky (meteors) circa 1800, physicians do not need to wash their hands between patients circa 1840, heavier than air flight is not possible circa 1900, Meteor Crater in Arizona was believed to be volcanic in origin circa 1960 (Eugene Shoemaker showed it was, in fact, caused by a meteor and the earth is pockmarked by such craters), black holes will accumulate mass forever circa 1980 (Stephen Hawking using quantum mechanics showed that black holes can evaporate), and so on. [33-35]

The result of the CIA controlled Robertson panel's recommendations was to quell active public research into UFOs and to suppress the amount of information available to the public. By December of 1953, it had become illegal for a member of the military to release any real information about UFOs to the public, and airline pilots were also under penalty of law not to release any information either. [36] At some point in 1954, Air Force Regulation 200-2

(AFR 200-2) took effect making all sightings submitted to the Air Force classified material, and prohibited the release of any useful information to the public unless the sighting could be positively identified. [37] This was to give the public impression that all UFOs could be identified. Thus, the military sightings and restricted airspace encounters were "born classified" as Stanton Friedman has put it, and any civilian sighting reported to the Air Force became classified. [38] This explains government investigators telling people not to talk about their sighting once it had been reported. This could also be the basis for the men in black mythology or fact-ology as the case may be. Project Blue Book had become an exercise in public relations sometime in the middle 1950s. The most serious government investigations were behind the scenes into high credibility, classified cases. Kevin Randle backs this up in his recent 2012 book *Reflections of a UFO Investigator*. [39] Although young at the time, Kevin Randle was one of the first UFO investigators to examine the files of Project Blue Book after they were declassified in 1975. The files showed that after the Ruppelt years, around 1953 and onward, the investigations were many times merely going through the motions to give the public the impression that thorough investigations were being done for all credible cases. So, we conclude the story of how science was manipulated by the CIA into killing open public research into UFOs, and how the era began in which UFOs became taboo in the eyes of most scientists.

As if the Robertson Panel's Report was not damaging enough, the contactee nonsense came to prominence in the mid-1950s. Probably the most damaging thing that hurt ufology mightily in the eyes of science were the so-called contactees that made worldwide headlines in the press. Led by George Adamski, this ridiculous nonsense has done almost irreparable damage to the respectability of ufology in some circles to this very day. [40] A number of the contactees apparently became quite wealthy through fees for speaking engagements, and there are some modern-day contactees such as Billy Meier that are following in Adamski's footsteps. To any objective observer, the contactees were conmen and con-women exploiting a situation. Some skeptics want to claim that alien abductees are just continuations of the contactee movement, but this does not hold up to logical investigation, as we will see later. The contactee movement was led by conmen and con-women into what was very much a sort of UFO religion around the contactees. Ufology was looking less and less respectable in the eyes of science, and the Air Force was also downplaying unexplained UFO reports and publicizing cases that could be explained. And then along came Jung's book.

In 1958, Swiss psychologist Carl Jung published *Flying Saucers: A Modern Myth of Things Seen in the Skies* in Europe, and in 1959 it was published in the US. [41] Suddenly the founder of analytical psychology seemed to be saying that UFOs were all in the mind of the observer, especially if one ignored the last chapter in Jung's book. This probably became the seed of the psychosocial movement in ufology in Europe. The last chapter threw some water on the rest of the book and by 1960, a year before his death, Jung admitted to being of several minds about UFOs. [42] He considered the purely psychological cases to be a minority, he mused that ball-lightning/atmospheric plasmas could be the cause of some UFO reports (some seven or eight years before Klass), and he concluded that radar showed some UFO reports to be real and their behavior indicated that they may be interplanetary. He seemed to embrace three of the possible explanations for UFOs that are on the scene today: psychosocial, atmospheric plasmas, and extraterrestrials. The skeptical literature seldom mentions anything about Jung other than the implication that all UFO encounters are in the mind, a premise that Jung actually rejected. Jung also severely criticized Menzel's mirage and reflected image explanations for UFOs. The full range of Jung's views about UFOs is seldom given in any publication, but the skeptical literature seems to be attached to the one premise that Jung flat out rejected: that UFOs are all in the mind. But overall, Jung's book contributed ammunition to those looking to discredit ufology, and the credibility of ufology was again dealt a body blow.

In summary, the CIA struck a mighty blow against the public respectability of ufology using the Robertson panel in 1953. While it was somewhat damaged in the eyes of science, ufology got up and staggered back to its feet. Then along came the lunacy of the contactee era that dealt the respectability of ufology another severe blow in the 1950s. By 1960 there was Jung's psycho-babble about UFOs seeming to far overshadow what he later concluded: that some UFOs were real and possibly not of this earth. Some years later, in 1968, the Condon report summary stated that UFOs offered nothing worthy of scientific study and subsequently the Air Force shut down Project Blue Book. This should have delivered the knockout punch to ufology, if the phenomena were totally baseless.

However, the 1970s happened and UFOs were everywhere worldwide, demanding to be recognized. Unfortunately, it did not help that the UFO cause was championed by the tabloid magazine, the *National Enquirer*, which further alienated the science establishment toward the subject. The subject of UFOs became taboo to most scientists, at least publically. If ever a phenomenon was screaming for formal scientific study, it was UFOs in the 1970s, but what was the scientific establishment's response: nothing, nada—unexplained phenomena do not exist.

CHAPTER 3

Astronomers and UFOs

"All truths are easy to understand once they are discovered; the point is to discover them."

-Galileo Galilei, (1564-1642), Italian natural Philosopher, Astronomer and Mathematician

Vallee was one of the first to point out that some official sources have misled the public by implying that astronomers do not see UFOs. [43] None other than Edmond Halley (English astronomer, mathematician, physicist, discoverer of Halley's comet, 1656-1742) himself had a UFO sighting for more than two hours that consisted of a lighted object so bright that one could read the printed page by its light in the year 1716.

The first photographs of what appeared to be spherical UFOs in the telescope (elongated disks due to the slow film were shown crossing in front of the sun) were made by an astronomer in Mexico in 1883, according to the rense.com website. [44] Hundreds of these objects trailing each other over two days were seen with all going in the same direction at about the same speed so that an explanation involving the breakup of a comet or meteor is possible. If just one had stopped and changed direction, then the comet or meteor explanation would be untenable. The Project Blue Book Files indicated that about 1% of all reports later classified as unknown came from amateur and professional astronomers, missile trackers, and surveyors.

In the 1950s, a survey by J. Allen Hynek of 45 fellow professional astronomers found that five reported UFO sightings. [45] Two of those in Hynek's survey were Clyde Tombaugh, the discoverer of Pluto, with six UFO sightings, and Lincoln LaPaz with two sightings. Tombaugh believed that UFOs were extraterrestrial in origin, and scientists that dismissed UFOs without study were being unscientific.

In the 1970s, Peter Sturrock (now emeritus professor of applied physics at Stanford University) conducted two large surveys of the American Institute of Aeronautics and Astronautics (AIAA) and the American Astronomical Society (AAS). About 5% of the members admitted that they'd had a UFO sighting, but refused to have their names used. [46] I would expect that the real number would probably be double or triple that due to the great stigma against admitting such things in mainstream scientific circles, but maybe not if the surveys were done anonymously. Sturrock, who has had a very distinguished scientific career, is one of the few truly mainstream scientists to openly and honestly look at the UFO question. In 1982, he helped to establish the Society for Scientific Exploration to explore subjects ignored by mainstream science. The *Journal of Scientific Exploration* has been published since 1987 by this organization.

There are many amateur astronomers looking at the skies as well, and an informal survey in 1980 of amateur astronomers found that about 24% reported seeing objects that they could not explain, even after their best efforts to do so. [47]

On the other hand, Carl Sagan (1934-1996), the famed astronomer and author of the masterpiece *Cosmos*, never endorsed any current UFO cases of the day, even though he wrote about UFOs all the time in a hypothetical sense, as in his 1963 article about direct contact among galactic civilizations by relativistic spaceflight. [48] Sagan thought about the possibility of life on other planets and what aliens would be like probably more than most other humans on the planet. However, he dismissed all UFO cases with this statement of unbelievable arrogance and contempt for the unwashed masses in *Cosmos* on page 292, calling all UFO reports "...the unsubstantiated testimony of one or two self-professed eyewitnesses." He seemed to be unaware of cases confirmed by eye witnesses at several different locations for the same sighting, telescopic, photographic (not all are fakes), TV crew videos, radar, radar-visual, radar-visual-radiation spike, and today FLIR (infrared camera images), or the reports from law enforcement, airline pilots, highly trained military personnel, and aeronautical specialists. He also seemed to be unaware of his fellow astronomers' sightings, or dismissed them as well.

At least Klass knew that as a debunker he would have to try to discredit radar confirmation of sightings. Sagan seems to have been either uninformed or did not want to know about the UFO cases that remained unexplained after thorough investigations. He did not seem to be interested in investigating UFO reports at all, as he seemed interested just in making sweeping statements dismissing them (S. Friedman calls them proclamations), which is almost beyond belief for someone that wrote about hypothetical UFOs all the time and lamented about where the aliens might be. Even Klass knew that some eyewitness accounts were fairly accurate when confirmed by several other witnesses, to the point that he had to come up with the ball-lightning/earth-light UFO hypothesis to explain some of them. Sagan dismissed UFOs in a very flippant and unconcerned manner, showing very little scientific curiosity about something he wrote about all the time. It is beyond rational comprehension except when one fact is considered. Sagan was one of the founders or proponents of SETI, which is the search for extraterrestrial intelligence through the detection of radio signals from space. To in any way admit that there might be something to the UFOs and there might possibly be some evidence suggesting that space aliens might be on the earth already would seriously undercut funding efforts.

Dr. Seth Shostak, who is the lead scientist with SETI today, of course, also has a huge conflict of interest on the subject of UFOs, which have the potential to steal his thunder. [49] UFOs have the potential to interfere with funding efforts for SETI. SETI was considered, at one time and maybe still is in some circles, to be nearly as far out of the scientific mainstream as studying UFOs. All areas of science are competing for funding in order for studies to continue. Far out requests have to struggle mightily to get funded. SETI is quite an exciting undertaking indeed to those in the general public who are interested in such things, but if the ETs are already here, then that undermines the funding efforts. I completely support the efforts of SETI, because to me it is good science.

However, Stanton Friedman totally disagrees with my position, and makes some valid points as to why SETI is a very long shot. There is no love lost between the SETI crowd and the UFO crowd. There are few who would want to straddle both worlds, but I believe there is merit in both. I hope they find a signal soon, but the odds are against something being found in the near term. As they get better and better equipment, and can rapidly process more and more signals, and different types of signals, they could find that ET signal hidden amongst a huge noise background. Shostak is negative about UFOs or the need to study them, but the conflict of interest is apparent. However, if SETI discovers a signal or NASA finds microbial life somewhere, it will be the biggest boon to ufology ever, because many people will realize that we are not alone in the galaxy or the universe, and they may rethink the UFO situation.

At this point I am quite confused about Sagan, who worked closely with NASA.

A number of former NASA personnel have made public statements confirming the existence of UFOs. Cady Coleman informed Mission Control during the shuttle mission STS-73 in1995 that a UFO was present. Maurice Chatelain, Edgar Mitchell, Scott Carpenter, and Gordon Cooper, among others, have made statements that UFOs are real and are probably extraterrestrial in nature. Paul Hill was a world renowned scientist in aerodynamics who worked for NASA and its predecessor for over 30 years. During his time at NASA, he had two different UFO sightings of his own, and served as an unofficial cataloguer of UFO reports received by NASA. Hill wrote a book about UFOs from the viewpoint of an aerodynamics specialist. The book was published posthumously in 1995, five years after his death. In it, he covers the physics of the reported UFO behavior and flight characteristics in a highly technical manner. [50] Richard F. Haines, a former NASA scientist, is now a ufologist, although he prefers to call it studying unidentified aerial phenomena, UAPs, reported by civilian and military pilots. [51] Did Sagan not have any contact with these individuals or hear about the UFOs following our spacecraft? Did he simply follow a confidentiality agreement not to disclose what he knew about classified material? Was he afraid that the general public was not ready for such a disclosure, or was he shielded from the UFO information? Was SETI a way for the general public to get used to the idea that we may not be alone?

I simply cannot believe that Sagan did not know the truth of the situation. A possible explanation comes from Friedman's 2008 book. In it, Freidman presents evidence to suggest that Dr. Donald Menzel, who was a public debunker of UFOs, privately was involved in a

TOP SECRET group known as Majestic-12 or MJ-12, a presidential advisory group that was trying to figure out where the UFOs came from. Of course this is disputed by skeptics and some ufologists, but Friedman found the documents that showed Menzel was secretly involved in several ways with our government. Menzel's secret involvement in a group like MJ-12 is not out of the realm of possibility. Menzel is purported to have speculated that the UFOs came from outside the solar system. Could this actually be true of Menzel? Could this be true of Sagan? Was Sagan possibly a public debunker, but a private insider with top secret information concerning UFOs that he could not disclose?

Dolan and Zabel, in their 2012 book, discussed Sagan and speculated in much the same way that I did above. [52] I wrote my passage several months before reading Dolan and Zabel, but they were the first in print with this speculation that I am aware of. They point out an episode of the NBC-TV series *Dark Skies* that aired for 18 episodes in 1996 and 97 and portrayed a Sagan-like character that had to choose between knowing the truth and not being allowed to publicly say anything about it, or to be excluded from knowing the truth but being free to discuss the issue in public. Zabel was one of the producers of *Dark Skies*, and has speculated about Sagan in this way for quite some time. However, I was unaware of it until I read their book in mid-2012.

Friedman and Sagan both attended the University of Chicago at the same time and had frequent discussions on opposite sides of the issue of UFOs over the years. However, Friedman does not bring up this possibility about Sagan, and there is no evidence to support it. If this is not true in any way, then I would otherwise have to conclude that a brilliant man and scientist examined the area of UFOs in a completely illogical manner and in a completely superficial manner, and made regal proclamations in exquisite prose dismissing UFOs without any serious investigation. I almost cannot bring myself to believe that.

John Alexander, in his 2011 book, points out that Sagan was an elitist that did not think the average Joe on the street should be allowed to discuss various scientific subjects. [53] In my view, he could have very well believed that the masses could not handle the truth about UFOs, or that the masses were such poor observers that their reports were useless, or he could have been under strict security clearance requirements to disclose nothing. He could even have been so determined to see that SETI was funded that he refused to look at the UFO data in any serious way. It may have been all of the above, or maybe I am totally wrong. The reader can see that I am quite conflicted about Sagan, who was probably the best communicator of science we have yet seen on the world stage.

CHAPTER 4

NASA, NORAD, NSA and Radar or the Lack Thereof

"Data is not information, information is not knowledge, knowledge is not understanding, understanding is not wisdom."

—*Clifford Stoll, U.S. astronomer and author*

The astute reader is probably wondering what about NASA (National Aeronautics and Space Administration), NORAD (North American Aerospace Defense Command), NSA (National Security Agency), and all other aerial data gathering groups in the National Security arena. If anybody on earth knows the truth about UFOs, then it would be NASA, NORAD, and NSA. NASA does not formally comment about UFOs, and sends out a reference to the Condon report when someone inquires about UFOs as if nothing had happened in the intervening 45 years. However, NASA seems to regularly bring up to the public the possibility of life elsewhere in the Milky Way galaxy and in the universe. One way was through the Mars meteorite announcement in 1996 that NASA scientists believe is evidence for the existence of microscopic life on Mars at some time in the past. [54] In reality, some personnel within NASA know all about UFOs and know that they are very real, but it is not an openly discussed subject and is officially not talked about. Unofficially, there are a number of astronauts and other former NASA personnel that have spoken publicly confirming their belief that UFOs are real, and others have been overheard in space announcing the presence of a UFO as mentioned earlier.

The average person probably doesn't think about it much, but should realize that we have satellites out in space watching our borders and the movement of everything that flies in our airspace. Reportedly, we were doing this by the late 1960s. In the 1988 book *Out There* by Howard Blum, the number of objects tracked in orbit around the earth was in the thousands. [55] The main theme of the book was aimed at trying to identify unknowns that would enter the airspace over the continental US from time to time. In Alexander's 2011 book, the known objects tracked daily as they orbit the earth was listed as about 9,000, and the known objects tracked at least once a month numbered 20,000. Off the record in Alexander's book, a young NORAD technician indicated that he tracked one or two UFOs a month. Blum's book detailed that unknowns or bogeys were encountered fairly frequently back then as well. The belief is that the extensive security network in place today watching the skies of North America sees UFOs nearly every day. Anything giving a radar return equivalent to a two-square meter metallic object can be tracked, according to Sturrock. [56] That represents an object about five feet by five feet square, and that was in1997. Today's capabilities may be even greater. Yes, NORAD knows more than most about UFOs, but they are not talking and they are exempt from the Freedom of Information Act. Alexander disclosed that higher ups in NORAD discourage UFO reports from the radar technicians. Alexander indicates that it is the same old fear of ridicule and being judged mentally unfit for their jobs that keeps many military personnel from sending UFO reports up through the channels. They soon learn not to report such things if they want to keep their careers on track, depending upon the viewpoint of their superiors. The NORAD technician noted that most times the UFOs were on a non-threatening track. Some jet interceptors might have been dispatched if the track was nearing a sensitive security area. Getting classified information about such encounters can be difficult or next to impossible.

Ground based radar cannot see aircraft flying close to the ground as well as the downward facing satellite radar can. Ground clutter becomes a problem for ground based radar in that case. For a number of decades now, air traffic control radar has had software that removes anything not acting as a normal aircraft so that a UFO can maneuver in and sit directly over a major airport undetected by radar. This seems to be what happened in the 2009 O'Hare Airport case in Chicago. [57] To put the radar data into perspective, consider that civil radar, such as air traffic control, has a primary radar and a secondary surveillance radar (beacon radar) that usually only follows aviation aircraft that have a transponder identifying signal. [58] All normal aircraft, passenger planes, helicopters, and military aircraft have transponders for identification. The use of transponders came about in WWII to determine friend from foe. Civil defense ground based radars attempt to identify everything in the air, whether the craft has a transponder or not, but may not see an aircraft flying below a certain altitude at a certain distance away due to ground clutter. There are ways to minimize ground clutter, but something flying very near tree-top level can be very difficult to distinguish.

The technology associated with radar is quite complicated. There are short-range radars and long-range radars. There are radars that operate at different wavelengths and scan rates with many technical subtleties. Sometimes the weather causes problems for a radar operator. The thing that debunkers bring up every time an unknown target shows up on radar is thermal or temperature inversions in the atmosphere. Temperature inversions involve a layer of warmer air sitting on a layer of cooler air at the earth's surface, and can cause what are known as "radar angels" on the radar scope. These are also known as anomalous propagation signals. These "radar angels" sometimes result from the radar beam being bent down toward the ground by atmospheric or tropospheric ducting, and being reflected by something on the ground. After the uproar in 1952, "radar angels" were studied carefully. A number of other phenomena caused by temperature inversions were identified. It was shown that a number of "radar angels" can be generated by vortices or unstable air pockets associated with temperature inversions. The radar beam is not picking up something on the ground, but is being weakly reflected by pockets of air with high refractive index gradients. This type of "radar angel" moves at the wind velocity, or at double the wind velocity at the altitude in the wind direction, and gives a weaker leading edge signal. Usually these "radar angels" are hazy or shadowy, and can be distinguished from normal radar returns of an aircraft, which give a strong leading edge signal. How can this fool a radar operator who has been trained to recognize it? An unknown target moving with the wind, or at double the velocity of the wind, with a hazy return signal looks nothing like the radar return from a hard surface aircraft moving with much higher velocity. Also, in general, the radar operator can calculate how severe the tropospheric ducting will be to determine if "radar angels" will be a problem.

In the 1950s and 60s, Menzel claimed that radar-visual UFO cases could be explained by temperature inversions with a mirage of the object on the ground showing up where the radar seems to show the object in the air. It was not a valid explanation, because visible light beams are not refracted in the atmosphere to the same extent that radio waves are. It was a completely impossible explanation for radar-visual cases where the object was sighted by observers on the ground and by pilots in the air. It was also an impossible explanation for ground radar and air radar in pilot visual cases.

Klass knew he had to attempt to debunk radar-visual cases, so he employed temperature inversions even when there were none, radar equipment malfunction, and radar operator error to debunk radar cases, but some of his explanations could only be believed by skeptics who forego critical evaluation of a fellow skeptic's explanation. Klass claimed that the famous Washington, DC case of July 1952 was caused by a temperature inversion, and claimed the Air Force said so as well. In the famous July 29, 1952 news conference, Major General John Stamford, Director of Intelligence, and Major General Roger Ramey, Director of Operations, indicated that the incident might have been caused by a temperature inversion, but this was before the incident had even been investigated. In fact, there was a weak temperature inversion, but it was not strong enough to cause the radar signals that were seen. Ruppelt himself did the calculations that showed only a weak effect, and the radar operators said they saw some very weak "radar angels" or anomalous propagation signals that were disregarded. Ruppelt disclosed in his book that one of the Air Force's radar experts, Captain Roy James, would label every case a temperature inversion if no one called him on it. He was with Ruppelt and others in Washington, DC when they were confronted by reporters before any investigation had been done. True to form, he blurted out that it could have been a temperature inversion instead of saying that the case was under investigation. To the press it seemed that the Air Force said that the case was due to a temperature inversion, but after thorough investigation the case went down as unexplained in Project Blue Book. Atmospheric physicist and radar expert Dr. James E. McDonald (1920-1971), who Freidman has called the greatest ufologist of all time, also evaluated the case, and concluded that the anomalous propagation signals were too weak to fool anyone. [59]

So, the press was bombarded with statements about temperature inversions, and Ruppelt even briefed President Truman that it could have been a temperature inversion before the investigation was begun. However, the Air Force never corrected their statements after the investigation officially listed the case as unknown or unexplained. It is clear that the Air Force was not being above board and was trying to mislead the public some six months before the Robertson Panel was convened by the CIA. Sometime during the July 1952 uproar, Lt. Colonel Moncel Monte, an Air Force public information officer let it be known that there was a standing order to shoot down UFOs, if they would not land after being signaled to do so. It became known from Air Force sources that a number of encounters had occurred, but the UFOs had always out maneuvered the Air Force interceptor jets and had flown away before they could be fired upon. Wow! What an admission! Unknown craft that outflew our best jet interceptors are out there! There appear to be hundreds of UFO-jet interceptor incidents whose details may never be known publically. I wonder why skeptics never bring up these incidents?

The Washington DC 1952 incident involved several radar-visual sightings, and the radar images of six or seven objects were sharp and distinct indicating the presence of hard surfaced aircraft of some type flying directly over the Capitol and near the White House. Radar operators can usually sort out the signals and figure out what is going on, but Ruppelt gave a case, also in the Washington, DC area, that fooled the radar operators. An unknown was spotted on radar that gave a strong clear signal and did not appear to be hazy or shadowy. Two jet interceptors were dispatched and they soon approached the target from the radar operator's vantage point. In fact, they were very near the target, but they saw nothing. Circling around several times, they realized that they were over the Potomac River, and on the river was a large tourist boat. They soon realized that the radar operator was seeing the tour boat. This implies that a "radar angel" that results from a vessel on the water may appear sharp compared to something on the ground, because there is no ground clutter associated with the signal. The fact that there was no visible image or mirage to be seen in the air where the anomalous radar signal indicated the object to be was another strike against Menzel's mirage explanation.

Most radar-visual cases occur at night, as do most UFO sightings. Skeptics just chuckle and say this is due to man's inherent fear of the nighttime coupled with our active imaginations, which turn ordinary things seen at night into the extraordinary. But detecting something unidentified on radar and visually seeing something at that same location in the sky is hard to explain no matter when it occurs. Ruppelt purposely pointed out at least two cases of daytime radar-visual cases in his book to illustrate that radar-visual cases occur in the daytime as well. In one case, the radar tower personnel realized that the unknown was close enough to be seen visually. They went outside and observed a dark, cigar-shaped craft flying at the position where the radar indicated the unknown target to be.

After all this talk about radar conformation of UFO sightings, some baffling cases are of apparent UFOs that did not produce a radar return. One of the most famous of these cases occurred in 1997 over Phoenix, AZ with hundreds, and possibly thousands, of witnesses who saw a very large V-shaped something with lights flying silently from Henderson, NV (near Las Vegas) to Phoenix. That is a distance of about 300 miles that was covered in about one hour. The Phoenix Lights case of March 13, 1997, was shown not to be flares that were dropped by the military on a training mission or to be due to the hoaxer which claimed to have released flares attached to helium balloons. Two attempts to reenact the hoax failed. [60] The scope of the phenomena was covered in Dr. Lynne D. Kitei's 2004 book (second edition in 2010) *The Phoenix Lights: A Skeptic's Discovery That We are Not Alone*. The book details that the case did not begin in 1997, but in 1995, and related phenomena was reported to occur periodically over at least the next decade. The photos, videos, and 35 mm film show that the Phoenix Lights emission spectrum did not match in any way that of flares, as analyzed by Jim Dilletoso and confirmed by others. The behavior of the lights did not match flares either. To me, the most startling revelation in the entire book was on page 227, where Dilletoso states that the only spectral match to the lights that they had found as of 2010 was with a 35 mm photograph of the light phenomena seen at Fatima, Portugal on November 13, 1917 (possibly he meant October 13, 1917). If this is true, it is truly earth-shattering evidence that I will discuss later.

Another such case where there was never any mention of radar was the 1951 case of the Lubbock Lights, and there must have been some radar installations within range. To me, this case has a lot in common with the Phoenix Lights. The Lubbock Lights were "explained" by Project Blue Book as migrating birds, but that explanation does not hold up on closer examination of the case. The Lubbock Lights were spotted by some Texas Technical College professors (now Texas Tech U.) about 9:10 p.m. on August 25, 1951, and were photographed later by a student, Carl Hart, Jr., who had heard about the sightings. [61] The professors and friends made some 14 or 15 sightings in all over a period of weeks, and determined the time of flight overhead and angle in the sky, and that there was not any sound. The exact details of time of flight and angle in the sky determined by the professors are seldom mentioned when the case is discussed, but should be center stage. The typical angle to the horizon was 50-60°, and the angular velocity was about 30° of arc per second. Covering 30° of arc in the sky per second means something is either flying very, very fast or very close to the observer. At that angular velocity a flying object spans from horizon to horizon in just six seconds. The professors never saw the blue-white (also referred to as green-white) lights in a V formation, but once saw them in a semicircle formation and mostly in random formations. The professors were unsure of the photographs, because of the V formation shown (some were of a double V sometimes flown by migrating birds on rare occasions). From the series of photographs taken showing 18 lights in a double V formation, Project Blue Book experts determined they were individual lights, as they seemed to move independently of each other in the series of photographs. From the published photographs one can see that seven or so of the lights were about twice the size of the other 11 lights, and the largest light was about four times bigger than the smallest light.

A number of people said what they were seeing were birds reflecting the newly installed, bright mercury vapor street lights, and Project Blue Book listed the case as solved by birds, but several things argue against that conclusion, and one thing completely disproves it. A strong argument against the bird proposal was that a professional photographer tried to replicate the photo with real water fowl reflecting the streetlights and got only very faint images that did not match. Secondly, the professor's reports of the occasions where the lights flew overhead in a random fashion and not in formation conflict with a migratory bird explanation. Thirdly, if it was birds or another natural occurrence, why did the phenomena never show up again in the area year after year?

Some years later, Menzel could not adequately duplicate the photographed lights with his optical effects, and ultimately ended up labeling the photographs a hoax, but, if a hoax, it is one that no one else has been able to duplicate. Whether the photographs are real or not (they appear to be real) has little bearing upon the professors' observations. The professors, all engineers or scientists, soon wanted to wash their hands of the entire affair, as they were hesitant to be linked with unexplained phenomena. A telegram was sent to Ruppelt indicating that they were embarrassed to say that they now believed they had observed a very commonly occurring natural phenomenon, but Ruppelt would not say what it was in the first edition of his book. Perhaps their explanation involved the night flying moths that Ruppelt said solved the case in the second edition of his book. I nearly rolled in the floor laughing when I read that passage. Fifteen or twenty moths flying in a nearly perfect semi-circle?!?! The professors' concerns show that they were human and they were worried about how they would appear to their peers. While I completely understand the situation only too well, their concerns are not germane to the case, but their observations are. The 50-60° elevation and the incredible speed of 30° of arc per second can be used to further test the migrating bird hypothesis and the night flying moth hypothesis.

I found a fact sheet from the Smithsonian National Zoological Park concerning neotropical migratory birds; the type of birds that migrate over the North American continent. [62] The lowest flying migratory birds are some waterfowl that fly as low as 200 feet. The fastest migratory fliers are also some waterfowl that fly at 50 mph. On occasion, some migratory birds may fly a little faster. The professors' observations could be duplicated by waterfowl flying at 65 mph (faster than typical) at an elevation of 50 feet (much lower than typical), but the birds would have been only about 60 feet away from the observers, and would have been easily recognized as birds. There was enough light and the professors would have been close enough to easily see the wings flapping (the same is true for the moths, which can fly up to 35 mph and would have only been 42 feet away). How fast would migratory birds have to be flying if they were flying at a 200-foot altitude (the lowest listed by the Smithsonian) to be consistent with the professors' observations? The answer is about 200 mph. This shows that the migratory bird hypothesis completely fails to explain the case. Project Blue Book technical personnel and the professors should have done the same "what if" evaluation, and should have seen how ridiculous the migratory bird or moth explanation was. Blue Book personnel noted that the professors were unable to determine the altitude of the lights and took that part of the investigation no further. They did not do any "what if" calculations, which would have quickly shown that birds were not the answer or moths either.

Swords and Powell, in their expansive 2012 book *UFOs and Government: A Historical Inquiry*, discussed the Lubbock Lights beginning on page 130. [63] They point out another professor of mathematics who was able, along with his father-in-law, to observe the lights from two different locations for the same flight, and that they were able to time the flight. From this information, Professor Underwood calculated the minimum altitude as 2,000 feet and the minimum speed as 700 mph. The professor wrote a letter to Project Blue Book giving his results, but his letter seems to have been ignored or lost in the shuffle.

In 1977, an analysis by a research group known as The Ground Saucer Watch located in Phoenix, AZ, validated the lights. The group had studied a number of other UFO photographs and dismissed them as clever hoaxes, but Hart's photographs passed their scrutiny of computer-aided tests. [64] They said the objects were large in size and a long way from the camera. That was my impression of the lights from just looking at the pictures online. It clearly was not due to migrating birds, contrary to what Project Blue Book concluded. Curiously, most of the accounts that I have seen about the Lubbock Lights do not mention radar, or the lack thereof, in the case directly, but there was speculation by Ruppelt about a radar hit of an object some distance away from Lubbock traveling at 900 mph that could have given rise to the Lubbock Lights and a somewhat similar case in Albuquerque, NM on the same night involving a large, delta wing type of craft with multiple lights. The Albuquerque witnesses thought the Hart photographs from Lubbock looked very much like what they saw. The timing and direction would have been correct for this to have occurred, and the Albuquerque lights/object also did not show on local radar. There seems to have been very similar, multiple light phenomena in the southwest region of the US some 46 years apart (Lubbock Aug. 25, 1951, Albuquerque Aug. 25, 1951, and Phoenix 1997). The phenomena seem not to appear very often, but when it does, it typically does not show on radar.

Another case with some similarities to the Phoenix Lights was the Stephenville, Texas case that occurred on January 8, 2008. Witnesses reported seeing something enormous in the sky passing overhead, but this time there were radar unknowns—several of them. Sometime later, Seth Shostak of SETI was on a segment of *Larry King Live* dealing with the case. He was opposing the view of some UFO researchers.[65] Shostak dismissed the extensive radar evidence of unknown targets with a statement that said of course there was radar data, because there were planes in the area at the time. This implies that the radar operator could not tell the difference between a plane, which has an identifying transponder signal, and an unknown target. He did finally say, "Let us review your radar evidence and see what it indicates," but not without a lot of verbal jousting. Arguing on national TV may be good for ratings, but it is not the essence of science. It was a mostly emotional dismissive response, not a scientific one. By the way, the radar data showed the 10 fighter jets sent to intercept the unknowns and the unknowns. The radar returns of the unknowns indicated them to be hard surface targets. One unknown was tracked over, or very near, then President Bush's Crawford, Texas ranch. [66]

During my search of the internet for UFO information related to my family's sightings, I discovered some supporting evidence concerning radar. I found a website dealing with the worldwide UFO flap of the 1970s. [67] The year 1973 had the most UFOs reported worldwide of any year to date. Something else quickly caught my eye. The website reported that the civil defense radar site in Columbia, Mississippi was one of the five or six most active radar sites in the world for radar hits of unknown targets during the 1970s. These unknown targets, of course, could have been UFOs. Columbia, MS is some 70 miles south of my family's farm. Now, I do not know what the capabilities were at the Columbia radar site (at what altitude a target would have to be to be seen at 70 miles), but some of our sightings ventured up high enough to be seen. I have not researched the date, time, or location of these unknown hits to see if there is even stronger evidence to support our story. In any event, with the level of activity that we saw there should have been a lot of radar hits of unknown targets in the 1970s at the Columbia site, and there were.

I was unaware of several of the major UFO cases in Mississippi until after I started writing this book and had begun a search of the prior literature. Searching the prior literature was something I always did in my research career, so I thought I had better at least find out if someone else had already written our story or something similar. I did not find another story of ongoing family sightings in Mississippi, but I did find two of the most well-known UFO reports in Mississippi history: the Flora case in 1977, and the Taylorsville case, also in 1977, that strongly tie in and support our story. Taylorsville is in the southeastern part of Smith County, so there was already some UFO activity reported in Smith County, Mississippi during the 1970's. I was totally unaware of these cases until August of 2011.

In the following chapters I will tell my family's story. It is a story a little different from the usual single UFO report. It is a story of ongoing sightings that span more than a decade, and the total number of sightings may be upwards of 40 in all. Not all sightings that I am presenting were in Smith County, Mississippi, but most were within a six-mile radius of my family's home. The vast majority of sightings were along a one-mile strip of country road. On several occasions, my paternal grandmother told us of an incident that happened to her and her children in the mid-1930s. The story was verified by my father and his brothers and sister (my uncles and aunt). They are all deceased now, and I am in my sixties. I am writing this to clear my conscience by ending my silence in this matter and telling the world what we have seen. I also wanted to preserve these observations for my daughters and for anyone else who is interested.

In discussing arguably the most famous UFO case in Mississippi history, the Pascagoula abduction case of October 11, 1973, Klass made some comments that were critical of Charles Hickson and Calvin Parker, because they waited two entire hours before trying to report what had happened to them. The absurdity in Klass's criticism is that this was one of the most promptly reported UFO abduction cases in history. What sort of comments would Klass make, if he were still with us, about someone waiting nearly 40 years to make their sightings public? Would I get the Klass "treatment," or be totally ignored? Oh well, better late than never.

Chapter 5

Central Mississippi Location of Smith County

"Why is it that UFOs are mostly reported by someone such as a farmer out in the boonies?"

—Ron LeLeux, Mayor, Sulphur, Louisiana 2000-2008 and Louisiana author

Before discussing the first unusual sighting (actually in the 1930s) that occurred in this area of rural Mississippi, let us put into perspective the location of Smith County, Mississippi, and the family farm. The map shows the location of Smith County relative to Jackson, the state capital, and the major highways crossing the state. Smith County lies about 34 miles east and a little south of Jackson. The county line lies some six miles or so south of Interstate Highway 20 (I-20), and the main state highways cutting through the northern part of the county are MS-13 and MS-35. The family farm site is barely within the county in the extreme northwest corner near the village of Polkville. To the west is Rankin County and to the north is Scott County. Polkville students were allowed to cross county lines at that time, so we actually attended school for a number of years in Morton in Scott County, which was closer than the other alternatives.

The second map shows a zoomed-in view of the family farm in the northwest corner of Smith County near Polkville at the time, but today it is within the village limits. The farm lies down Bradshaw Road. According to family lore, the Bradshaw family, my maternal grandmother's family, settled in the area in the second half of the1800s after discovering gold in California, and had become quite prominent in the area by 1900. Also according to family lore, the governor of Mississippi stayed with the family on one occasion when a flood caused a reroute in his travel plans. There are a number of Bradshaws who are my second and third cousins, and so on, currently living on this road today. There are also several more houses on the road now than in the 1970s.

My uncle, Arlon Palmer, built a house in 1961 across the road from our house, and we lived in the house off and on for a number of years, as will be discussed a little later. Just down Bradshaw Road and over the hill on White Road is my grandparents' (the Penningtons—Roy and Lottie) former place some 550 yards by the twists and turns of the road from our house. This, then, is the unlikely layout for an incredible sequence of events that happened primarily from 1970 into the mid-1980s. However, there were some sightings of unexplained phenomena before the 1970s, and I learned during the course of writing this book of some recent sightings that occurred in 2011 and 2012. The next two chapters will discuss the earlier sightings before 1970. The first of these occurred in the 1930s, and was quite mystifying for those members of my family who were involved.

Smith County Mississippi
Germantown
Southaven
Holly Springs
Senatobia
New Albany
Oxford
Tupelo
Batesville
Clarksdale
Water Valley
Brinkley
Marianna
Helena
Forrest City
Cleveland
Grenada
Aberdeen
Monticello
Indianola
Greenwood
Starkville
Columbus
Lake Village
Greenville
Belzoni
Louisville
Kosciusko
Yazoo City
Philadelphia
Canton
Ross Barnett Res
Forest
Meridian
Vicksburg
Jackson
Smith Co
MISSISSIPPI
MS-13
MS-35
Polkville
Smith Co.
MS-28
Taylorsville
Hazlehurst
Magee
Collins
Laurel
Brookhaven
Monticello
Hattiesburg
McComb
Columbia
Bogalusa
Prichard
Picayune
Gulfport
Biloxi
Pascagoula
Baton Rouge
Slidell
Courtesy of Univ. of Texas Libraries-Public Domain Maps

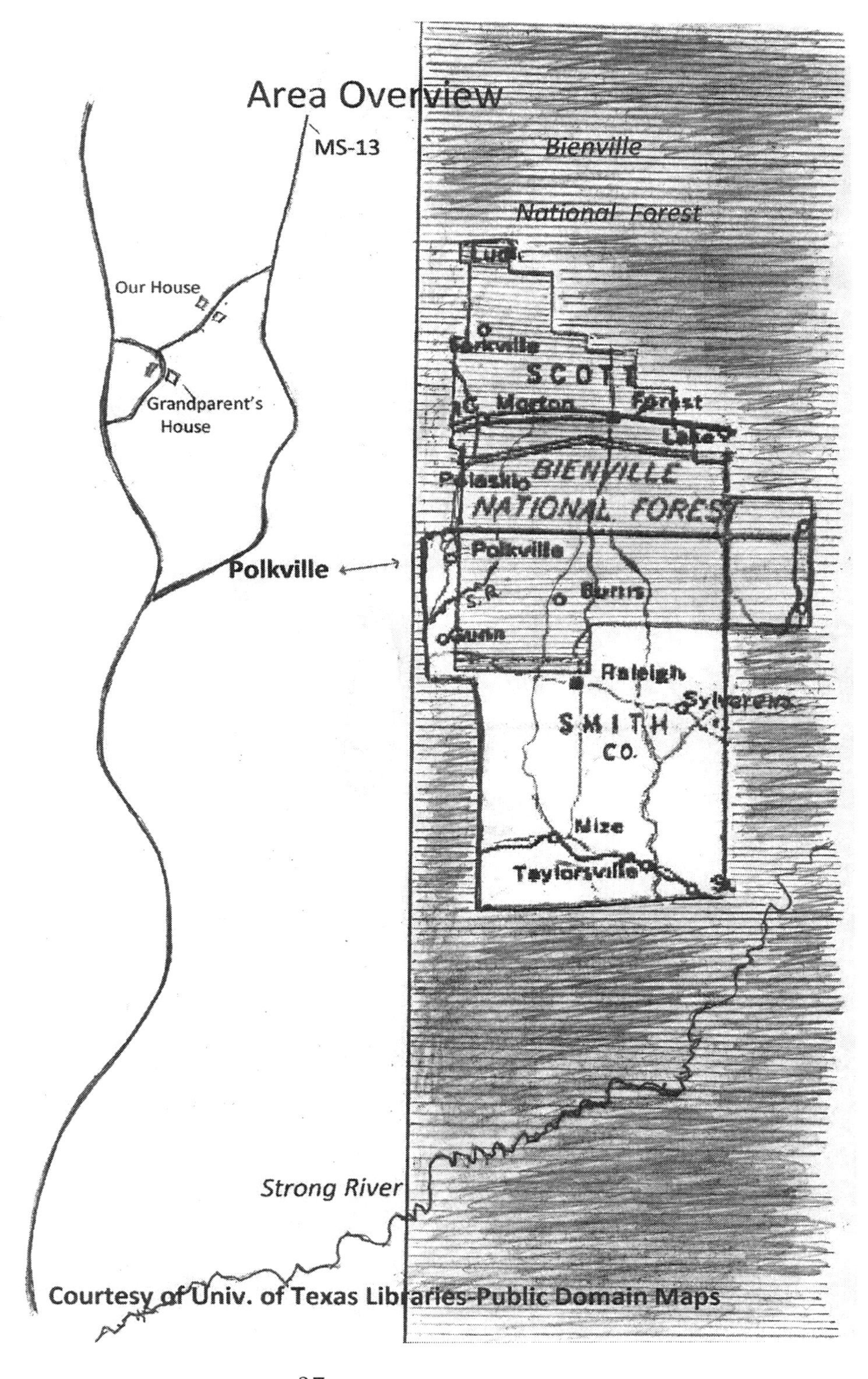

Courtesy of Univ. of Texas Libraries-Public Domain Maps

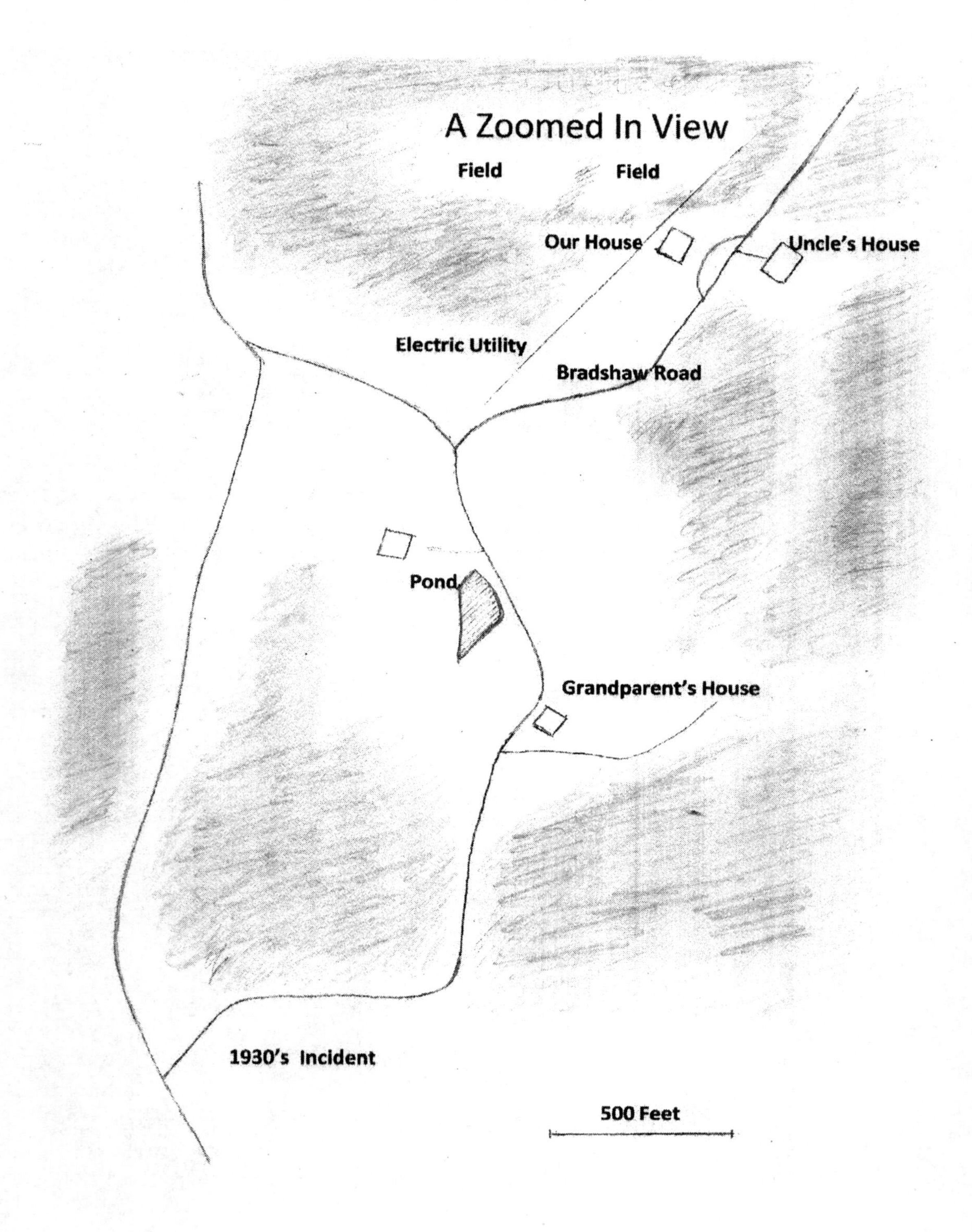
A Zoomed In View
Field
Field
Our House
Uncle's House
Electric Utility
Bradshaw Road
Pond
Grandparent's House
1930's Incident
500 Feet

CHAPTER 6

A Spotlight from the Sky

"The most beautiful thing we can experience is the mysterious. It is the source of all true art and all science. He to whom this emotion is a stranger, who can no longer pause to wonder and stand rapt in awe, is as good as dead: his eyes are closed."

—*Albert Einstein*

Carl Sagan, in *Cosmos* on page 303, told of an account of the first encounter of a Native Alaskan group with French explorers in 1786. [68] The Native Alaskan group was the Tlingit tribe that had no written language. The story was passed down by word of mouth for about 100 years before it was told by Cowee, a chief of the Tlingit, to a Canadian anthropologist. The French had a written account of the meeting and interaction, which overall remained friendly and ended on good terms. The account of the Tlingit was basically quite accurate, although they called the large French ship a giant canoe with white wings. The story shows how a very strange encounter with only eyewitness accounts can remain basically accurate even after having been passed down by word of mouth over several generations.

In much the same way, my grandmother, Lottie Pennington, would tell her grandchildren stories of true events that had happened in her life to pass the time, to entertain us, and to possibly pass some significant events down through the family history. She was a deeply religious person and frowned upon anyone misrepresenting anything. She told us, for example, about the time when she was about 12 years old and someone had hidden in the barn waiting for her to come home from school. Her parents had to be somewhere else that day, or were working in a field far away from the house. I do not remember that detail exactly. When she got home and was about to go into the house, she saw the barn door move and open to a slight crack, and then close quickly. In that moment of terrible realization, she knew someone was there watching. She instantly bolted for the road and began running as fast as she could for the neighbor's house about half mile away. The next thing she heard besides her own feet hitting the ground was the sound of someone else's feet hitting the ground running behind her. Closer and closer the sound of running steps came as she was literally running for her life for all she knew. Then when she had just about given up hope as the sound of running steps got closer and closer, and seemed to be right behind her by only a few feet, the neighbor's house came into view. The sound of the running steps behind her started to slowly fade and then stopped. She never looked back, and ran on to the neighbor's house to safety.

She also told us about pranksters that would come around people's houses at night out in the country when the man of the house was away. They would hit the house with a chunk of wood or a stick, first one side and then another. On one occasion, she opened the front door slightly and held the door as my father, then a boy, fired a shotgun into the night in order to scare the pranksters away. They remained terrified all night long anyway.

There were several other stories that I do not remember very well after all these years, but there is one story I remember very well, because it was so incredible then and it is still incredible today. Here it is.

On a farm in rural Mississippi in the mid-1930s, there was a period of time, maybe in early July, when the crops were laid by. That meant that the crops had come up, had been worked with a hoe to get rid of most of the weeds, and any last application of fertilizer had been made. There was less work to be done for a while until harvesting began, and people visited neighbors a little more than during the busier earlier summer. This was during the Great Depression, and many people in the area did not have automobiles. Telephone service, once available in the area before the depression, was no longer available. If you wanted to talk to someone, you had to go to their house or they had to come to yours. Most people walked or rode a horse to go to a neighbor's house, usually not more than a mile or two away. Most families had only one or two horses, and usually had wagons for harvesting in the fields, but most did not have a nice buggy to travel in. A few had cars, but most did not. Most just walked for 30 minutes to an hour to a neighbor's house, visited for an hour or two, and walked back home.

As previously indicated, my grandparents' house was located about 550 yards down the road from my parents' house shown in the figure, except that my parents' house had not been built at that time. One summer night my grandmother decided to visit her sister, who lived about a mile and a half away. My father, who would have been a teenager, his two brothers (one older and one younger), and his sister (the youngest), went with my grandmother on the visit. They had eaten their evening meal early, around 5 or 5:30 p.m., and walked to her sister's house for a visit, getting there sometime around 6 to 6:30 p.m. They visited until after dark, which would have been a little before eight in the days before daylight savings time.

After their visit, they began their walk home at around 8:30 p.m., expecting to be home by nine, or shortly thereafter. It was dark as they walked along when they noticed up ahead at some distance—maybe more than a quarter mile away—that there seemed to be some sort of light in the road or near the road. The road made some twists and turns, and came to a straight section where they saw that the entire width of the road was lit up.

As they approached the light, they realized that there was a shaft of light some 30 or 40 feet in diameter coming down from the sky and illuminating the entire width of the road and ditches on either side. They cautiously approached closer and closer to the shaft of light. It was eerily quiet. They looked up to try to see the source of the light, but could not make out any details—just a shaft of light as far up as they could see. They were very hesitant to walk through the light, so they started looking for a way around the lighted area. They found a barbed-wire fence with a thicket was on one side and a briar patch was on the other side. There was no way around the light without a lot of effort. They noticed that the shaft of light was not wandering around or moving in any way. The light was steady and was lighting up the same area of the road with no drift or sway. I questioned my father about this on several occasions. He said the light was steady with no movement, just as if a mounted spotlight was being shined down onto the ground. They were in a state of confusion and disbelief. Still not wanting to go through the light, they would put a hand or foot into the light and draw it back. My father told me that inside the lighted area it was as bright as daylight.

Finally, after great hesitation and seeing no way around it in the dark, they hurried through the light. They seemed to be okay as they came out on the other side. They walked on, but turned and looked at the shaft of light every few seconds for a while. One of them looked back after they had gone a little less than a quarter mile and the light just blinked out and was gone, as if someone had flipped a light switch. There was no electricity in their area at that time—electricity came sometime after World War II.

They were totally mystified by what could have produced the light. My grandmother simply told the story to show how strange things can happen in this world, things with no explanation then or now. We all listened, spellbound by the story, and wondered what could possibly produce a light like that in the middle of rural Mississippi in the 1930s. Reproducing the shaft of light incident today would require mounting a 30-foot diameter spotlight to a tower built over the road and connecting it to an electrical source, something that would take quite a bit of time and trouble. A hovering helicopter with a spotlight might be able to produce a reasonable shaft of light, but not that large of a diameter, and the noise and wind would ruin the effect. Although the first helicopters were being developed at that time, such helicopters were not around in the mid-1930s. It might be possible for a dirigible with a huge, battery powered spotlight to have been tied off over the road securely enough to shine the spotlight down without a noticeable swaying motion, but who would go to such trouble to do this as a prank, even if it were possible to do, for an audience of a limited number of people. Tying the rigging would have been tedious, and the tied-off rigging would have been visible, as would the huge size of the dirigible airship.

This incident is not without historical precedent. As I was watching a History Channel program on UFOs earlier in 2011, they had a program on that was a repeat and had aired a few years earlier. [69] It was called "Russia's Roswell," and the main part of the program was about an air battle between a Soviet fighter plane and a UFO. At some point in that program, they talked about a case in Russia in 1892 where a UFO with a large spotlight had hovered in a Russian town shining the beam on the ground. The spotlight was seen by people on the

ground, but the UFO was seen clearly by people in the hills surrounding the town as it hovered and shined the spotlight down.

UFOs are not some new phenomena that occurred after World War II, as many people seem to think. My literature search found that UFOs had been reported at least as far back as 2,340 years ago with none other than Alexander the Great himself, and have been reported occasionally down through the intervening centuries in secular history. [70] They go back even further, if you consider religious documents such as the Bible and ancient texts from India and elsewhere. Some historical researchers, notably Thomas E. Bullard and Jerome Clark, say the documentation of UFOs further back than 300 years is poor, according to the website *www.hyper.net/ufo/summary.html*, and I would agree. [71] A UFO report would not make it into any documented form unless witnessed by a person of some importance or rank, so the reporting rate would be many times lower than even the low modern rate.

Other cases of UFOs with powerful spotlights were reported over Bridgewater, Massachusetts in 1908, in 1910 over Huntsville, Alabama and Chattanooga, Tennessee, and in1913 over Milwaukee and Sheboygan, Wisconsin, according to the primary source "A Century of UFOs," *UFO Roundup*, Vol. 4, No. 36, December, 1999, found on the website www.*zetatalk.com/theworld/tworx134* . [72] My grandmother had never heard of such things in any case. Such reports usually make the local papers and are talked about for a while in a local area, but are promptly forgotten by most people in a few years.

Based upon the communications of the times, my belief is that there were probably many more cases of a spotlight shining down from the sky along rural roads throughout the country during the early part of the twentieth century that were never reported. Most would have been talked about among family and friends, but probably never even reported to a local paper. The sightings just become family stories about strange events.

This incident that was so shocking to my father's family is completely unexplainable as a natural phenomenon or manmade technology. It shows on its own, without any other reports, that there is something unknown out there, something that is unknown to mainstream science, something that is an unexplained phenomenon. Perhaps from having heard this story from a fairly early age, I was more open to the possibility of there being evidence of unknown, unexplained things out there in this world.

Chapter 7

What Was the Red Light?

"Even a minor event in the life of a child is an event of that child's world and thus a world event."

—Gaston Bachelard (1884-1962), French philosopher and poet.

This sighting goes back to 1963, and may or may not be related to the later sightings in the 1970s and 80s. As stated earlier, my uncle had built a new house across the road from our house. He built it to retire in someday, but for the present he was working in New Orleans. He allowed us to move into the house for the time being. Our house faced east and his west (see the map layout in the satellite view preceding this chapter). There were fields about 200 feet to the north of our house that we had grown cotton in at one time, and later corn. Sometimes various other crops were grown as well. The trees and bushes beside and surrounding the field on the house side were smaller and much more sparse than they are today. There were taller trees and woods to the west and north of the field.

In the 1960s in the rural South, people spent more time outside when the weather was nice than they seem to today, especially in the early evening after dark. Our family would frequently take a stroll at this time of day around the circular driveway or up the road a little way, or down the road and then make our way back to the house. We sometimes sat on the porch and talked a little while. We were outside a lot at night in the evening compared with most people today, who are inside in an air conditioned environment.

When we were living in our house and started outside for a walk, we were facing east toward the pasture and woods behind my uncle's house. In none of our strolls out and about that originated from our house would you be looking straight at the fields that were near the house for any amount of time. This is because the fields were off to the side, and the lower field was offset back and not in line with the house. When we moved to my uncle's house, our nightly strolls began by walking down the 90-foot or so driveway that looked out over the fields. Looking straight ahead, the fields would be in full view, especially the lower field, until we turned north or south at the road.

On one of these strolls in 1963, not long after we had moved across the road into my uncle's house, Clark, my brother, spotted a red light hovering out over the lower field. He was seven years old at the time. I was 15 and my sister, June, was 16. My parents, my sister, and myself were occupied discussing something else and we said something to the effect that it was just an airplane, so don't worry about it. We brushed it off as nothing unusual, hardly even looked, and went on talking. (Maybe we were in training to be editors of scientific journals). Coming back from our walk, the field would be to our side and to our backs, and we would have had to purposefully turn around to look back at the field. None of us turned around to look, but Clark did. He saw that the red light was still hovering over the field, but he did not say anything since we had so thoroughly ignored him the first time. The sight terrified him, but he still kept quiet. It terrified him so badly that he had nightmares about it later, and it remains one of his most vivid childhood memories. Many years later, I learned in March of 2013 while reading Project Blue Book unexplained case reports, that red lights were frequently associated with unexplained UFO reports from Mississippi in the early 1960s.

Even a seven-year-old child realizes that a hovering light is a very unusual sight, and one that hovers in the same spot for upwards of ten or fifteen minutes is very puzzling and can be terrifying, especially to a child. I don't know why we ignored him so completely, since red could mean that something was on fire. I think someone quickly glanced, saw that it looked like the red light on an airplane, and did not think any more about it. Although a plane that close to us should have made some noise, but it did not. We were just not expecting to see anything unusual, and were preoccupied discussing something else, so it did not register with us, but it did with him. Even though UFOs had been in the news since the late 1940s, we just did not pay much attention to it. After all, what are the odds that we would ever see anything out of the ordinary in our quite corner of the world?

Klass made a big deal about people going out to look for UFOs once someone had reported them in a given area. He implied that most people would let their imaginations run wild and report ordinary celestial objects, such as the moon or Venus, as UFOs. Klass believed that people would be seeing UFOs everywhere when in fact there were none to be seen. It's true that some people might allow themselves to get carried away, but most down-to-earth, honest, level-headed people will not. In fact, in the community where I grew up, most people only talked with family or good friends about the unusual things that they had seen. I have got to believe that this is generally true in many areas of the country. This fits in with the fact that most UFO encounters go unreported. Why? Most people do not want to be ridiculed about what they've seen or experienced, or told they did not see what they actually did see.

One thing I discovered during my research was that there is a disconnect concerning the percent of UFO sightings that are reported. The number 10 percent comes up frequently as the percentage of sightings that are reported, but this is much, much too high based upon the percent of the US population who claim to have seen a UFO in their lifetimes—about 5%. I found a website that indicated there had been 150,000 documented UFO reports in the entire world over the last 65 years. [73] If 50,000 of those were made in the US, then we can make a reasonable estimate of the percentage of sightings that are reported and formally documented in this country. Over the past 65 years, I estimate that 480 million individuals have

Red Light

lived in the US. I also found that the average number of people that witness the same UFO sighting is about 2.6 per sighting, or about the same as witness a bank robbery. [74] Based upon these numbers, the total number of sighting events would be about 480,000,000 x 0.05/2.6 or about 9.2 million sightings over 65 years. The percentage of reported sightings would be (50,000/9,200,000) x 100 or 0.54% rather than 10%, or about 19 times lower than commonly assumed. Another way to say it would be that if only one quarter of one percent of the population actually have seen a UFO, or what they thought was a UFO, then the reporting rate would be about 10 percent. Using the often quoted figure that 95% of all reports can be explained, and with a total of 50,000 reports, then that would indicate a grand total of 2,500 unexplained UFO reports for the US over the last 65 years. While the reporting rate is quite small, the total number of unexplained cases is fairly large, because the accumulation time has been about 65 years. Of course, the point can be made that the unexplained cases really average 20% or more (Battelle Institute -*Project Blue Book Special Report Number 14*) and that the total unexplained US cases may be as high as 10,000 or more after 65 years.

How many people have totally ignored a sighting of something unusual as we did on that night in 1963? It probably numbers in the millions. We were completely oblivious, because in our mind set at the time such things only happened somewhere else to someone else and not to us.

CHAPTER 8

The Endless Looping Light

"Extraterrestrial intelligence will be elegant, complex, internally consistent and utterly alien."

—*Carl Sagan in Cosmos, P. 296*

In June 1970, I had just graduated from college and was getting ready to get married and later to enter the University of Southern Mississippi graduate school working on a higher degree in chemistry. My family had moved to Morton about 12 miles north on MS Highway 13 during my undergraduate years, but had just moved back to my uncle's house temporarily while our house across the road was being renovated. My uncle was working as a civilian contractor in Vietnam at the time. Our house was almost ready to move back into, and we made frequent trips back and forth between the houses.

One night about 9:15 p.m., my father, mother, Clark and I were strolling to our house down my uncle's driveway. My sister was not with us on this walk, as she had gotten married a few years before and moved to Pearl, MS near Jackson. About halfway down we looked up to see what was apparently an "airplane" at high altitude flying from west to east almost directly over our location. The "plane" was not directly overhead, which would be at a 90° angle to the horizon, but was near an elevation angle of 80° to 85° to the horizon. You had to look up at a pretty steep angle, but not directly overhead. The significance of the angle is that the spot on the ground directly underneath the "plane" could be a quarter mile to two miles north of our location, depending upon the plane's altitude. The "plane" was really just a steady white light. No blinking lights could be seen. Normally, a plane has a red blinking light on its left side and a green blinking light on the right side. There are white lights as well for collision avoidance and landing. Sometimes with an airplane at a high altitude of 25,000 to 35,000 feet, it might be difficult to tell the color of the lights. Usually one can see some indication of the blinking lights, but there was none on this occasion. The light was nearly as small as the background stars, but about two or three times brighter and traveling from west to east. The light appeared brighter than the light seen for most planes at great altitude, which is not much brighter than the background stars. The light was not a celestial body. It was moving. How anyone under any circumstances can think that Venus or Jupiter are moving is completely beyond my comprehension. I know that there is the so-called auto-kinetic effect that skeptics love to evoke [75], but stare as I might on a dark night I have never experienced it myself when looking at stars in the nighttime sky. I think there were too many visual clues around, such as tall trees, utility poles, etc., to act as visual points of reference.

So on our way down to the other house, we saw what we took to be a plane at a high altitude going in an easterly direction. Viewing time was a glance upward for two or three seconds. We hardly stopped walking to glance up at it. No big deal. On our way back to my uncle's house about 15 or 20 minutes later, we saw what appeared to be the exact same "airplane" going in the same direction in just about the same location in the sky. Somebody said it was weird to see two airplanes in the nearly exact position on our trip down and back. At that point, I stopped and started watching the "plane" more closely just out of curiosity. Everyone else continued on to the house and went inside. There was a tall pine tree beside the driveway to act as a point of reference. It soon became apparent that the light was not an airplane. Now up to this point, I had not thought anything about the fact that there was no noise. A plane that was apparently that far away cannot be heard over background noise, but there was little background noise that night out in the country. Probably atmospheric conditions have to be just right to hear a plane at that apparent altitude, although both Clark and myself have heard planes at fairly high altitudes before out there, just a faint sound that you have to listen very carefully to hear.

It soon became very clear that the light was not an airplane when it stopped dead still as it approached another light of the same magnitude that was hovering motionless in the sky. I had not paid much attention the other light before, as it was not moving. The two lights seemed to touch one another briefly. This seemed much too close, for what I realized now might be helicopters, to approach one another. Then the light made an ever so slight move to the side and began flying in the opposite direction without any obvious turning maneuver. It was reversing the path we had seen before. It was now flying in a westerly direction. It was then that I noticed the third light. The third light was hovering motionless to the west. The flying light continued until it stopped right beside the third light. In a kind of mirror image instant replay, they seemed to touch. Again, the light made a very slight move to the side and began to fly to the eastern hovering light without any obvious turning maneuver. It only seemed to stay next to the hovering lights a few seconds, the elapsed time might have been five or six seconds or so. The flight back and forth between the two hovering lights took no more than 30 seconds one way. When I stretched my arm with pointed finger and followed the path of the light, the distance traced out was about three feet, more or less. That is about three feet at seven feet and four inches elevation (the end of my pointed finger). This would allow me to do various "what if" calculations, but I suppose the correct way to express the pathway the light took is to say it traveled an approximate 25-30° arc in the overhead sky out of the 180° arc from horizon to horizon.

Now I believe that after we saw the light the first time as we walked down to the other house, the light had continued to travel back and forth for the entire 15 to 20 minutes we were down there. I do not know how long the light continued to fly in this way. I watched two complete loops and the start of a third. After that I went into the house and got ready for bed, as I had a physically demanding summer job that required me to get up by 6 a.m. In hindsight, I wish that I had watched the looping light all night, if need be, to see what eventually happened. I still cannot find a reasonable explanation of why any conventional aircraft would maneuver in this manner, or that any can actually maneuver in this way. Again, I repeat that there was no noise. If this were three normal helicopters carrying out this maneuver, we should have heard some noise, a lot of noise.

I really did not know what to think of the behavior of the light and just left it as puzzling until one day I sat down and did some calculations. The maximum altitude for a helicopter is about 18,000 feet, which is far lower than the maximum altitude for commercial jets at about 40,000 ft. From my observations, I can calculate the speed of the flying light and the distance between the two stationary lights, if I assume they were at an altitude of 14,000 feet, a little below the max altitude. To get the distance between the stationary lights, simply take the three-foot distance my finger traced and multiply it by 14,000 feet divided by 7.33 feet. The answer is 5,730 feet, or about 1.08 mile, and since this distance was covered in 30 seconds max, then the average apparent ground speed would average about 130 mph. However, the trip started at a speed of zero mph and ended at zero mph. An average of 130 mph could be achieved by starting at zero and ending at 260 mph, but it ended at zero mph. This means that for the light to be a helicopter at 14,000 feet, it could average 130 mph by accelerating to 260 mph at the halfway point and then decelerating from the halfway point to zero as it came to the opposite stationary light. This is just beyond the calculated speed capability of a helicopter type of craft, which is about 250 mph, although it is well within the capability of tilt-rotor aircraft (helicopter-airplane hybrids) that were developed sometime later in the 1980s. The fixed rotor helicopter speed record was set in 1986 by the Westland Lynx in the UK at about 249 mph. [76]

The light did not appear to accelerate and decelerate noticeably. It appeared to travel at a steady speed from one stationary light to the other stationary light, and to stop abruptly. The light appeared to attain a cruising speed at about a second from disengaging from the hovering lights, and to maintain this steady speed until a second before stopping beside the hovering lights. The light performed in a way that appeared to be beyond the capability of any known helicopter of the time, but let us examine that closer.

The skeptic might rightly say that I had assumed the lights were higher in the sky than they actually were; that if I chose an altitude of 3,500 feet, then the helicopter only has to accelerate to 65 mph at the mid-point of its trip, and that would be an easy cruising speed for most helicopters. My response to that is a 3,500 foot altitude with a viewing angle of 80° would place the lights' location about 1/8 mile north of my location and 3,500 feet up. Three normal helicopters that close should have created lots of noise, and there should have been

blinking lights to be seen as required by the Federal Aviation Administration, FAA, but special operation helicopters may not have these lights.

An engineer friend of mine in the 1980s mentioned, upon hearing the details of some of our sightings, that possibly we saw some of the noise reduced helicopters that the military has. He indicated that lower noise helicopters had been developed and had been used in Vietnam in the 1970s. A normal helicopter generates about 100 decibels (db) of sound at a distance of 100 feet. While noise suppression is quite complicated, let us use the estimate of 80 db at 100 feet as the noise level of a noise dampened helicopter. This represents a rather large difference in sound levels in actual operation. The equations below can be used to estimate the relationship of sound intensity levels in db and distance.

Dis2/Dis1 = SQRT(10exp((db1-db2)/10))

Dis2 = Dis1*SQRT(10exp((db1-db2)/10))

By rearranging the equation, we get the equation for calculating the sound intensity db level at a second distance if we know the dbs at the first distance.

db2 = db1- 10*log($(Dis2/Dis1)^2$)

For the sound intensity level of a normal helicopter to fall to 80 db, an observer has to be about 1,000 feet away, or 10 times further than for a noise dampened helicopter, which is also known as a "silent" helicopter. The "silent" helicopter's sound level would fall to about 50 db at about two-thirds of a mile away. A normal helicopter would need to be over six miles away for the sound to fall to 50 db according to the calculations. A sound level of 50 db can be heard if one is in a very quiet place, but not otherwise. Of course, the human ear only hears sounds in the frequency range of about 20 to 20,000 Hz, and moving the frequency of sound outside this range will result in a noise that is silent to human ears. Humans generally perceive higher frequencies to be louder than lower frequencies at the same decibel level, but lower frequency sounds tend to travel further than high frequency sounds.

As stated earlier, my engineer friend suggested that some of our sightings could have been due to the flyby of "silent" helicopters. However, most of our sightings were fairly up close and personal, and it can be shown in most cases that the light or object approached to within a few hundred feet of the observer where even the "silent" helicopters can be readily heard.

How do these sound levels relate to this night in the summer of 1970? The night was very quiet with a completely clear sky, and the ambient sound level was approaching 40 db. A silent helicopter would have been just detectable up to about one mile away under normal observing conditions. Three "silent" helicopters without proper navigation lights that were hovering and flying at 7,000 feet, ¼ mile north of my location, could appear to be noiseless.

A problem comes up with the acceleration rate of helicopters from a hovering maneuver. The actual acceleration rate varies with airspeed in a complicated manner, but tends to average about 6-8 mph per second each second, or 6-8 mph/s^2 from zero to maximum airspeed, usually about 180-220 mph. At 7,000 feet altitude, the helicopter would need to average 65 mph to cover just over a half mile in 30 seconds, but has to be at zero mph at the end of the 30 seconds. It can do this by accelerating to 130 mph at the mid-way point and then decelerating to zero at the end. This would average 65 mph. The problem is that at the halfway point in time, the helicopter would only be at about100 mph and would need to keep accelerating to complete the course. It would need to accelerate for at least 22 seconds, reaching about 150 mph, and then decelerate to a complete stop in the remaining eight seconds. A helicopter making this maneuver would appear quite jerky, and would be unable to completely stop and hover only eight seconds after reaching 150 mph. The helicopter would have to make a rapid deceleration maneuver by pitching up and then arcing down. The flying light was observed to be rock steady as it traversed from hovering light to hovering light. Also, the close in maneuvers for a helicopter would look considerably different from what was observed, and would involve a turning maneuver. In addition, if we were to assume that it was a military training exercise of some type, the question arises as to why a military maneuver of this type would be made over a civilian area far from a military base? While the area has a low population density and the Bienville National Forest is only a mile or so away, the maneuver was at least partly over a civilian area.

My perception was that the lights were very high in the sky at perhaps 35,000 feet or higher. The nearly instant attainment of cruising speed, the nearly instant stopping, the extremely close approach involving the apparent touching or docking of the flying craft and the hovering craft, and the lack of a slowing or a turning maneuver argue against even a silent helicopter explanation. An airplane was ruled out long ago. Also, the skyhooks and whatever other balloons that can be out there cannot traverse back and forth across the sky in that manner. We are left with no convincing conventional explanation for so simple a sighting. The question is why would any craft from any source be making this continuous looping maneuver? The behavior itself seems to defy logic. I can only come up with two possibilities: pilot training or recreation. A military training exercise set up over a civilian area is extremely unlikely. Over flight of civilian areas will be necessary to get from one military base to another, but setting up a training exercise over a civilian area is against regulations.

I can just hear an official government explanation that I witnessed an inflight refueling maneuver of some sort, but that would be another ridiculous explanation. No planes or helicopters would approach each other so closely repeatedly. Once they had maneuvered into position, they would stay there until the refueling was completed. If refueling, the two aircraft would have been seen once and, depending upon altitude and the viewing location, would have been gone in 30 seconds to a minute or so. On the other hand, maybe I witnessed Mars and Venus in a mating ritual, which makes about as much sense as a lot of the "official" explanations do to someone familiar with the facts. My question is, can anyone explain so simple a sighting?

Note that if the light was at 35,000 feet, then an average speed of 350 mph would be needed to travel between the two stationary lights, and for any type of conventional aircraft would require a mid-point speed of 700 mph with deceleration to zero. That would break the sound barrier at this altitude and would be pulling a continuous 2 Gs for the transit back and forth. The apparent behavior of the object was that after about one second it was at cruising speed, which would give a G force of almost 16 Gs, or 8 Gs if spread over two seconds. This would be survivable perhaps, but would be quite an unpleasant experience for human pilots to repeatedly perform, as they can withstand 9 Gs for only short periods of time. The main component of the G forces was in the horizontal direction, so it would not be as likely to cause someone to blackout as a vertical force would.

Stanton Friedman has pointed out that the instant accelerations reported for a number of UFOs may not be as detrimental to humans or living beings as commonly believed. He offers some convincing evidence based on rocket sled testing and a rocket sled accident that humans can withstand some very high G forces for a very brief period of time. [38] Humans seem able to survive on the order of 42 Gs for 0.75 of a second, 30 Gs for one second, or 10 Gs for 3 seconds. Friedman's point was that a flying craft doing some of these reported high G force accelerations does not necessarily have to be flown by robots or remote control.

As an aside, about a week or 10 days after sighting the looping light, I had a dream about a flying saucer landing in the front yard of my uncle's house. It was clearly a dream and we all had a good laugh about it the next day when I related it to my family. The dream seems to have come from my subconscious mind telling me that a possible explanation for what I saw was a UFO. Up until that point, I had not considered seeing a UFO as even a remote possibility. In my way of thinking at the time, seeing a UFO was too improbable an occurrence to consider. This dream was akin to a Woody Allen type of absurd dream sequence.

Chapter 9

Something Was Watching Over Us

"If an advanced civilization were to arrive in our solar system, there would be nothing we could do about it. Their science and technology would be far beyond ours."

—*Carl Sagan in Cosmos*

For a long time, we thought that the incident described in this chapter was the first incident that was definitely the phenomena we came to call "The Light," but it was not, as we found out later. The incident in this chapter happened in November of 1972. The exact date of this incident is lost to the fog of time, but the incident very probably happened around November 23 or 24, 1972 (the Thursday and Friday of the Thanksgiving holiday). Having time off from work and school after the typical start of hunting season and the meteorological records all point to those dates. Buford Jones (my sister's husband) and Clark were out hunting in the woods to the west of the lower field. They had stayed until nearly dark and had gotten lost on their way out of the woods. It was after nightfall and they were still trying to find their way out of the woods. The night was clear and crisp with little wind. Finally, they saw car lights on a nearby road and knew the direction to take to get back to our house. They had entered the lower field and had gotten to a place where a pond would later be dug and appears on the satellite view today. They were not going particularly fast in the dark and they were tired. For some reason, Clark had an uneasy feeling. He sensed that something was not right and looked up above their heads. He instinctively grabbed Buford's shoulder and said, "What the heck is that?" Buford looked up at that instant. They both got the shock of their lives! There above them was something huge blocking out the starry sky, and it appeared to be close, very close. It had a large, bright white light on the bottom of it that blocked getting a clear view of the top part of the object, just as the details of a car cannot be seen clearly on a dark night when one is looking into the headlights. The uneasy feeling Clark had earlier may have been that the ground and surroundings were much brighter than they should have been.

At that instant, Clark raised his gun and pointed it at the object out of a defensive reflex to the startling situation. Buford said, "Boy, don't shoot at that thing. Don't even point your gun at it. Are you crazy!?" Several thoughts flashed into Buford's mind at that instant. One was that something like that might have superior fire power, and also there was the very practical aspect that the bullets could bounce off and hit them. Either way they were distracted for a few seconds, and the object, which had been hovering silently above them, began to move silently and in a few seconds moved to a position about 450-500 feet away near the road, indicating a speed of 40-50 mph while moving. The object hovered motionlessly at the second position at an estimated 75 to 100 feet above the ground. The top of the object was hard to make out once it had moved to the second location, as the bright light on the bottom made it difficult to see any details of the top, which appeared as just a large outline. Clark and Buford's description of the bottom of the object was consistent. Both said that most of the bottom was giving off light. They described an area maybe 15 or 20 feet across that was not perfectly rounded, but more oblong where the light seemed to be coming out of it. When seen from a distance, the light did not look as big as from directly underneath. It was believed that the bottom came down to a cone, and the very bottom of the cone was what was seen at a distance. From directly underneath at an estimated elevation of 75 to 100 feet, the lighted part appeared quite large.

The diagram shows the object's path that night in November of 1972. The object stopped abruptly near the road and hovered motionlessly there for several seconds before the light turned from brilliant white to red. The red light then blinked out a few seconds later. The object was difficult to see at that point in the darkness, as the moon was not out. (Moonrise on Nov. 23 was about 8:00 p.m. and about 9:00 p.m. on Nov. 24. Due to the tree line along the horizon, the moon would not be noticeable until at least another hour later). Clark had been totally mesmerized by the object as he wondered what in the world could they be seeing. Buford said a little before that, "If that light goes out, I am getting out of here." Clark found himself coming back to his senses and realized that Buford was running to the house as fast as he could go. He later said, "There I was, a frightened 16-year-old, and Buford never said anything when he actually did take off. He just took off faster than I thought he could move." Buford, 28 at the time, later said, "Yeah that was faster than I thought I could move, too." Other recollections from the moment were that the upper part of the object was 30-50 feet across with a generally round shape. Clark later said, "I could have hit it with a rock, it was so close." At its second position, the object was only several hundred feet away, and yet the upper structure the light was attached to was difficult to see due to the bright white light and a dark night. The inability to see the top of the object when the bright light was on was a constant for a number of the later sightings.

Startled and badly frightened, both ran on to the house in a state of disbelief. At the house, everyone was already anxious because Buford and Clark were getting back so late, but now both were excitedly trying to tell what had just happened to them. Buford was badly panicked and wanted to call the Jackson airport (28 miles away to the west and slightly north) or the Smith County Sheriff to report what had happened. My father said, "No! We do not want to do that. They'll think we are crazy! What you saw was something to do with the Air Force or military. You didn't see anything weird." Philip J. Klass could not have said it better. At first we thought my father's reaction to the incident was completely understandable. We thought at the time that he had been keeping up with military capabilities, and that he just figured Buford and Clark had seen some kind of training flight. However, training flights over civilian areas are strictly forbidden. He was a veteran of the Army Air Force in WWII before it was later split off from the Army to become the Air Force in late 1947. He would have been well aware that a training flight or the testing of new aircraft was seldom conducted over civilian areas. He had this reaction to the incident even though he and his family had quite the unexplainable sighting in the mid-1930s. One could say that the 1930s incident was of a somewhat different nature, but was it really? To us, he did not seem to connect this current incident and the spotlight incident decades before, even though both involved a bright light shining down upon several people and whatever was behind the light was not really visible. He was adamant that no report of the incident was to be made to any type of official agency. We found out later that there was a little bit more to the story concerning my father than what he was letting on. There was a lot more actually. Both my father and my mother would gradually change their points of view as the seemingly incredible events continued to unfold over the years. When one's view of reality is being challenged right before their eyes, most of us do not handle it that well.

The debunkers' retort to this sighting would probably go something like this: What Buford and Clark saw, in their view, was a large weather balloon, perhaps something like a skyhook balloon, with a small leak. The "balloon" with the right sized leak would have descended to near the ground to a point of neutral buoyancy instead of rising up and bursting in the upper atmosphere. It just happened to descend to a point where it seemed to stop directly above them. The bright light was the moonlight reflecting on the metallic sensors trailing underneath the balloon (even though the moon was not out), or possibly it was equipped with a light. The wind blew it from its first position to its second position (even though there was little wind). The appearance of the light turning red was due to balloon fabric getting wrapped around the sensors or around the onboard light. This is the way debunking goes, and it requires a good imagination. Sometimes the debunkers are right, but not in this case.

Weather balloons do not fly faster than the wind, which this object did, as the night was almost devoid of wind, and the brightness and size of the light could not be due to a reflection or a lighted balloon. If this were the only sighting, then the debunkers might be able to hold their ground however concocted their position might be. No remains of a weather balloon or a much larger skyhook balloon were ever found in the nearby area. If a skyhook balloon were down that low already, then it should have gotten entangled in some of the taller trees that lay in the path ahead. Most weather/skyhook balloons rise up until they burst in the upper atmosphere, and their remains then fall to earth like any other object falling from thousands of feet in the sky.

DoN't shoot at That THing, Boy!
What THE!!!

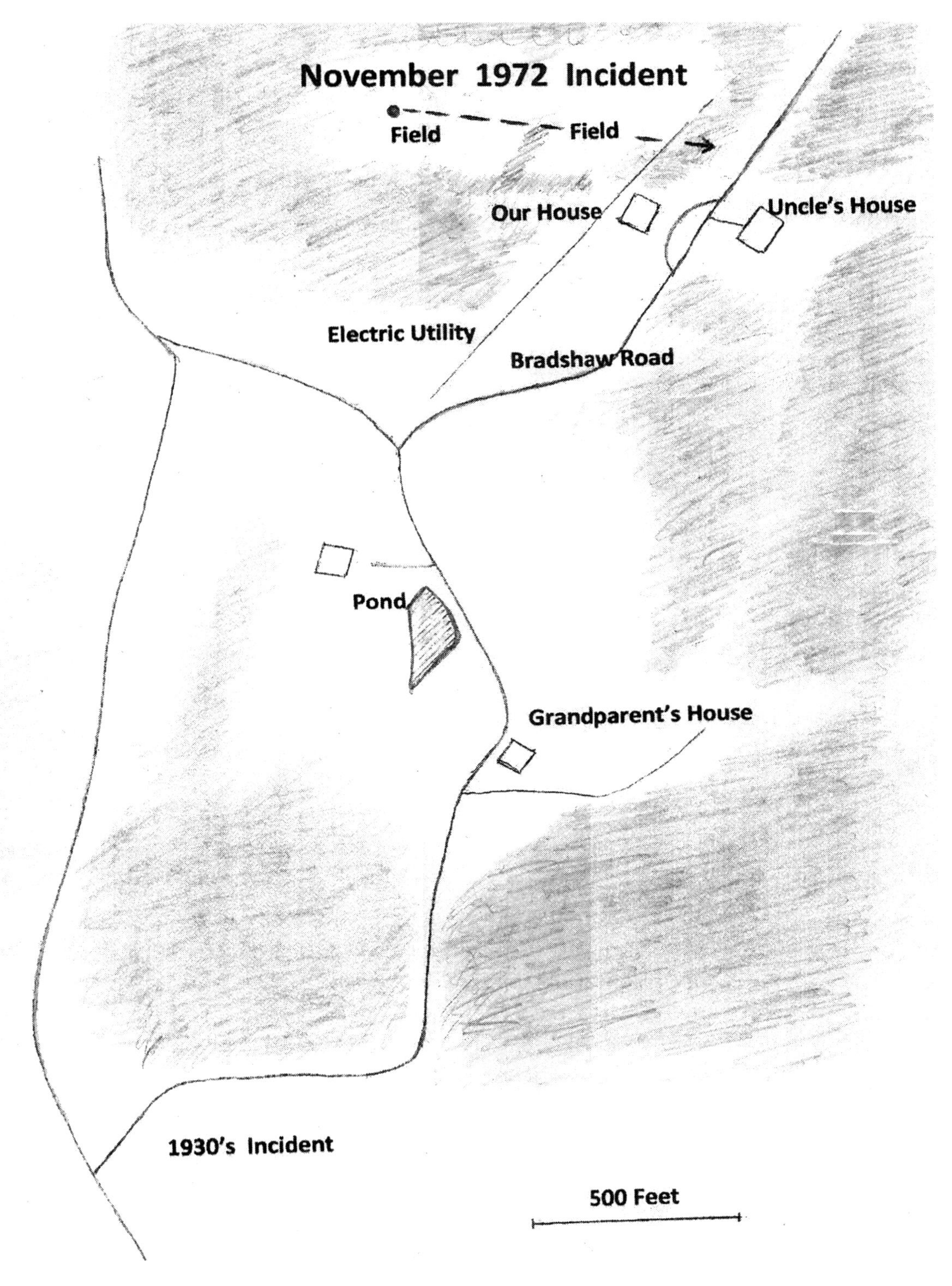
November 1972 Incident
Field
Field
Our House
Uncle's House
Electric Utility
Bradshaw Road
Pond
Grandparent's House
1930's Incident
500 Feet

Another favorite of debunkers are Chinese lanterns. Chinese lanterns are small hot air balloons and have been responsible for many UFO reports. I helped make and launch Chinese lanterns on several occasions when I was an undergraduate at East Central Junior College in Decatur, MS. It is clear from his book that John Alexander would admonish me greatly as a hoaxer, and for engaging in something that could start a fire and be quite dangerous to others. We were in it for the excitement and not thinking about the danger of starting a brush fire. (In my initial draft of the chapter, I told how we built them, but I am leaving this out so as not to encourage such stunts.) A Chinese lantern is held up until the hot air gives it lift, and then it is off just like a hot air balloon. It then drifts with the wind. To an observer they are an erratic and strange light in the nighttime sky. The light flickers with the flames, so a flickering light traveling at about the speed and direction of the wind are dead giveaways for Chinese lanterns, but they do appear quite unusual in the nighttime sky. The object hovering over Buford and Clark was not a Chinese lantern. The upper part of the object would have been illuminated by the flame, as light usually appears to be coming out of the top or sides of a Chinese lantern. The actual object flew faster than the wind, as it was a very still night. The light on the object was an intense white like a bright fluorescent light with sometimes a very faint bluish cast to it. The light itself seen from underneath appeared to be as large as maybe 15 or 20 feet in diameter and very bright.

In most of UFO history, a particular witness usually only has one or two encounters in their entire life. We had no way of knowing that our sightings were just beginning. After each particular encounter early on we did not think about or expect to ever see anything unusual again. The persistence and the improbable nature of the subsequent events never cease to amaze me. Clearly, there was something systematic and regular going on, and I will speculate on that in later chapters.

Both the ufologist and the skeptic become suspicious when someone says they have had numerous UFO sightings, but what are you supposed to do? Look away and say, I better not look because I will lose my credibility if I see more than one or two of these things? That is not going to happen. If some unusual object or light comes flying by, you are going to watch it if it happens only on one occasion or two dozen separate occasions. In Ruppelt's book no one suggested the personnel at several top-secret laboratories had a repeater problem. UFO sightings were so frequent at some government laboratories and other installations that scientists and technicians started a sky watch club to unofficially keep track of them. UFOs were seen on a number of separate occasions flying over the government facilities dealing with nuclear materials in New Mexico and Oak Ridge, Tennessee. Contrary to the Mogul balloon and skyhook balloon hypothesis of skeptics, what the labs were seeing was not balloons. Ruppelt specifically mentioned that his team had access to all the secret balloon launch records, and ruled balloon launches out as the source of the UFOs flying over the nuclear labs.

Chapter 10

The Maneuvering Light

"To myself I am only a child playing on the beach, while vast oceans of truth lie undiscovered before me."

—*Sir Isaac Newton (1642-1727), physicist, mathematician, astronomer, natural philosopher, alchemist, and theologian*

This is the only sighting where I recorded the date, time, etc., as it was my first direct sighting of what we later came to call "The Light," and it made a very dramatic impact upon me. I never expected to see something a second time, but I was to have four sightings in all. I feel somewhat remiss about not having exact dates for all the other sightings, but I never thought that I would see anything similar again, or be writing a book about these incidents or reporting them to the world.

It was Sunday, March 18, 1973 at about 7:35 p.m. It was a few weeks before daylight savings time was to take effect, so it had been dark about an hour. (Sunset had been about 6:15 with the end of twilight around 6:40, meaning that it was dark by 6:40 p.m.). I was putting our suitcases and whatever else into the trunk of our car. My wife and I had been visiting our parents over the weekend, which is something we did every second or third week. I was in graduate school at the University of Southern Mississippi in Hattiesburg at the time and we got away when we could. The drive to Hattiesburg took slightly less than two hours, so we would leave by eight and be back just before ten.

In our everyday lives, most people notice and comment about the weather a lot, as it affects every aspect of human existence from what activities we can do to our mental attitudes. When I stepped outside with the suitcases, I noticed how still the night was with no perceived wind. Unusual for March, I thought, which gets a lot of windy days. Later, a check of the meteorological conditions recorded at the Jackson airport showed an average wind speed for that date as 4.3 mph with a maximum gust of 8.9 mph.

When I closed the car trunk I was facing south, since the car was facing south, but I turned toward my uncle's house rather than toward my parents' house. If I had turned toward my parents' house, which would have been the natural way to turn with my back to my uncle's house, I may not have seen anything at all that night. As I turned, a large bright white light at a low altitude was moving from right to left across my field of view. A few seconds before, a view of the light would have been blocked by my uncle's house and trees. The light was huge compared to any light I had seen before flying in the sky at night. The best estimate I can give is that it appeared as big as a one-foot diameter street light seen at a distance of 200 feet. A one-foot diameter light seen at 100 feet appears to be approximately the size of the moon, so the light appeared to be about one-half the diameter of the moon. It was many more times larger than the planet Venus, and brighter. The light seemed to be just above the tree tops and on a path at an angle to the north. It was totally silent and moving quite slow for an object in powered flight that appeared to be so close to me. My estimate at the time was 30-50 mph, which is near the stall speed of most private prop airplanes. It was moving toward the upper field, and I saw it pass behind a tree in the foreground at about the 35 foot level. Within about 10 seconds or slightly less, it was out over the upper field where it just stopped still in flight. There was no slight sway as you might see in a helicopter—nothing—just instant dead still hovering silently over the field. I stared at the hovering light for a few seconds, maybe six, maybe eight seconds, and then I hurried into the house to get everyone to come outside to see this light. It may have taken 20 or 30 seconds to get everyone outside, but I really do not know for sure. Five of us (my parents, my brother Clark, my wife Esther, and I) were outside looking at this large white light that was not blinking, just hovering motionless. The object was at a low angle to the horizon, as we only had to look straight ahead without raising our gaze much at all to see the light over the field. After a short time with all of us watching, say 10 seconds or so, it started to move to apparently retrace its path. A curious thing was that the light gave no indication that the object was turning around. The light did not waver or temporarily dim or blink, or give some indication of turning around as you might see for a helicopter turning around to go in the opposite direction. Only if the light were located at the bottom center or top center of the object, and was symmetrical in all directions, could an object turn without any change in the appearance of the light. However, it is hard to believe that any flying machine can turn in that manner without a slight wobble, dip, or wavering move. We all got a feeling of weirdness about the light's appearance and the way it moved. By describing these details it is hoped that you, the reader, can begin to understand just how strange the light appeared at times.

As it began to move, I heard nothing, but my mother said she heard or felt a very faint humming or buzzing sound. The light proceeded to more or less retrace its path, heading now from left to right in our field of view at about the same speed as it had before. The object had been about due north of us as it was hovering. As it retraced its path, it was moving in a southeast direction to our right. We all stared in silence, as within about six to eight seconds it had passed silently behind my uncle's house due east of our position. The object had covered half the angular distance from horizon to horizon in 10 seconds or less. That is 90 degrees of the 180 degrees from horizon to horizon from directly north of our position to directly east of our position in 10 seconds or less. This meant that the object was fairly close to us just as it had appeared to be (within a certain range of uncertainty). A faraway object to the north would have had to have an extremely large light and be able to accelerate at a tremendous rate to give the same impression.

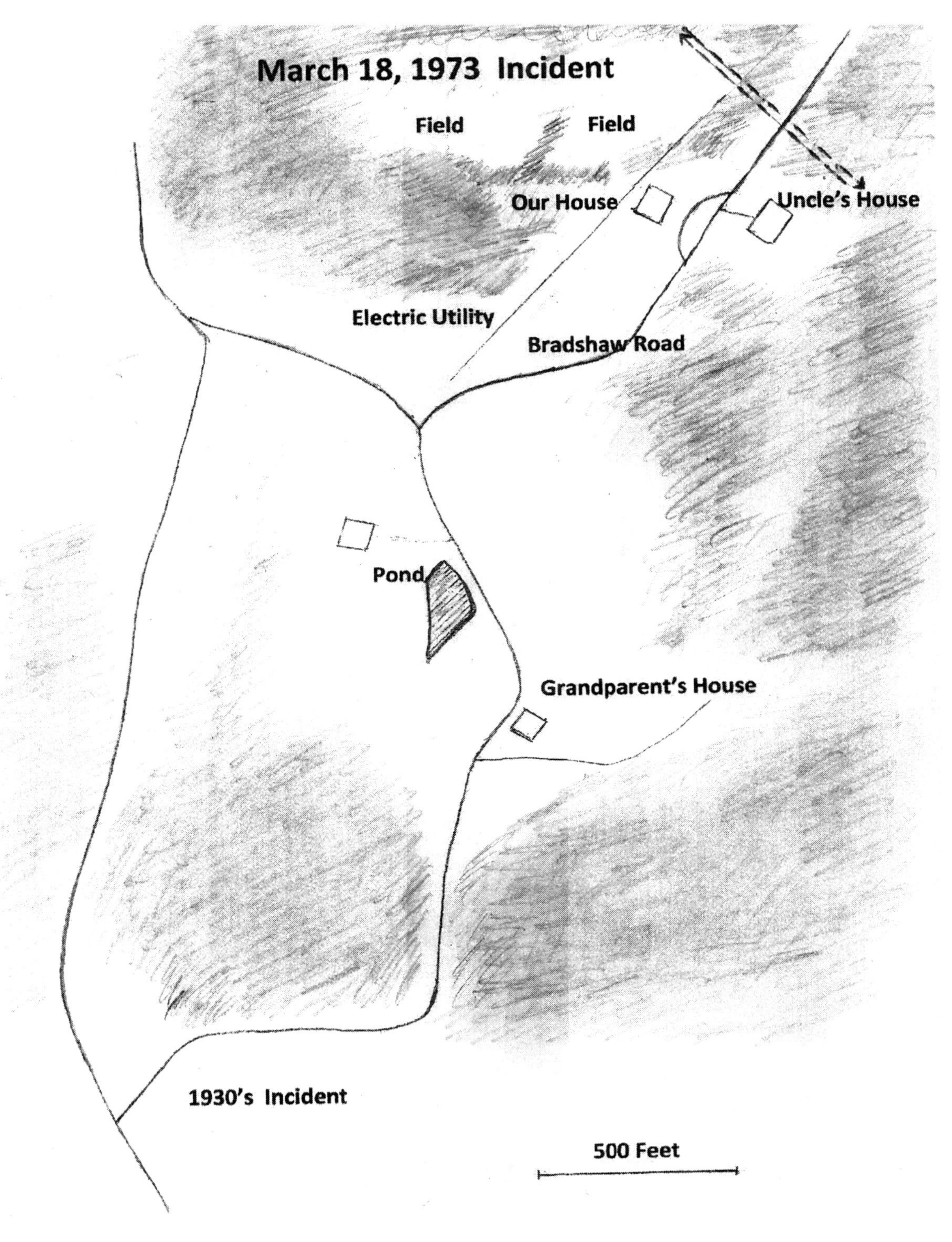
March 18, 1973 Incident
Field
Field
Our House
Uncle's House
Electric Utility
Bradshaw Road
Pond
Grandparent's House
1930's Incident
500 Feet

Powerline
March 18, 1973

This is one of the sightings for which my engineer friend suggested the possibility of a "silent" helicopter and frankly, that was a very reasonable explanation to put forward. Calculations based on the acceleration performance limitations for helicopters (about 8 mph/s^2 average acceleration up to 100 mph) reveals that for the light to have been any type of helicopter, even today, it could not have been more than 600 feet away, and a "silent" helicopter would have produced at least 64 db of noise at that distance, which would have been easily heard. Suppose it was a top-secret "silent" helicopter capable of twice the acceleration rate. In that case it could have been as far away as 1,200 feet. The sound level would have been about 58 db, which could have been heard, as no one was saying anything and it was a still night. We were simply staring in silence at something quite unusual. If the light had actually been one mile away, then the light would have been 20–30 feet in diameter and able to accelerate from a hovering position to 1,080 mph in 10 seconds. The object would have broken the sound barrier and pulled about 5 Gs all the while. This acceleration rate from hovering is completely impossible for a man-made helicopter of any type, and a Harrier Jet could not pull this off either, as its top speed is limited to less than 800 mph. The object also did not display the hovering and turning maneuvers required of any known helicopter type of aircraft. The key point here is that to evoke a top-secret military aircraft as the explanation, it would have needed to have performance capabilities that have not been attained even today, some 40 years later, for any type of known aircraft.

Fortunately, there are enough landmarks in this sighting to make a decent estimate of the object's flight path, speed, direction and altitude. The distance to the tree that the object went behind at an apparent height of approximately 35 feet was measured to be 180 feet. From this, the angle to the horizon can be calculated to be 9°-11°. I also know from this that the object could not have been any closer at any time than 200 feet. In crossing the road, power line, and tallest trees around, the object could not, at any time, have been lower in altitude than about 60 feet. The approximate path the object took that night can be determined. This estimate is given in the diagram following this chapter. Notice that the upper field where the light hovered was where the lighted object in the November 1972 sighting was seen just four months before.

My estimates are that the light was closest to me when I first saw it at a distance of 250-350 feet away. The object moved a minimum of 420 feet in 8-10 seconds for a speed range of 29-35 mph (remember the maximum wind speed recorded that day was about 9 mph). When to our north, my estimate places the object at 460-600 feet away at an altitude of 90 to 120 feet. The actual size of the light itself is estimated to be 2-3 feet in diameter. As mentioned earlier, if the light were further away than I calculated, then the speed, altitude, and size of the light would all be greater. The direction of the object's first path was SE to NW, and the second path was NW to SE. I did not realize until I looked at satellite maps that the road by our house is angled somewhat from NE to SW so that anything crossing that section of road squared up with the road is on nearly a NW-SE track, and this object's track appeared angled to the north when coming from an easterly direction, so the track was very near a true NW-SE track.

Now Klass railed about people saying a light or object seemed to interact with them when he believed that it was natural phenomena or an airplane and could not be interacting with ground observers. But I have got to confess, this light at times gave the feeling that it was interacting with us, such as the November 1972 sighting. Again on that night in March 1973, it would have been easy to imagine that it saw me and stopped to take a closer look, and then, once I brought a crowd outside, it decided to leave. That is the feeling I got at the time, but that is purely speculation.

Clearly this sighting was of a lighted object that appeared to be under intelligent control and flying faster than the fastest wind speed on that date by at least a factor of three or four. Also it reversed direction and retraced its original path after hovering for a minute or so. All of this shows that it was not any sort of wind-blown object or ball-lightning. The real key to this sighting is that it is extremely improbable that natural phenomena or wind-driven objects could appear to retrace the same flight path. It is also impossible for such objects to be flying faster than the wind speed. There was nothing about its behavior that indicated a randomly drifting object or ball of energy. This appeared to be a real object flying around silently with either a pilot or under remote control. Several of the aspects of the object's movements cannot be replicated by any known aircraft, even today—being able to go forward or in reverse without any sway or wobble, and showing no indication of turning around and, of course, flying silently at close range and being able to hover completely motionless. To explain this sighting, one has to come up with something that can fly from a dead still hovering position at least 600 feet north of an observer to a position at least 300 feet east of an observer in no more than 10 seconds time, and do it noiselessly or with only a faint humming noise.

If Klass could not debunk a sighting with the eyewitness's stated information, he would ignore certain details of a sighting until he could fit it into a conventional explanation and, if that were not possible, he attacked the credibility of the witnesses or called them outright hoaxers. He could not conceive that sometimes the incident could be due to something that is not natural or manmade, something unexplainable.

My father's reaction to this sighting seemed to me to be one of bewilderment more than anything else. He was cautious at this point, but he knew he had witnessed something strange. I did not understand until later that my father was wrestling with integrating these sightings into his world view of reality, and that there was already much more to the story.

Two of my other sightings of "The Light" were over or near the upper field. Clark was with me both times, and both times we were out in the summer between nine and eleven looking for the light. Now, this in itself is incredible. These sightings were after 1974 in the mid-1970s when I was home maybe three times a year. We went out looking for "The Light" and, although we had to wait an hour or more, "The Light" eventually showed up. Maybe that would not be a surprise to Klass. He said if people went outside, then they would see something, although nearly every time he told them they had seen the planet Venus or Jupiter, or ball-lightning or earth-lights, or a meteor or an airplane coming in for a landing, and many times he would be correct.

However, sometimes the official or skeptical explanations of certain sightings get so twisted that they strike me as humorous. This leads me to my first UFO joke based upon the Groucho Marx joke, "Last night I shot an elephant wearing my pajamas." Pause. "How he got into my pajamas, I will never know." [77] My joke goes like this, "Last night I saw a UFO wearing my pajamas. The experts tell me it was the planet Venus." Pause. "How the planet Venus got into my pajamas, I will never know." Okay, Okay, Jay Leno will not be calling anytime soon. The joke makes about as much sense as the typical official or skeptical explanation does for many sightings. However, a lot of people actually believe these explanations no matter how ludicrous they are.

Skeptics and debunkers latch onto any explanation no matter how much the explanation deviates from the facts of a case, because of their beliefs concerning UFOs and unexplained phenomena. This is completely normal human behavior. We all look for explanations that justify our beliefs, whether these explanations are correct or not. I have always been a little bit different in that I try to find out the facts before forming an opinion. I try to avoid any belief-driven conclusions in matters of science. I try to avoid belief or disbelief by trying to assign probabilities to certain things, and trying not to allow myself to have absolute acceptance or rejection of an idea or a concept without thorough investigation. It is not an easy way to view a world that pressures you toward absolute acceptance or rejection of certain things, even though these things are many times too complicated to view in a simplistic way.

Returning to our sightings on a summer night when we had been outside from about 9:00 p.m. until about 10:30 p.m., Clark and I were about ready to go inside. We both had powerful flashlights. We were going to try to shine the flashlights on the object, if it came along, to see more clearly what was behind the light. We were standing in the driveway slowly scanning the sky in all directions. I was facing north, the direction of the upper field, when I caught a glimpse of a light a little to my left through the trees and underbrush in the foreground. We darted to the entrance to the field in time to see "The Light," large and bright, moving at the same steady pace we usually saw. It was just below the upper field flying very low at that point, and appeared to be flying over the far edge of the field in no hurry. It was a large white light slightly smaller than half the size of the moon and very bright. We decided to try to intercept it as it crossed the road rather than trying to run through the field that was planted with crops at the time. We turned around and ran the 200 feet or so back to the driveway and headed up the road. After going maybe 100 feet, we were just in time to see it crossing the road. Measuring the distances later, it was easy to see that "The Light" had moved about 700-900 feet in the time that we had moved 300 feet running at 3/4 speed. We ran up the road to a point maybe another 300 feet up the road where we viewed the light directly from behind as it was flying steadily at an estimated 50 feet above the treetops about one third of a mile away. We misjudged our interception maneuver rather badly. Again, the entire incident occurred in total silence. There was no sound whatsoever, even though we clearly approached to within 400 feet of the light. The speed of the light was calculated to be in the 30-50 mph range as it crossed the road about 130 yards beyond the entrance to our driveway.

"The Light" was seen flying over the upper field on a number of occasions over the years. One night my father was out at the edge of the field and the ground lit up. He looked up to see a large white light on its slow and steady pace crossing the field. He just watched it as it silently crossed in nearly the same location we had seen before. We got the idea that it flew over the fields quite often. It was the same silent, bright white light not blinking not varying in any way, and appearing to be the largest thing in the sky, sometimes appearing nearly as large as the moon. It became a staple of conversation as the question was often asked, "Have you seen 'The Light' lately?"

We found out later that my father had seen "The Light" crossing the fields as early as 1970. He had been the first to see it, in fact. We do not know how many times he had seen it before we had our sightings. What we do know is that he was struggling mightily with what "The Light" could be. He was hoping to put his sightings aside, but our sightings brought him back into the middle of it. At first, and for a number of years, he could not accept that there could be an unknown phenomena going on. Later, he would say to Buford and Clark that even if it were what they thought it to be, no one would believe us if we told them. I was not home more than two or three times a year, but I did notice that he did not want to talk about it, and only reluctantly could I get him to talk about what he had seen during the middle 1970s. My father had accepted, just as Klass and mainstream science, that UFOs were just a perception problem, but eventually you have to come to grips with what your lying eyes are seeing. Hardly anyone's perception could be so distorted as to get the essential elements of the sightings wrong.

On another of my visits, Clark and I were out looking again, and again it appeared. We were ready for it, or so we thought. Clark had some binoculars in the hope that we could get a better view. However, this time it flew out over the woods to the north of the upper field and, before Clark could get his binoculars focused, it appeared to land in the woods and disappear for a minute or two. Finally, I spotted it shining between some trees. It could be seen through the vegetation clearly, appearing to have landed in the woods. It stayed for a few minutes and took off. We never got a good look with the binoculars. The next day, Clark looked around in the woods for the spot where it appeared to land, but could not find anything. Upon exploring the area, he did not see any clearing that a large object could have landed in. So the mystery continued. Maybe the light appears to be much the same, but there are objects of different sizes flying around with the same type of light. Only an object no bigger than about five feet in diameter, or 10 feet at the maximum, could have come down and landed or hovered in that wooded area without breaking some limbs or disturbing some vegetation.

Zoom

By 1976, some other people in the community had seen "The Light" and some of its strange behavior. We learned some years later that my father had talked to several people that had seen a large, bright white light fly into the edge of their fields near their house and just hover there, or they saw it slowly crossing their fields. They thought it was strange and wondered what it was, but nobody ever reported it as far as we know.

Buford Jones related a story from his boyhood that he did not think much about then, but now realizes that his parents saw a bright white light crossing their fields routinely. His parents would sit outside on the porch at night, as did many people at the time, to allow the house to cool off in the summer before retiring for the night. He remembers hearing them talking about a silent white light flying across their fields noiselessly on several occasions. It could have been as early as the mid-1950s when they first saw it. This was when his family lived on a farm near Laurel, Mississippi, about 50 miles to the southeast of my family's farm.

Others in the community saw "The Light" in the 1970s as well. Clark worked in the shirt factory that was in Polkville at that time in the middle 1970s, and one day overheard some of his co-workers who were friends with each other talking about a strange light. It seems that on this occasion eight or ten friends were camping out on the banks of Strong River just a few miles east of Polkville. After dark in the early evening, a silent white light flew over at low altitude and circled around. The light persistently flew around and hovered for a while. The campers got the feeling that the light was watching them. At some point, they all got creeped out, jumped into their vehicles and left. This shows that people with normal reactions tried to get away from the strange light, but Clark and I were trying to catch the darn thing. I do not know what that says about us.

My wife recently reminded me, after this section had been written for several months, about the time "The Light" appeared to follow us on our way back to Hattiesburg one night. It seemed to follow us for 40 to 50 miles, but I did not stop or even roll down the window to be sure it was not a reflection of some type from within the car or a true mirage, although the distance of 50 miles and changes in viewing angle tends to rule out a mirage or reflection. We did not want to stop and get out of the car, but that is exactly what we would have had to do in order to confirm that the light was an object moving independently of us. We only saw something like this on one occasion. I am reasonably sure that it was "The Light," but we did not take the necessary actions to absolutely verify that it was.

Nearing the completion of this book, another family member told me out of the blue about "The Light" following their car one night. I had not questioned them or told them anything about our experience. In their case, they observed a large white light that seemed to be following their car. They stopped the car and got out to observe the light, which also stopped and hovered, moving ever so slowly. After hovering for a short time, the light took off at a fast clip and flew away. By getting out of the car, they confirmed that the light was not a reflection or a refraction (a mirage), but that it was a real flying object.

Every so often I also hear of other members of the Polkville community that have seen "The Light" flying by at night. Most describe a large, bright white light flying silently not far above the treetops. There are probably many family stories similar to ours in the area that have never been reported. I have not actively gone out into the community seeking out stories as of yet.

Chapter 11

Night Lights at the Community Dump

"It is a common experience that a problem difficult at night is resolved in the morning after the committee of sleep has worked on it."

—*John Steinbeck, (1902-1968), Pulitzer and Nobel Prize Winning Author, The Grapes of Wrath, Of Mice and Men*

This chapter almost did not make it into the book. Only sightings that could not be explained by conventional means were under consideration. There could be no borderline incident of maybe something unusual that was seen, or maybe what was seen was just an airplane. We had discussed the subject as the sightings continued back in the 1970s so that we would be slow to say something was a definite UFO unless it met certain criteria. It must display a behavior of some type that was completely unexplainable by conventional means. Clark did not want things added to the book that could be easily dismissed or criticized as being something else. When discussing my plans to begin the book, I got conflicting statements from my mother and my brother about one particular incident.

My mother said that she saw what looked to her to be a flying saucer hovering out over a field with lighting coming from the inside. She said, "It is the most vivid memory that I have of all the sightings back in the 70s," pretty strong stuff indeed. When I mentioned the sighting to Clark, he surprised me greatly by saying we should probably leave that sighting out of the book. I was very puzzled, but I could not write about an incident where two witnesses were in such disagreement. Disappointed, I listened as Clark further said that he had seen the lights and they were strange. It appeared to him to be an object hovering just above the treetops at about 1/2 mile away that did not move while they were at the dump, but he was busy unloading the back of the pickup truck and could only look a few seconds at a time. When he was through unloading the trash at the dump, my father was in a hurry and left immediately. Clark would have stayed to investigate, or would have driven closer and confirmed what they were seeing, but was not able to do so. He wanted to be absolutely certain that the lights were not from an airplane a long distance away flying directly toward them. The landing airplane illusion can fool an inexperienced observer, as critics can say it was just an airplane seen at a greater distance away than was perceived. This can be a great out for debunkers, because they can always say with a degree of correctness that you cannot judge how far away the object was, so your observation must be wrong. However, some people can be fairly accurate in their distance judgments, even if it is not clear exactly how they can do this. Because he was not able to do his normal investigation and get a much better handle on the sighting, Clark did not feel comfortable including the incident. I did not press the issue any further. The last thing we wanted was to include something that could be easily explained as a conventional object.

A few weeks after I had given my family a rough draft of the family experiences, Clark called me to say that he and my mother had decided that the sighting at the community dump should be included. After reading the other chapters about the family sightings, Clark had a somewhat restless night where his mind kept thinking about the sighting at the dump. The next morning he came to the conclusion that there were enough details to the sighting to determine that it was not an airplane. I am glad that he did, because once I heard the details, I knew the possibility of it being an airplane was quite remote.

The incident happened sometime in 1975 when Clark was just 19 years old. He had gone with my father and mother to a community dumpster that they frequently used. It was already a little after dark when they arrived. They were in the family pickup truck and Clark had the honors of unloading the trash. They saw the apparently hovering lights as they were nearing the dump. The lights were seen above the treetops, but were quite low on the horizon. In later years, Clark saw several airplanes at varying distances away from the dump location, and all were at a much greater angle in the sky than the lights on this particular night. The Jackson airport lay in a direction behind the lights 25 miles or so west of their location. Clark had to immediately swing into action and begin unloading the back of the pickup. He could only see the hovering lights as he was going and coming unloading the truck. He could pause for a few seconds at a time, but had to complete the unloading task. My mother and father sat in the truck with the truck engine idling.

This was another frustration to Clark, as he would have killed the engine and listened for any noise, such as that of an airplane, but my father was not about to participate in anything of the sort. Finally, I began to fully understand the situation. My father, at that point in time, had admitted to the family that we (and he) had seen UFOs and things that could not be explained, but he did not want it known outside of the family. He did not want people to think we were crazy. He did not encourage anything associated with the phenomena, or any pursuit or investigation. He was not about to wait around at the dump to further observe the object, or cooperate in any way along those lines, such as driving down a side road to try to determine the exact nature and location of the object. Clark was stalemated and frustrated by the situation, and had just preferred to put it out of his mind since he was not able to investigate it in a logical manner.

My mother, on the other hand, was viewing the object while seated inside the truck. The front windshield acted as a frame around the object. Any change in its apparent position would have been easily noticed and the time that she had for viewing the object was about 90 seconds. That is much longer than one of our typical sightings, which could last 30–40 seconds before "The Light" would be obstructed by the trees. Today everything is so overgrown compared to the 1970s that seeing the light from the yard would require it to fly very near or almost directly overhead. So her viewing time was quite long compared to one of our typical sightings. The other factor is that the cab of the truck limits someone of normal height to less than a 20° viewing angle relative to the horizon. If you lean forward, you can see a greater angle upwards, but she was sitting normally and looking out forward at the object. My estimate of her viewing angle is 15° based upon a test I did sitting on the passenger side of my Ford Explorer. Sitting normally, I was able to see the top of a 30-foot tree that I walked off to be about 100 feet away. With my eyes at approximately five feet above ground level, that gives a viewing angle of 15°. If I am off in my estimates a little, it does not impact the outcome of the conclusions that can be drawn from the calculations. In other words, the viewing angle can be 12-20° and not have much impact on the final conclusions. To see an object out of the front windshield, the viewing angle has to be less than 20° above the horizon unless one is leaning forward, and she was not. As stated above, the frame of the cab of the truck serves as a reference point as to whether or not the object moved during those ninety seconds. According to my mother, the object did not move and did not change its position in any way during that time. Clark has the same recollection, but he was constantly moving himself during those 90 seconds.

Based upon a viewing angle of 15°, the altitude of the object is calculated in the table below for a given distance away.

Distance	Altitude	@ A Viewing Angle of 15°
1/4 mile	341 feet	
1/2	683	
1	1,365	
5	6,825	
10	13,650	

Clark believes the object was hovering about 1/2 mile away, so if he is correct, then the altitude of the object would be 683 feet if viewed at 15° to the horizon, or 455 feet viewed at a 10° angle. An object viewed at 1/2 mile away as just above the treetops can be 450-700 feet in the air. That is the nature of distance perception. Faraway objects can be much higher in altitude than we perceive, but the altitude, or apparent altitude, is just one factor in a sighting, and it turns out not the most important in this case.

The appearance of the object, according to my mother, was that of a hovering flying saucer with light appearing to come from the interior. The outline of the saucer shape was very faint. There was not a bright white light on the bottom to obscure the top of the object. The lighting was definitely from the top. Even though she described the outline of the saucer

shape as very faint, she believes that is what it was. Clark could not stand and stare at the object, as he was in constant motion and could only stand still and look for a few seconds at a time. He does not recall the faint saucer shape. What he saw was *four* horizontal white lights close together "parked in the sky," in his own words. Four horizontal white lights were seen without the bright white light on the bottom. These lights were distinctly visible as separate individual lights, and were steady and not blinking. When he told me that, I knew that I had never seen an airplane present four horizontal white lights as it came in for a landing. I had seen as many as three white lights on an airplane coming in to land at the Lake Charles, Louisiana airport, but this was not as viewed from the front. It was viewed from the side at a distance of about 1/2 mile away, as judged by the size of the plane. I also saw the green and red blinking lights on the plane during that observation.

The typical illusion of hovering for a plane coming in for a landing involves the forward landing light or lights, and it can appear to be a hovering UFO for 30-50 seconds, and at times perhaps longer. The illusion can be powerful because of the descending angle that a landing plane follows as it gets closer to the observer and does not cause the observer's viewing angle to change very much as the plane approaches. Contrast this to a plane flying at constant altitude where the viewing angle constantly changes as the plane approaches the observer. However, if one looks long enough at a descending plane, usually for 30 seconds or so, you can see a wobble or a slight course correction. Keep looking and the forward light becomes noticeably larger, and the plane may eventually fly over your position. A commercial jet coming in on a descending flight path at a reduced speed of 200 mph or more can easily cover three miles in one minute. Once it covers 50% of its distance away when the observer first saw it, then the light will appear to have doubled in size. The illusion of a hovering UFO then gives way to this is an airplane coming in for a landing, especially once the observer sees the green and red blinking lights as the plane approaches.

As we continued to discuss the sighting, Clark gave me the information that conclusively disproves the airplane landing illusion. He said that the four lights could be covered by the end of his thumb. As you stand and line up your thumb with an object in the sky at a 10-20° viewing angle to the horizon, your thumb will be approximately two feet from your eye for an average size individual. Armed with this information, one can calculate how far away the lights would be if they were from an airplane. The key is the calculation of the apparent size of the object, as the distance away is varied. I used 1/2 the width of his thumb as what it took to cover the lights from view in order to make it easier to support the landing airplane illusion. The 1/2 width of thumb measurement was taken as 0.375 inches. Using this, the distance across the lights can be calculated at various distances from the observer.

Length spanned by the four lights as a function of distance away.

Distance	Width from Light 1 to Light 4
1/4 Mile	21 Feet
1/2	41
1	82
5	410
10	820

Now let us suppose what they saw was a Boeing 747 on a landing flight path. Since the wingspan of the largest 747 is about 225 feet, the distance away could not be more than about 2.75 miles. Such a plane coming in on a landing trajectory would have covered that 2.75 miles in less than a minute. It would have overflown the dump site before they ever left. Finding a place to land a 747 on that flight path would have been the tricky part. Placing a plane at two miles away would mean it had a wingspan of 160 feet, and it would have arrived at the dump site just as my family was departing. Likewise, if it were a plane with an 80-foot wingspan, it would have been about one mile away and it would have overflown the dump site in about 45 seconds. Placing a plane at 1/2 mile away would mean it had a wingspan of 40 feet, and if flying at 80–100 mph it would have overflown the dump site in 20 or 25 seconds. Again, all the planes would have to have been on a descending flight path to appear to be hovering. This what if evaluation leads to the conclusion that the four horizontal white lights could not have been an airplane, as does the absence of red and green blinking lights.

For our distance estimates to be realistic we have to ask, does the distance and width estimates violate the distance resolution of the human eye for distinguishing separate lights, or how far apart do the lights have to be to be seen as separate at a given distance? At ½ mile the lights have to be at least about 20 inches apart to be seen as separate lights according to the website *www.ehow.com/about_6603780_limit-resolution-human-eye_.html* . [78] This would mean a 10-inch separation at ¼ mile and a 40-inch separation at one mile would be required as the minimal distance apart to distinguish the lights as separate lights, and so on in that way. Our estimates for a reasonable size of the lights put the lights about 10 feet apart or greater at ½ mile, so no, the distance and object width estimates do not violate the human eye's spatial separation limitations.

Let us consider the possibility of a helicopter being present. No sound was heard over the sound of the idling truck, and a standard helicopter at 1/2 mile away could have been easily heard over the sound of the idling truck. Typical noise levels for a helicopter are about 100 decibels at 100 feet away. Plugging into the equation in Chapter 8 gives a noise level of about 72 db at ½-mile distance and about 66 db at a one-mile distance. This should be easy to hear over the approximate 55 db maximum noise level for the idling pickup truck with a standard gasoline engine and a working muffler when a person is standing about 15 or 20 feet away. A noise dampened helicopter, which had been developed by this time, would produce about 55 db at a ½-mile distance and would be difficult to hear near the idling truck. A standard helicopter would have to be well over a mile away to be that quiet.

I have never seen a helicopter present four horizontal white lights as its only lights. The fact that the light appeared "parked in the sky" with no notice of slight swinging side to side does not fit with a helicopter hovering for that long. The fact that my father barely acknowledged the sighting and would not allow any further investigation hindered obtaining enough information to accurately determine the distance the hovering lights were from the dump site. Clark feels comfortable with a distance of 1/2 mile away, which would make the object at least 40 feet in diameter, or 40 feet across the lights. Placing the object closer than 1/2 mile means my family could have heard some noise even if it were a noise dampened helicopter, and placing it a mile away leads to a helicopter over eighty feet long for the four white lights seen from the side.

There is at least one military helicopter nearly 100 feet long and that is the Chinook (CH-47D). [79] The possibility that such a military aircraft would be hovering out over a field in an isolated rural area is a very remote possibility. Also, remember that I used one-half of a thumb width in my calculations. If I used the full thumb width it would make the width of the lights 160 feet at a one-mile distance, and there would be little possibility of the lights being due to a Chinook helicopter. However, there is an uncertainty in Clark's estimate, as there is in any measurement no matter how carefully done, and enough uncertainty in this case so that it could have only taken one-half a thumb width to cover the lights. What can be said is that the helicopter explanation is not very likely. I will wager that a Chinook helicopter seen from the side at night does not display four horizontal white lights, but I have not tracked down this detail. There should also have been some red and green, blinking navigation lights for a helicopter according to FAA regulations. Helicopters used by the Drug Enforcement Agency, DEA, and other agencies such as border patrol and the FBI, are known to operate without the required running lights, but would they display four white lights instead? [80] Not very likely. Clark knew it was no use in trying to get my father to drive toward the object or try to triangulate the position of the lights. Without doing that, a definitive determination of the position of the lights could not be done, so that leaves us to consider what else the lights could be. One possibility that would be pushed by skeptics is that these lights could have been a mirage of some type.

For the lights to have been a mirage there would have to have been a temperature inversion (warm layer of air sitting over a cooler layer) and some source lights some distance away capable of producing the image of four lights in a horizontal row fairly close to one another.

This is not a common image seen on the ground—four round lights equally spaced in a straight line. The type of mirage seen in the air is called a superior mirage, but the situation with mirages is hardly simple with the atmosphere acting as a prism, a weak lens, a strong lens, a mirror, or a series of lenses, depending upon the exact conditions. The most thorough explanations that I found concerning mirages were on the website ***http://mintaka.sdsu.edu/GF/explain/atmos_refr/phenomena.html*** with copyrighted information by Andrew T. Young. One thing that may argue against a mirage is that the lights were seen on approach to the dump, and when traveling away from the dump they were at an angle that was to the side of the image. The lights were seen at an appreciable angle from the head-on view at the dump-site. I do know that optical phenomena can be very dependent upon the viewing angle in a number of cases. I am not that familiar with mirages. An airplane explanation is completely ruled out, a helicopter explanation is very improbable, but a mirage or some sort of optical explanation remains a small finite possibility. If the lights lifted up, maneuvered around, and flew away, then the mirage explanation would be impossible. Regardless of the mirage possibility, this sighting has been included because of the family dynamic that it revealed at that point in time in 1975, and the dynamic revealed today as we consider those past sightings. There is also the possibility that it was a hovering UFO. My mother still quietly maintains that she saw a hovering flying saucer that night, and that this is her most vivid memory of all the sightings during the 1970s. I also became aware while doing research for this book that this sighting had more than a little similarity to two well documented sightings in Mississippi during the 1970s time period involving hovering UFOs with horizontal bands of lights.

Chapter 12

Collision Course?

"Never act until you have clearly answered the question: What happens if I do nothing?"

—*Robert Brault, Freelance Writer*

This incident happened sometime around 1976, most likely in the fall, as we did come home for a visit around that time, it would have been dark around 6:30 p.m., and it is sometimes warm enough to sit on the porch for a while before going inside for the night. I was visiting my grandparents. We were sitting on the front porch, as was customary when visiting for a little while. My grandparents would usually have their evening meal between five and six, and went to bed between eight and nine every night. We were visiting on the front porch around 6:30 to 7:00 p.m. I was sitting facing out toward the pond that was in the hollow in front and below their house. My grandparents' house sat on a ridge looking out to another ridge to the west that was some 700 feet away. A hollow lies between the ridges, and the ground drops down some 30-40 feet or more below the ridge lines. A pond lies at the bottom of the hollow, and another uncle's house was near the pond at a little higher elevation. My grandparents were sitting at the end of the porch and were turned slightly facing me, as I was them.

We were talking back and forth, but I cannot remember about what. Probably we were talking small talk about what I had been doing and about what they had been doing when suddenly I noticed a white light in the distance over the far ridge. It was probably about 1,000 feet away at that point. I thought to myself, *That looks like 'The Light.' Is it hovering? No, it is moving toward us.* At that point I was trying to remain engaged in conversation with my grandparents and keep an eye on "The Light" at the same time. I never was any good at multitasking. It appeared to be moving directly toward us, but how could I tell that? When a plane with its forward landing light is flying at a descending angle and directly toward an observer, it can appear to be a hovering UFO. This has been the cause of many a UFO report over the years: a plane a mile or two away flying steady and directly at an observer as it descends to make a landing. The angle to the horizon for a descending airplane does not change dramatically in contrast to the constantly changing angle for an approaching plane flying at constant altitude. The illusion can be convincing for a while. Both Clark and I had seen planes in this way and had independently become aware of this illusion, so our guidelines for calling something a possible UFO are that the light must display some other behavior that shows it is not a plane or a helicopter in order to be designated as a possible UFO.

I could tell that "The Light" was flying directly at us, because it was changing altitude and showing slight deviations in motion. It was definitely not hovering. When it came over the ridge, it was 75 to 100 feet above, but as it flew on and the elevation of the ground sloped down and the trees thinned out, it dropped down in altitude as if following the ground. I observed this as I was trying to remain engaged in our conversation, glancing back and forth between the light and my grandparents, unsure of what I was seeing. My grandparents were not looking out over the hollow and had not noticed the light flying in our direction. My grandfather had his back to the light and a porch post obstructed my grandmother's view somewhat. When I glanced back at "The Light" over the lowest part of the hollow, I realized that it was flying at about my eye level and on a collision course with us. I was beginning to become concerned. I wanted to yell something like, "Hit the deck! A UFO is about to hit us?" My grandparents would have probably thought I had gone completely crazy at that point. I had never discussed "The Light" with them, and I did not know if my father had either.

Just a few seconds before it would be too late to do something, "The Light" raised up some 50 or 60 feet on the same general course, I have to assume, as I lost sight of it due to the porch overhang. I believe it rose up and continued on the same course that would carry it directly over or near my grandparents' house. We heard absolutely nothing. There was no sound whatsoever. From the time that I first saw it to the time it got to my grandparents' house I would estimate as well over 10 seconds, maybe 15 seconds. The speed would then be about 45-50 mph. I could not make out any structure that the light was attached to, but I was in panic mode and not in observation mode.

The lack of any sound (to my ears at least) was a characteristic of "The Light." I have noticed different aircraft that fly over or near my present home can be heard quite clearly or even noisily inside the house, even though my house is well insulated. There is not much debunking room for a flying object going faster than the current wind speed (the wind was not noticeable, which usually means 10 mph or less) that is silent and is maintaining a certain altitude above the changing terrain. No, it was not a cruise missile or a UAV, unmanned aerial vehicle, although it was flying in a similar manner. Predator drones did not come into being until the mid-1990s. [81] Both of these, cruise missiles and UAVs can fly low, but both are noisy enough to be heard at close range.

The drawing shows the course "The Light" flew that night. It turns out that this was one of its frequent flight paths. Clark used to park just off SCR 141 and watch for "The Light" to fly over the first ridge, as it would be visible longer and easier to see than when it flew near our house. He did this when he was trying to convince some skeptical friends of the existence of "The Light." He made believers out of a number of them that something unusual was flying by.

One of my daughters asked me why I did not point out the light to my grandparents. I do not really know, except that I was uncertain of what I was seeing until it got close enough, and then I was mesmerized and not thinking. I hesitated and then it lifted up and went over the house. I guess I could not believe that it would fly directly at us. We could have gotten off the porch and hurried into the yard to see it going over the trees behind my grandparents' house, but I just sat there in an indecisive state of mind, which I sometimes fall into and forget about anything around me. It was a surrealistic moment and I was caught off guard for a few seconds.

One day I was looking at satellite maps and came to realize that a flying object that stayed on the flight path over my grandparents' house would fly directly over Polkville. If it kept going on the same trajectory, it would fly very near Taylorsville and eventually very near to Laurel.

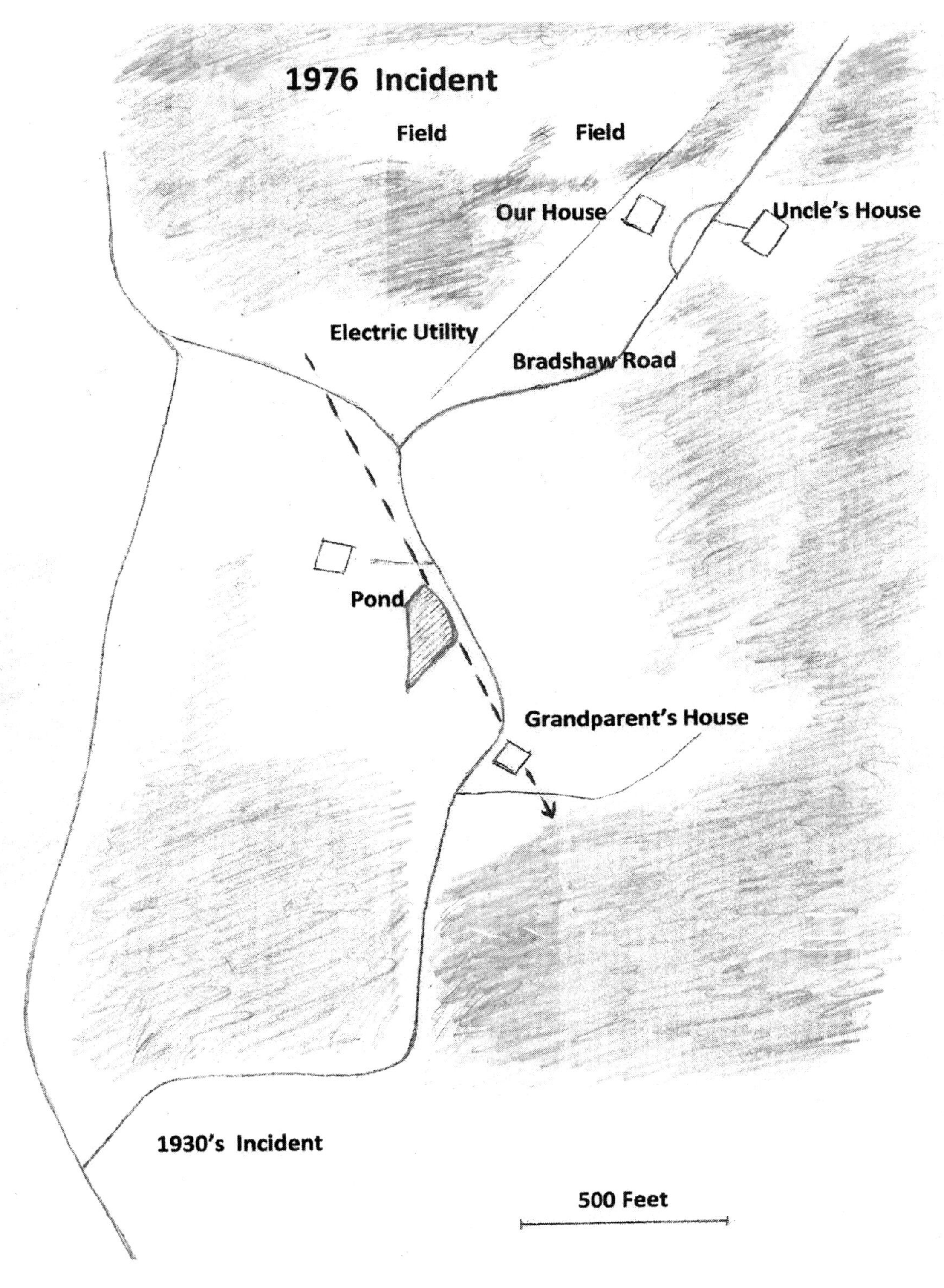
1976 Incident
Field
Field
Our House
Uncle's House
Electric Utility
Bradshaw Road
Pond
Grandparent's House
1930's Incident
500 Feet

pond

Chapter 13

The "Light" and the Airplane

"...There is another and more humiliating possibility - alien intelligences so superior to us and so indifferent to us as to be almost unaware of us. They do not even covet the surface of the planet where we live - they live in the stratosphere. We do not know whether they evolved here or elsewhere - will never know... Some few of them might study us casually - or might not."

—*Robert A. Heinlein, (Science Fiction Writer) Grumbles from the Grave.*

One night in 1978, Clark decided to go outside and look for "The Light." After some time he noticed a bright white light hovering motionless over the woods behind and to the south of our uncle's property. He had not noticed it at first, as he was looking for a moving light, but it was too bright and too large to be a star or a planet. He said to himself that this must be "The Light" because of its size. It was probably a quarter mile away or more, and 300 to 600 feet high. Of course, in a Klassian analysis of this sighting, the light would have to have been Venus or another planet or a mirage, as no other possibility could exist. Anything that happened after that would have been Clark's wild imagination gone berserk. Clark watched it closely to be sure it was not an airplane flying directly at him in a descending flight path from a mile or more away. He did not want be fooled by that particular illusion that can look very much like a hovering UFO as discussed in the previous chapters.

It just hung there in the night sky and looked too large to be any star or planet, and showed no detectable movement for a period of time. He was sure that it was "The Light." Clark started to walk toward it as it hovered south of our uncle's house perhaps a quarter mile away more or less. He crossed the road into our uncle's yard and had gone less than 50 yards when it made a slight move down and toward Clark's position. He stopped and watched it closer at that point. Then, it started to move again in a slow zigzag motion sideways and upward. That move was followed by a somewhat faster zigzag motion, higher still, and then zoom it was gone. It was gone in a snap of the fingers. He stared for a few seconds and headed back across the road. Once back in our yard with a total elapsed time of 20-30 seconds or so, Clark heard an airplane coming from the west from behind our house. A small plane flying maybe 1000-2000 feet altitude flew almost directly over the house and directly over where the "Light" had been hovering. This was a further indication of what appeared to be intelligent control over "The Light." It behaved as if it were well aware of other aircraft approaching its position and it left. Now, it is a little disconcerting that an unknown object apparently controlled by an unknown intelligence can just hang around in the sky and be watching us or our neighbors, or who knows what. It could move to a position where it looked like a background star or planet, and not be easily detected visually, but for what purpose we do not know. Clark believes he has seen it hovering before, but unless it moves as it did in this incident, or is so close that the size of the light is too large to be any star or planet, he could not be completely sure it was "The Light."

This sighting temps the skeptic to think of the auto-kinetic effect where staring at a stationary light for a long time gives the illusion that a star or celestial light is moving erratically. However, there were trees acting as points of reference in any view in any direction when standing in the yard of my parents' house. A point of reference in the view prevents the auto-kinetic effect from happening. The only view where there may not be a tree to be seen is if one looks directly overhead, but who can crane their neck like that for very long. The other point is that the light left a little before the airplane arrived, leaving just a dark sky at its former position. The light also appeared larger than any star by a factor of three or four times, but was smaller than what we typically saw when we saw "The Light" crossing the field. It was an individual sighting, and Clark is not prone to any psycho-illogical nonsense so favored by skeptics and debunkers, but neither are most people.

CHAPTER 14

The Madison, Mississippi Incident

"It's like deja vu all over again."

—*Lawrence Peter (Yogi) Berra, Baseball Great*

Another family member was involved in a close encounter, but his identity has to remain undisclosed. Most of us do some questionable things sometimes; things that we may come to regret later. Sometimes we drive too fast in order to get somewhere on time, or run under that yellow light a little too late. Most of us push the boundaries at times when motivated by the need to succeed in one way or the other. However, sometimes the boundaries are pushed a little too far. In the fall of 1974, one family member, along with three others, was in the woods near Madison, Mississippi at night spotlighting for deer. (Madison is north and slightly east of Jackson near I-59 North some 32 miles in a straight line from my family's farm in Smith County). The desire to get that trophy deer can drive some to push way beyond the boundaries allowed by legal statute. The penalty for getting caught by the game warden would be quite serious. Loss of guns, hunting gear, possible loss of vehicles, stiff fines, and possible jail time were nothing to regard lightly. Our hunters knew this, but the desire to get a trophy deer overrides reason sometimes.

Our four hunters were equipped with lights fitted on their caps so that they could hold the deer in the spotlight while keeping their hands free to aim and fire a gun. The saying that someone looks like a deer in headlights is not a complement. It means you look like you do not have a clue as to what is going on. Their spotlights were on. They were actively pursuing their prey, scanning for eyes shining back in the darkness, when they saw a bright light flying low in the distance. Somebody said, "It's the game warden in a helicopter with a searchlight." They jumped like they themselves had been shot. They put out their lights and knocked the cap off of one of the guys who did not quite understand and did not react quickly enough. They glanced at the "helicopter," which then seemed about 100-150 yards away, and in a blink of an eye it jumped directly over them in a motion that is difficult to describe. Our witness said, "I don't know how to describe the way it moved. It was over there a hundred yards away and then it was over here just like that in an instant." They dropped to the ground in response to the object suddenly being so oppressively close above them, and they probably had subconsciously continued to react as if it were the game warden. They were giving up, surrendering to this sudden intrusion. At nearly the same instant, they realized that it was not a helicopter or the game warden. They looked up in confusion as this unknown intruder stopped and hovered directly over them without a sound. The bottom light now appeared much larger than when seen from a distance. No one could get a clear view of the top of the object, which was just a dark outline. It was at treetop height, and the trees in that area at the time were barely 40 feet tall.

The object above our group of hunters was completely silent and there was no perception of air movement such as a helicopter would produce. It appeared for all intents and purposes to be the same or an exact copy of the Nov. 1972 object. The craft had a lighted bottom that was estimated to have a slightly oblong, lighted surface 15 feet across when seen from below, but when seen at a distance the light appeared to be smaller than would be expected, but still large for an aircraft. They looked at the craft in total disbelief and thought about the unreal nature of the situation. The top of the craft was hard to see clearly because of the bright undercarriage light, even when staring intently at it. Staring up in that moment, they could hardly believe their eyes with hearts pumping with surprise, fear, and amazement all at the same time. The craft hovered noiselessly over them for "less than 10 seconds" and then shot straight up and out of sight. Our witness said, "I have never seen anything move that fast." Of course, none of the hunters were going to report the sighting due to the incriminating circumstances.

Unbelievably, here was another case in the same family of a close encounter where a UFO hovered at no more than 50 to 100 feet in altitude above two or more people. The common denominator of the two close encounters was that those involved were outside in the woods or fields at night. Klass was right; if you are outside at night, you will see some things. It occurred to me later that this would make a great ad for PETA. Even UFOs are for the ethical treatment of animals. It seemed as if the lighted object was giving the hunters a dose of their own medicine.

The behavior of the craft strongly indicates that its occupants or its controllers saw the hunters and wanted a closer look. I cannot come to any other conclusion. I am sorry Mr. Klass, but the UFO appeared to be interacting with the people involved. The maneuvers the craft displayed are beyond anything the military had then or has now. The silent flying, hovering noiselessly and motionlessly, the abrupt quick movement, and then accelerating at high speed straight upward in a manner that could create enough G force to stun or kill a human, are all beyond the known and possible behavior of manmade aircraft.

Chapter 15

By the Light of the Moon

"Every once in a blue moon, something new comes along that scrambles your preconceptions"

—*Thinkexist.com -anonymous, female*

This sighting occurred sometime in 1978. My parents had, by this time, come to the realization that something quite unusual and unexplainable was going on. My father's view had started to change from what it had been after the incident in November of 1972. My parents were, quite understandably, somewhat concerned that this unknown phenomena was regularly seen so close to our house. My father was still conflicted about the explanation for all of these sightings, but by this time they had told my uncle across the street what we had seen. He reacted a little bit miffed at first. He said, "You mean to say that you were seeing some strange lights flying around for a few years before you decided to tell me anything?" Well, in defense of my parents, it is not a subject you just bring up out of the blue by saying something like, "By the way, we have been seeing these lights we believe are UFOs flying around and a lot of them have been going right beside and around your house. Just thought you might want to know."

Things continued on in a somewhat tentative state of mind for my parents, particularly for my father, until the incident I am about to describe happened. Even though my father had perhaps 15 or 20 sightings of "The Light" by this time, he was still hesitant about the UFO explanation. Undeniably, my father realized that something very unusual was going on, something that just did not fit into his previous paradigm. Sometimes when one is least expecting it, a situation becomes clearer, much clearer.

One evening not long after dark, my parents were driving south on MS-13 headed from Morton and going back home. At a short distance from the intersection known as the Cracker Station intersection, my parents saw what looked like "The Light" off to their left flying low and in a direction that would bring it across the highway. The flying white light looked as if it would, if it kept flying the same course, cross the highway at some short distance in front of my parents' car, which was then traveling at about 50 mph in a generally southern direction.

As they first glimpsed it off to the left of the road, it was probably more than 1,000 feet away and maybe 300 to 400 feet to the left of the highway. The moon was up and quite bright so that the object was backlit by the moon and the upper part of the object could be seen clearly. The path the object took was not to cross the road in a perpendicular manner.

Instead, it took an angular path. As a result, they were able to view the object a little longer and at a closer distance. In fact, it ended up flying directly over my parents' car at a low altitude of perhaps 50-60 feet. My parents were able to see the object clearly as it angled toward them and the road. It was a surrealistic scene, the object, backlit by the moon and clearly visible, grew larger and larger until it seemed to fill the windshield of my parents' car just before it passed overhead. For the first time, the upper part of the object was clearly visible. There had been no sound that could be heard above the normal road noise, unlike a crop duster plane that can be heard distinctly in the same situation. The total viewing time was 7-8 seconds. The time that it was at a distance of about 300 feet away and closing, and perfectly lit up for viewing, was perhaps three or four seconds. That may not seem like much, but it is.

A person can see a lot in just one to one and a half seconds of viewing time, much less seven seconds. I demonstrated this to myself viewing bridges and overpasses for seven seconds while riding as a passenger in a car. I also tried a test using my house and the trees and shrubs beside my driveway. By standing in the woods behind my house and looking back at the street, there is a 35-foot wide viewing slot to view traffic coming down the street at a distance of 200 feet. Now, 35 feet may sound pretty wide, but it only represents about one and one half seconds of viewing time for a car going by my house at 30 mph (the speed limit is 25, but few people drive that slowly). I found that in one and a half seconds I can recognize my neighbors and their cars, and most makes of the cars and trucks that are going by. Of course, these are familiar things I see every day. Something unfamiliar would take longer to mentally process, but six or seven seconds is long enough to get a fairly good idea of what you are seeing. You can tell an airplane from a helicopter, or a bird, or a balloon, or a kite, or a UFO.

As confirmation of this, tests of visual perception have shown that a familiar object can be identified in 1/10 of a second or less for a normal person, and the decision to take action or not can be made in 3/10-5/10 of a second. [82] Viewing an unfamiliar object for 7+ seconds is long enough to process what you are seeing, even at night. If the object is coming directly toward you and is getting larger and clearer all the time, then the observer can get a fairly accurate idea of what they are seeing. Of course, my mother as the passenger in the car, had a longer viewing time because the driver will likely be glancing back and forth at the road and the object.

My father, who was driving, described what he saw as a flying water tower with less height than width, and a top and bottom section joining in the middle. He placed the size at 30 or 40 feet across. My mother, the passenger, described it as two half rings joined with a wider band at the middle, and with a clear canopy on top that had a dark shadow in it suggesting a pilot, but she could not make out any more than that. In other words, they saw a type of classic metallic disc or saucer UFO. We finally had a view of the top of the object seen by two witnesses. The structure behind "The Light" had finally been revealed more fully.

My parents started referring to it as "that spaceship" to each other and to Clark. My father had finally come to accept that a UFO type object or objects under control of an unknown intelligence were flying around. This was after nearly a decade of sightings. I sometimes wonder if I could have put Klass in my father's shoes, and if somehow he could have seen what my father saw, if he then would have finally changed his mind. There is no way to know. I have to believe that most people would have finally come around to face the reality of the situation, even most scientists and most skeptics, but who knows. One thing is for sure. When you see "The Light" for yourself and what it is attached to, those standard explanations of it being the planet Venus or Jupiter, or a mirage or ball-lightning, or an earth-light or a weather balloon, or a Chinese lantern or a meteor, or a military exercise or an advertising plane/banner, or a spotlight/searchlight, are simply wrong. Of course, it would be possible to explain away an individual sighting here and there, if there were only a few sightings. Since there were so many sightings, because of the range of the sightings and the different sightings had elements that tied them together, they are, in their totality, difficult to dismiss in any objective manner. How can anyone deny that some sort of unexplained phenomena was occurring?

Chapter 16

A Landing in the Field?

"You can't depend on your judgment when your imagination is out of focus."

—*Mark Twain's Notebook (Harper and Brothers, 1935), p. 344.*

Sometime in the early 1980s, my father went outside one night and was walking around in the front yard and driveway. When he glanced out toward the upper field, he saw what looked like a bright light on or near the ground, and it looked as if it were close and not very far out into the field. He moved a little closer and saw what looked like a substantial metallic structure around the light, and it also looked as if there was some type of movement around it. He basically panicked.

He rushed inside the house and said to my mother, "Get Clark! I think that spaceship that has been flying around all the time has landed out in the upper field."

Now here is one of the more incredulous parts of this story. The recollection is that it was a Friday or Saturday night, and Clark had already gone to bed by 9:30 p.m. He recalls being worn out (probably from staying up all night the night before). Needless to say, his usual bedtime on the weekends was in the wee hours of the morning. Putting aside that almost unbelievable detail, Clark was half asleep and half awake, and heard my father when he came into the house. This immediately roused him up, and after collecting himself for a minute, he was excited about the turn of events. His excitement was mixed with some fear. As he decided to go investigate whatever might be out there, he carried a shotgun with him. (Clark was the natural choice to go investigate being strong, fast, and agile with excellent eyesight).

My parents were standing on the porch as he was getting ready to confront the unknown. They were both saying things like, "Be careful, son; watch yourself. Don't get too close to it. If it looks like trouble, get back here quickly," and so on. My father said, "If there's any trouble, I'll call the law." Clark headed out slowly, moving closer and watching. My parents were watching his every move. He crept closer to the edge of the yard looking at the light and the shadowy movements around it. He confirmed to himself that it was definitely on the ground, and he crept closer still as he entered the path to the field that is surrounded with small trees and some vegetation. My parents watched and waited and listened. Clark was almost out of their sight as he stepped a little bit closer, and then a thought hit him. He had a pretty good idea of what he was seeing. A smile crossed his face. Another step or two and he confirmed exactly what it was.

Now, you would have had to have known my father to fully appreciate what Clark did next. Clark (and I) had learned our lessons well. At that point Clark let out a yell of distress and fired the gun into the air. After that Clark says everything is a blur. Pandemonium broke loose. My mother screamed and my father yelled, "That's it! I'm calling the law!"

With my parents going into hysterics, Clark came hurriedly scurrying back doubled over with laughter, as he shouted broken up between laughing, "It's nothing (uncontrolled laughing)! It's nothing! It's the pig (uncontrolled laughing), the pig feeder!"

My parents said, "What?"

Clark again: "It's the pig feeder that we just got a few days ago."

As it turned out, my father had decided to go into raising hogs because the market price had become very attractive. A few days before, Clark had placed the new, shiny, metal-surfaced pig feeder in the edge of the upper field until they could move it out to where it would be placed for growing hogs. It had slipped my father's mind momentarily when he had been outside and saw the apparent light on the ground in the field. A few years before, my uncle across the road had put in a yard light atop a post. The yard light was reflecting on the pig feeder and some of the intervening tree branches were casting shadows that moved with even a slight wind. So the light and the shadowy movements were explained. However, my parents did not appreciate the prank that Clark had pulled on them that night, although it was very much like the pranks my father used to pull when the opportunity presented itself. Clark had learned from the master.

One thing that struck me when I first heard the pig feeder story was how routine the sightings of "The Light" had become, and it surprised me how my father's view of things had changed from November of 1972. The explanation that it was all a military operation had changed to "that spaceship we see all the time." Indeed, what could explain seeing "The Light" so frequently?

Chapter 17

What are the Odds?

"The theory of probabilities is at bottom nothing but common sense reduced to calculus."

—Pierre-Simon, Marquis de Laplace (1749-1827), French mathematician and astronomer, from Théorie analytique des probabilités, 1820

Lots of ufologists and all skeptics tend to dismiss someone who claims to have seen too many UFOs over the years. For skeptics, even seeing one is too many. From a purely statistical point of view, the average person cannot expect to randomly see a UFO in their lifetime. It was usually considered to be a completely random occurrence. By the same logic, after a person has had one UFO sighting, the possibility of seeing a second is virtually nil. Even so, claiming to have had two separate UFO sightings may not immediately destroy your credibility with researchers, but more than that and you become a "UFO buff," a "Repeater Problem."

So after you have seen a UFO on two separate occasions and another one comes by, what are you supposed to do? Look away to keep from destroying your credibility? No, you are going to watch the darn thing and if it comes by again, you will watch it again. The point is that it is not always a random event, and clearly, in my family's case, the UFOs were flying close by and frequently. We were outside a lot and we saw them. After we saw them so frequently just by chance, we sometimes stayed outside longer looking for them and indeed, on some instances, eventually saw them fly by. I think the investigators are wrong on this point. Their assumption that all UFO sightings have to be random, once-in-a-lifetime occurrences just does not square up with experience. If the UFOs are flying by near you repeatedly, how can you not see them on multiple occasions? I think the researchers have basically missed a common phenomenon and have written off as "UFO buffs" or "Repeaters" some people with completely legitimate multiple sightings. No one is going to look away under such circumstances.

In any case, just about everyone would conclude that to see " The Light" so often, maybe five or six times a year at the peak, means that it was coming by very frequently. Also, about half the time that we went looking for "The Light" we saw it. It all points to a high frequency indeed. It does seem that some version of "The Light" was flying by quite frequently. Remember that the episode where it appeared to land in the woods could not have been possible for an object 20-40 feet across, so there is evidence of more than one size object flying around.

What do we know about the time of night that we observed "The Light?" We know that many times it was observed one or two hours after darkness fell. In the summer, that is nearly 11:00 p.m. In the winter, it's about 8:00 p.m. The rest of the year, it's about 9:00 p.m. The phenomena we called "The Light" did not seem to have a set time and the times we saw it, with only a few exceptions, were when we were outside for other reasons. Sometimes when Clark was out looking for it, he saw it around midnight or later. And no, it was not due to the auto-kinetic effect of staring at a dark sky until you see the stars move. Our observations were made with trees around. Trees could be seen from the corner of your eye standing at any point in our yard. Trees provide the eye with a reference point so that the auto-kinetic illusion does not occur. I have tried to see the auto-kinetic effect by staring at a particular star or Venus on several occasions and failed every time. No moving light, no nothing.

What do we know about the location or range of the flight path? "The Light" was seen crossing the upper field frequently, but Clark would stake out the flight path that went over my grandparents' house when he brought friends out. "The Light" could be observed longer on that path than by our house, because it rose up to go over the ridge. Also, my parents saw it about four or five miles north on Highway 13 as it was backlit by the moon. We know "The Light" ranged at least five miles from our house, and that there was probably more than one lighted object flying through the area.

The factors in figuring a probability for seeing "The Light" have a pretty wide range, but certain assumptions can be made. Consider that if "The Light" flew by in view every night and we were out scanning the sky all night every night, then we would see it 365 days a year. How can we tweak these factors to reach a frequency of seeing it six times a year? Well, to assume that the flyby frequency was six times a year would not work, because we were not out every night looking. If it came by every day within view of our house, then the frequency factor would be once a day or 365 days a year.

If the object my parents saw flew by during the daylight hours, then it could potentially generate dozens of sightings every day in the community. This would have quickly become known all over the area and since there were no reports or community buzz about daylight sightings, we will assume it only flew at night. Let us say that year round we have an average of about 12 hours of darkness. Our chance of seeing it when we are only out an average of 20 minutes a night is 1/(12/.33) or 1/36. This is arrived at by dividing 12 hours by 0.333 hours (20 minutes). Of course, we were sometimes out an hour in the summer and maybe only a few minutes in the winter. There may be a better way to account for the differing time factor, but let's go with an average of 20 minutes a night year round. Our assumptions are:

"The Light" came within sight of our house everyday:

Frequency Factor = Freq. F = 365 days/year

The time of night it flew by was completely random.
It flew only at night with an average night length of 12 hours.
Our average time outside at night is to be 0.33 hours:

Time Factor = TF = 0.33/12 = 1/36

Sightings/ Year = Freq. F x TF = 365 x 1/36 = 10.1 times/year

If we look at it another way, the time that we go outside is not random. It is within about the first two hours of darkness on average. This is the only time that we could observe a flyby of the light. One sixth (1/6) of the time, just by chance, the object would fly by in that two-hour time period if we keep the object's flyby time random. There are 120 minutes in the time period and we are outside on average 20 of those minutes. This is another random 1/6 factor for our equation. So again, the time factor for what faction of time we could see the light is 1/6 x 1/6 =1/36 either way of looking at it, as long as the object's flyby time is totally random.

Another factor we have to consider is that the object can fly by on its NW-SE track either north or south of the house. Unless we were out scanning the sky for the light, we could easily miss it if we were looking in the opposite direction, because it is generally only visible for a maximum of 30-40 seconds during flyby. This gives an observer's orientation factor of one-half chance of looking in the correct direction in order to be able to see the object. Adding this factor to the calculation gives a calculated sightings per year of 5.1. Our equation becomes:

Sightings/yr = Freq. F x TF x Observer Orientation Factor (OOF)
Sightings/yr = Freq. F x TF x OOF
Sightings/Year = 365 x 1/36 x ½ = 5.1

That is close enough to what was actually seen. Every factor was allowed to be random except the frequency factor, which was set at a nightly flyby (365 nights/year). We could randomize the days it flew by and hold something else to be non-random, such as the time of night or hours after darkness fell. We could arrive at five or six times a year in that way as well. The point is that some major part of our equation has to be held non-random or the sightings drop down to less than once in a lifetime.

Suppose a UFO flew by our house in easy viewing range once a year just by random chance. Our equation gives the sightings per year of 0.0139, or one sighting in 72 years just by random chance.

Sightings/yr = 1 x 1/36 x ½ = 0.0139

Now consider that if a UFO were to fly randomly from NW-SE all the way across Mississippi every night, then one would come close enough to our house to see it about once every two years. Our sightings-per-year equation gives a value of 0.0069 sightings per year, which is the equivalent of one sighting on average every 144 years. If we were out less than 20 minutes a night, then the odds would be even lower. If you are not out looking, you will not see anything.

Many professional astronomers have not seen that many UFOs because they tend to program the telescopes to scan a certain small section of sky at a certain time, and they are not scanning the skies all night as many amateur astronomers do. In addition, a nearby moving object zips through the visual field of a telescope so quickly that it can hardly be seen.

Could "The Light" have flown by in view of our house more than once a day? The times that we waited for it to come by and it did, indicate that the flyby frequency had to be high. Could it have flown by only in the first four hours of darkness? Possibly, as that would be by ten in the winter and one in the morning in the summer with an average of 11:30 p.m. year round. Holding this condition constant, we can again arrive at a six-times-a-year sighting frequency by manipulating the other variables. The main point is that these sightings were completely non-random. The phenomena behind the sightings were regular and repeating.

For comparison consider that at my current house there is a flight from Dallas, TX to Lake Charles, LA that flies nearby every night shortly before nine. That is a frequency factor of 365 days a year. Sometimes it flies directly overhead, but most times it is within one-half mile on either side of the house. I am outside on a random basis nearly every night, but usually for not more than five or ten minutes, except on rare occasions. In the summer, I sometimes run or walk through the neighborhood during the 8:30-9:30 time period to avoid the heat of the sun, but this is highly variable. In the past year, I recall seeing the plane three or four times, and hearing the plane go directly overhead about the same number of times. That seems to be a very low number of sightings for a flight that comes by every night in a tight time window that usually does not vary by more than 20 minutes, but it shows that very frequent, regularly occurring events may only be observed a few times a year if one is not consciously trying to go outside and watch for them.

What can explain such an apparent frequency of flybys of "The Light?" Well, it was clearly shown that the flybys were not random, but were occurring with some frequency, and this lasted more than a decade and may still be going on. If the flybys were totally random in every way, our equation shows that we would not expect to have ever had more than one, or at most two, sightings in our lifetimes, and possibly none. There was something completely non-random going on. Now, I know the active and quite clever minds of the debunkers who read this are going to either dismiss it as too incredible or say that we were seeing some kind of scheduled airline flight whose main collision avoidance light was bent by temperature inversion (mirage) in the atmosphere, or some other highly improbable explanation such as earth-lights. I would say that these particular explanations are so improbable as to be almost impossible, except that to be fair we do have to look into the earth-light explanation for some of the sightings. It could not explain the close encounters, the object backlit by the moon, the sightings where rock-steady hovering for some duration of time was seen followed by flight or very rapid flight, and the sightings where "The Light" maneuvers around in an apparently intelligent manner or appears to land, but it would give a way out to those that cannot accept there can be unexplained phenomena out there. Their worldview and their psyche could remain intact. They would not have to open their minds to the unexplained reality before them.

After some contemplation, I have come to the conclusion that many of our sightings were of the smaller object or objects. The landing in the woods incident clearly showed that the bright white light could be associated with a smaller object. The incident on my grandparents' porch is another situation where I should have seen some kind of outline if it were a larger object flying over. Even the bright light is not enough to obscure the outline of an object 30 to 40 feet in diameter going directly overhead at close range. However, the brightness of the light can completely obscure an object of five feet, or maybe even ten feet in diameter. I have to conclude that a remotely controlled object programmed to maintain a certain distance above whatever it was flying over flew silently over my grandparents' house that night. The apparent distance or height was 50 or 60 feet above whatever was beneath it: ground, trees or a house. The small size of the object in this case, and the closeness to the ground, meant that such an object might be difficult to detect on any radar, even a downward facing satellite radar where it might not draw much attention at such a slow air speed in an area considered non-threatening.

Is there a precedent for such a small UFO? There are reports of small metallic disks no larger than five feet in diameter from the historical record. I was once watching a program discussing crop circles when a video was shown of a small metallic disk. Though I have not researched crop circles to any extent, I find them interesting, even if it eventually turns out that all were made by humans. The more elaborate crop circles are quite incredible. A surveillance video camera overlooking a field from a nearby ridge recorded a small metallic disk flying around in the daylight. The location was in the UK, and the program was on sometime in the late 1990s or early 2000s. I do not know about the chain of custody of the video. Debunkers will claim that the video was faked in some way by altering the original. I do not think that it was faked, but I cannot provide evidence that it was not, and I cannot even reference the exact program it was on, but I know it was on TV because I watched the program.

Another reference to a smaller UFO is the December 1980 case of a US airbase in England called the Rendlesham Forest case. [83] The case spanned four days, or rather nights, around an airbase in the Suffolk, England area. What was not known to the public at the time was that nuclear warheads were located at the base in strict violation with treaties that we had signed with the Soviet Union. Just another instance where UFOs have checked out nuclear facilities, ours and theirs, all over the world, and no one has been able to do anything about it. [84] One incident in this case involved a smaller trapezoidal UFO that landed in the woods and was reported to have been actually touched by one of the airmen in the search party. Imprints on the ground and residual elevated levels of radioactivity were found at the site. Debunkers who learned of a lighthouse in the area claimed that the airmen at the base had become confused by the lighthouse beacon reflecting off of a farmhouse window. They claimed this without ever going to the site and getting actual locations and angles. A check of this on a History Channel program dealing with the case showed the debunker explanation to be nonsense. The lighthouse beacon pointed in the wrong direction and never pointed in a direction that could have caused any confusion. At least one of the formerly outspoken debunkers admitted so on the program, seemingly in amazement that a completely ad hoc, off the wall, knee-jerk, no thought, no actual investigation explanation for dismissing a UFO case could be wrong. What a shock; go figure.

I recently discovered that the prominent ufologist Jacques Vallee has the idea that the Rendelsham Forest incident was staged by military intelligence to test the base personnel. If that was the case, the main personnel involved, Jim Pennison and Colonel Charles Halt, do not appear to be aware of it, as they have appeared on several History Channel programs and other programs discussing the case without being aware of such an exercise. Such an exercise seems to be quite farfetched to me, and the technology involved seems beyond our technical capabilities. In other words, I do not think our military had the technology in 1980, or has it today, to stage such a test.

Returning to my family's sightings, a case like the Smith County UFOs is where real UFO research could be useful. Cameras could be set up that transmitted their images directly to a recording device untouched by human hands where the images could be analyzed, but not altered. Sensors that recorded the intensity and wavelengths of light coming from UFOs, their altitude, speed, and size could be set up, but all of this costs money, and takes resources and personnel. The scientists at the 1997 UFO physical evidence conference organized by Peter Sturrock [56] recommended as much for gathering physical evidence from UFO sightings. Ruppelt, as head of Project Blue Book at the time, had an even grander plan in place to study UFOs at the end of 1952 that was shot down by the findings of the Robertson Panel and the failure of some of the equipment that had already been placed in trial operation. However, since the UFOs dictate where and when they show up, and not the scientist, an effort like this could be very frustrating with years of effort that give no results. This is one reason that there have been only a few attempts at systematic study of UFOs where one sets up sophisticated equipment and waits for a UFO to come by.

However, in a case like the Smith County UFOs where objects flew by frequently and clearly were not random, a monitoring effort might have generated some results. There were clearly phenomena that were occurring repeatedly over time. We were just observers of the phenomena. We did not imagine the phenomena; we did not spin ourselves around until dizzy and look up to see the planets and the stars spinning around. We were not victims of the phenomenon known as the auto-kinetic effect where someone stares at a stationary planet or star against a dark background for so long that it appears to move. How anyone can be fooled by this is beyond me, because you can create your own point of reference by aligning the star or planet with your pointed finger and noting the angle of your outstretched arm. It will only move the very slow way that a star or the moon crosses the sky due to the earth's rotation. No, we were just minding our own business and the phenomena came to us, crossing our points of reference in tens of seconds.

The question sometimes comes up about why we did not try to photograph the lights. Part of the answer is in a preceding paragraph and part of the answer was due to my mistaken assumption that no useful information could be obtained using a simple 35 mm camera or a home movie camera. It turns out that 35 mm color film photos and 16 mm home movie film can be used to estimate a light's spectrum and power output. Photographic negatives are harder to tamper with to alter an image than the digital age images of today, and photographic images of lights at night are much truer images of the lights than what one gets with the typical cell phones of today. Never knowing exactly when the light phenomena would appear, thinking that we needed special cameras equipped with a grating to separate the various wavelengths of light and record their intensities, and that the cameras needed to be in a triangulated configuration, it seemed that great expense and effort would be needed trying to record something that may never appear again.

The 1997 conference on UFO physical evidence led to a report recommending study at two sites in the world where unexplained lights occurred frequently: Hessdalen, Norway and Marfa, Texas. In fact, some study had been attempted at Hessdalen, and the scientific panel had some excellent recommendations for further study. One recommendation was to use steady fixed cameras triangulated so that the position, speed and direction of an object or light can be determined unequivocally. As of this writing, I have not heard of a summary report or an updated report for the explanation of these lights. I did find that the CUFOS (Center for UFO Studies) monitors an ongoing project at Hessdalen. There is much speculation about these lights that I will get into later.

As for my family, by the mid-1980s Clark was no longer living at home with more responsibilities and less time to be out looking. Since I only came home two or three times a year with children of my own, and friends and families to visit, observation time was limited. We were just not there to look at the night sky as much at that point, and no one was outside as much at night after the mid-1980s. Clark did move back to the area in the 1990s and he and his son Chase watched for "The Light" on many occasions, but saw nothing.

Early in the course of my research, I ran across a report that similar light phenomena have continued to the present day in another Mississippi location not that far away from the Polkville community. About a year later, I learned that my niece, Tiffany, Clark's daughter, had a sighting of "The Light" sometime in April of 2011. Tiffany currently lives near the family farm. This is some 27 years since anyone in the family had reported a sighting, even though Clark and his son Chase have watched the skies out there on some occasions in the interim. This now means that four generations have seen some unexplained light phenomena near the family farm over a period of 75+ years. Tiffany's sighting was in many ways similar to the March 1973 sighting, which she had never heard described before. Her sighting involved a large light that hovered twice and changed direction twice.

While watching TV one night, she saw light coming in the rear picture window of her home. She looked out the window to see a large round light with an orange tint that appeared to be just above the treetops. She estimated that it could not be more than a quarter mile away. She described a streetlight-sized orange light (the streetlight seen at 100-200 feet) moving steadily from East to West. After a few seconds, the light stopped in mid-flight, hovered briefly for a few seconds, and then reversed course as it retraced its path a short distance before stopping again for a brief time. It then reversed direction again as it resumed its original course and headed out of sight to the west. The entire incident occurred without any noise that could be heard inside. The light was so large that it rivaled that of a full moon. There was no turning maneuver, no change in altitude. It was as if the light were sliding forward and backward on an invisible rail. The only difference in this light and what we saw in the 1970s and 80s seems to be the faint orange tint. She later told her husband, "I just saw my first UFO."

Then about a year later, sometime in April of 2012, Tiffany and two of her friends had another unusual sighting. They were outside on the front porch enjoying the evening when one of her friends shouted, "What's that?" Tiffany looked up to see an intense blue light burst so intense that she had to look away. At first they thought the light was a police spotlight, but immediately realized that the light was coming from just above the treetops. The light quickly faded and a smaller lighted object seemed to float down toward the ground in the distance. This secondary light phenomenon had a white bottom light and a blue top light as it slowly descended to the ground some distance away. There was no sound associated with any of the light phenomena. They ran inside and described what they had seen to their husbands and boyfriends, who promptly dismissed what they had seen as a shooting star (meteor). The peculiar thing about a meteor explanation is that the entire light phenomena were stationary and not moving. The only way that a meteor can give this impression is if it is traveling directly toward the observer or observers. However, the dual white and blue lighted object that descended slowly to the ground after the intense blue light faded out is at odds with a

meteor, which would be traveling at tens of thousands of miles per hour. The fragments from an exploding meteor would maintain a very significant velocity until impacting the ground.

I thought for a while that possibly an exploding meteor explanation could be made for this sighting, but I no longer think so. The widely reported Russian meteor case of early 2013 and its accompanying video showed a worldwide audience what an exploding meteor can do. A very substantial and damaging shock wave was produced along with the accompanying sound. There was no sound and no shock wave associated with the blue light burst. The blue light burst occurred near the spot in the 1970s and 80s where we saw a number of lights and objects crossing the road. It seems like quite a coincidence to have very unusual light phenomena occur near the same location time and time again and there not be a connection between them.

Chapter 18

The Mississippi UFO Experience

"Anyone who is going to make anything out of history will, sooner or later, have to do most of the work himself. He will have to read, and consider, and reconsider, and then read some more."

—Geoffrey Barraclough, (1908-1984) British Historian,

"You must always know the past, for there is no real Was, there is only Is."

—William Faulkner, Novelist, (1897-1962), Oxford, Mississippi

There was a UFO conference held in Mississippi in April of 2011. I found out about it while doing an internet search in August of 2011. One of the websites listed the best known UFO sightings for the state and I realized I had only heard of two of them before. [85] Nearly a year and a half later I came across some websites which listed the Project Blue Book Unknowns/Unexplained cases covering the years 1947-1969. [86A] There were about ten sightings listed as unknowns for the State of Mississippi from 1952 through 1967. There was a heavy concentration of unexplained sightings, seven in all, along the Mississippi Gulf Coast not far from Keesler AFB near Biloxi covering the years 1952 to 1965. Several of the sightings were made by base personnel. The first noticeable trend was that the preponderance of unexplained sightings for the state seemed to occur from central to south Mississippi. The PBB unexplained sighting listed for Greenville in 1953 was the farthest north, which is perhaps 30 miles or so north of an imaginary line cutting the state in half. Three unexplained sightings were from the central part of the state along, or a little south of, interstate highway 20 and, as indicated earlier, seven were along the Gulf Coast from Gulfport to Pascagoula. One thing struck me right away as I was reading the sighting reports from the late 1950s to the early 1960s. Three of the cases involved the sighting of objects with red lights similar to Clark's sighting in 1963 that the rest of us completely ignored.

Some of the more well-known sightings for Mississippi are listed below in chronological order. Only two of these six sightings were listed in Project Blue Book, because PBB closed its doors for good on January 30, 1970. Four of the well-known sightings occurred after that date. They occurred in the turbulent UFO decade of the 1970s, and one of them achieved worldwide fame.

1957 Mississippi Gulf Coast and Flyover of Four States. The RB-47 case of July 17, 1957, is a very famous case involving the most sophisticated electronic intelligence gear in the

US Air Force at the time. A radar type signal was detected coming from a UFO that followed the airplane for over 700 miles above Mississippi, Louisiana, Texas and Oklahoma. At one point the UFO was seen by radar in the air and on the ground, and visually appeared as a single, bright white light as it flew under the plane at a speed unmatched by anything the crew had ever seen. Klass claimed that the whole thing was a comedy of errors and that one of the most highly trained crews in the Air Force had mistaken a commercial airliner for a UFO, a very unlikely or even preposterous hypothesis. The object was tracked on civil defense ground radar, two forms of aircraft radar, and seen visually all at the same time. All forms of observation placed it near the same location. It disappeared from all four views simultaneously. There were other cases of this type that were reported in the mid to late 1950s where UFOs were reported to send out radar signals.

1967 Meridian. The Meridian case of 1967 was listed as unexplained in Project Blue Book (case 11869, July 10, 1967). The witness, Phillip Lanning, was a highly trained, former military officer whose car died just as he saw a large metallic object that "looked like a cymbal on a drum set and was a dirty metallic gray in color on the underside." The object flew away and his car restarted. This is very similar to a series of four or more reports in one night in the Levelland, Texas area ten years earlier in 1957 that were never satisfactorily explained.

The explanation for the Levelland incident, put forward by Ruppelt in the added last chapter to the second edition of his book in 1960, was St. Elmo's Fire. Now, that explanation is so preposterous that again I cannot help but laugh. Most witnesses described a large, 100-foot object on or near the ground. When the object lifted up and flew away, their cars were able to be started. Similar phenomena were seen in nearby New Mexico around that same time. I believe that Ruppelt, who was still involved with the Air Force censors, was pressured in the second edition into attempting to debunk some famous cases, and that he put in the most preposterous explanations he could think of so that the astute reader could see what he was forced to do. Maybe I am correct; maybe I am wrong. In the Meridian case, the witness was highly trained, unimpeachable, and described exactly what he saw. Skeptics and debunkers try to dismiss the many military and airline pilot sightings with a sweeping generalization fallacy that highly trained military personnel are no better observers than anyone else, which, if true, would mean all of their training was for naught.

1973 Pascagoula. The Pascagoula abduction case occurred on Thursday October 11, 1973. The story is as follows: Two men were fishing on the Pascagoula River at night when a craft some 30-40 feet across and 8-10 feet high with a blue light hovered on the riverbank 30-60 feet behind them just a few feet off the ground. They became aware of its presence due to a buzzing sound that caused them to turn around. A door opened on the craft and three strange occupants came out to abduct Charles Hickson and Calvin Parker. The men were taken inside and given separate examinations, then they were released in a state of panic. The humanoid beings that Hickson and Parker described were quite strange and not often reported before. They later came to be called the "mummy" or Michelin man type of alien with the distinguishing feature of claw-like or mitten hands with a thumb. Hickson later thought these beings could have been robots.

Although Hickson and Parker remembered almost everything that happened from their conscious memory, they had some difficulty remembering exactly how the craft had left. Sometime much later under hypnosis, Hickson stated that the craft left by shooting straight up in a flash of light. [87] They struggled with whom to report this to, or if they should report it at all, as they felt that no one would believe them. After calling Kessler Air Base and being told that the Air Force did not investigate such things anymore, the men called the Jackson County Sheriff's Office before briefly stopping by the local newspaper that was closed, and finally were escorted to the sheriff's office to report what had happened. The sheriff and deputies questioned them separately. The sheriff's interrogators tried every interrogation trick they could think of and could not break Hickson and Parker's story. Finally, they were left alone in a room and were secretly tape recorded. The sheriff was sure that they would give everything away as a hoax. The transcript of the tape reads like the truth as much as anything I have ever read. The audiotape itself is said to be even more moving. They were desperately trying to understand what had happened to them and why it had happened to them. Hickson later passed a lie detector test (both had volunteered to take lie detector tests, but only Hickson ended up taking the test because at the time Parker had suffered a nervous breakdown), with the examiner exclaiming about Charles Hickson, "Hell, he's telling the truth!"

Klass found fault with:

(1) the two-hour time delay in reporting the incident (Klass was simply being absurd about this point, as I hope even the most avid debunker can see. This is one of the most punctually reported abductions in UFO history). That Klass even brought this up shows his fanatical mindset.

(2) the credentials of the lie detector examiner even though he worked for Pemberton Detective Agency and had conducted over 500 lie detector tests (similar to if my employers had been upset because I was not Linus Pauling, a Nobel Prize winning chemist). When a UFO witness passed a lie detector test, and most did, Klass always found fault with the test or examiner, and those wishing to discredit the case lapped it up.

(3) Charles Hixson's past promotional practices as a supervisor at a shipbuilding facility (accusations that were disputed by Hickson).

A reporter, Joe Eszterhas of *Rolling Stone* magazine (later a scriptwriter in Hollywood) also wrote perhaps the best investigative article about the Pascagoula abduction at the time, appearing in the January 17, 1974 issue and entitled "Claw Men From the Outer Space" (a very earthy, vulgar beyond the pale, raunchy, irreverent, but even handed article telling it like it was in a very graphic way about the Mississippi Gulf Coast and the Pascagoula area as it existed at the time). [88] Many do not like the Eszterhas article, and it is too vile for my tastes, but it treated Hickson and Parker fairly.

A young reporter for the *Mississippi Press-Register*, Murphy Givens, knew that the abduction site was within viewing distance of a 24/7 drawbridge operator's booth and a 24/7 railroad bridge operator's booth. Also, although the abduction site was a little over a mile away from the Litton-Ingalls shipyard, he knew that there were manned security cameras around the perimeter of the shipyard looking out at all times.

Eszterhas indicated that Givens contacted the booth operators and the security camera operators to see if they had seen anything. They said that they had not. No further investigation was done, according to the article. If there was security camera tape or film footage for around the time of the abduction (9:00 p.m. to 10:00 p.m.), there is no indication that it was ever examined. It is important to note that the cameras appear to have been manned live, and perhaps recordings were not being made at that time. Some sloppy accounts of the story imply or state that the craft hovered over the water of the Pascagoula River, in which case the booth operators and the security cameras could have seen it, but Hickson's and Parker's account clearly states that the craft hovered in a "clearing" behind them making it more difficult, if not impossible, to see with the cameras unless it could be seen coming or going.

In the 1983 book, *UFO Contact at Pascagoula* by Charles Hickson and William Mendez, the abduction site is discussed in detail. The clearing was a break in the ten-foot high river grass behind the fishing spot. The fact that toll booth operators and drawbridge operators did not see anything was directly addressed. A UFO investigation group, The International UFO Registry, shined spotlights from the abduction sight directly upon the booths in question at night, and none of the booth operators reported seeing anything. There had been an accident in 1975 in which an 18-wheeler had crashed through the guardrail a few hundred feet from the operators' shack during the night and it was not noticed until morning. The booth operators might as well have been comatose. The question of whether or not someone in a motor vehicle could have seen anything was also addressed and the answer was yes, the abduction site could be seen for two seconds if the occupants in the car were looking in the right direction. Most of the time on the bridge, the site was blocked by a guard rail. The cameras at Litton-Ingalls shipyard were a mile away and could not have seen anything during the abduction. The best time to see the craft was when it was coming or going, but that was probably less than a second, at least in the going, and maybe 30 seconds of being within view as it settled into the clearing. It could have easily been missed. However, three separate phone calls came into the sheriff's office concerning a blue light near the river and around the docks area. One was a very credible witness, a former city councilman, with a lot of credibility at the sheriff's office, and another was a former military pilot. [89]

The Jackson County Sheriff's Department personnel that laughed when first confronted with Hickson and Parker's story, came to believe them, or at least believed them to be totally sincere. Ufologists Dr. J. Allen Hynek and Dr. James Harder (University of California at Berkeley, hydraulic engineering professor), came down to investigate and attempted to hypnotize Hickson to get more details, but had to stop when he became too frightened and agitated to continue. Some ufologists back away from the Pascagoula case because they say hypnosis was involved. Their reactions are extremely puzzling, because two witnesses gave the basic account and the vast majority of the details of the case before any hypnosis was attempted. Both Harder and Hynek became convinced that something very unusual had happened to Hickson and Parker. Hickson and Parker have never deviated from their story. Charles Hickson died at the age of 80 in September 2011.

One last point about Charles Hickson, he and four or five family members had a sighting of "The Light" on Mother's Day 1974, some seven months after the abduction incident. It was very clearly "The Light," as it is called by my family, and in *UFO Contact at Pascagoula* they also called it "The Light." (I ordered this book and read it in February of 2012 after deciding to write about UFO abductions). It happened late at night as they were traveling by motorcar from Hickson's parents' home near Saundersville, MS back to Gautier. They were on Highway 67 in an isolated area in the edge of the Desoto National Forest when an object with a large bright light on the bottom of it started following their car. The object got ahead of them, crossed the road in front of them from left to right, and hovered just above the treetops of 10-15 foot pine saplings. Coming over the road into position, the bright light lit up the road and trees underneath it. The object itself could not be seen, just the large bright light underneath it. Sound familiar?

The Hickson family stopped their car to observe "The Light" as it hovered at about 100 yards away. Here is where the object's behavior changed. The bright bottom light was dimmed or shut off, and small porthole or window lights could be seen around the middle of the object similar to what my brother and mother described at the community dump sighting. The full size of the object could be seen for the first time. It was large at maybe 100 feet across. This sighting and two others created a credibility problem for Hickson, because UFOs were thought to be randomly flying about willy-nilly without any rhyme or reason.

One person having two sightings or more within seven or eight months caused what ufologists called at that time the "repeater" problem and debunkers called "UFO buffs." If they have not already abandoned this myopic notion, it is time to now. If UFOs are real and involve real live occupants from some unknown origin, then their number one activity is very likely checking out the most advanced residents of terra firma. That is the most logical hypothesis for their repeat activity. If an observer lives near someone or something of interest to the UFO, or if the observer themselves is of interest to the UFO, then there will be opportunities for multiple sightings. Conversely, if the UFO is some type of as yet unexplained natural phenomena, and the conditions are right, then the same person may witness it a number of times. What's so difficult to understand about that?

The Mother's Day 1974 sighting was witnessed by five or six people. One was Hickson's wife, Blauche, who was in hysterics, as she was probably more disturbed by the abduction account than anyone other than Calvin Parker. One note about the case must be included. Hickson believed that the UFO occupants communicated telepathically with him on three separate occasions after the original abduction event. He was alone on two of these occasions and the message was about him being chosen and something important would happen or be revealed later, but on this Mother's Day sighting the message was "this is not the time," perhaps due to the number of people present or the very upset reaction of his wife. Of course skeptics will say that these communications were just imagined by Hickson and will imply he was crazy all along, although a battery of psychological tests over the years showed him and Calvin Parker to be in the "normal range" in every way.

Two well-known Mississippi cases occurred in 1977: the Flora sighting on February 10 and the Taylorsville sighting on May 25. I believe both of these sightings are related to my family's sightings.

1977 Flora. The Flora sighting involved observation by sheriff's deputies and other witnesses, according to the account in the Feb. 17, 1977 edition of the *Madison County Herald* in Canton, MS that was found on the website: *www.latest-ufo-sightings.net/2011/04/1977-flora-mississippi-ufo.html* [90]. This case is somewhat reminiscent of the Father Gill case in Papua, New Guinea in 1959. [91] The total number of witnesses in the Flora case is said to be as many as 24 in all, and many of them were law enforcement officers. A hovering object was seen by Deputy Kenneth Creel from his patrol car. It was seen out beyond a field near Flora at first, but it slowly moved closer until it was hovering directly over the patrol car at an estimated height of 20 to 30 feet off the ground. There was no bottom light; instead there was a soft row of lights around the middle of the round object that Creel said looked like a child's top. The lights seemed to shine through small windows that circled around the craft. The lights would change colors at times, appearing blue, red, green and other colors. The object hovered directly over the patrol car for about a minute before Creel backed up about 100 yards. At that point, Highway Patrolman Louis Younger drove to within a short distance of the hovering object and confirmed the description given by Creel. Other witnesses who saw the UFO from a distance included Deputy Charles Bowering and Highway Patrolman Joe Chandler. The object was reported to hover there for about 45 minutes. As it started to move, the police followed the object until it became obscured by the tree line.

The Flora case involves a UFO hovering closer and longer in the presence of credible witnesses than any case I have ever heard of, except the Father Gill case in Papua, New Guinea in 1959. Another witness that day in Flora was Mr. Pat Frascogna, who was so taken with the incident that he has become a local UFO researcher. Sightings near Flora have been reported all the way up until the present time, according to Frascogna. It is (or was) a town secret according to him, very much like the situation in Polkville. He has been researching the sightings and is working on a book about the subject. He spoke at the Mississippi UFO conference back in April of 2011, which I only found out about after I decided to write this book in August of 2011. I firmly believe that the Flora UFO and the Madison UFO, seen by a family member only a few miles away from Flora, are either one and the same or very closely related objects. Returning to the description of the object as looking like a child's top, the object described by my parents as it flew directly over them at low altitude can easily be said to resemble one. Again, I believe that what my parents saw, and the Flora and Madison objects, were one and the same, or very closely related objects.

1977 Taylorsville. May 25, 1977 was the date of the Taylorsville, Mississippi case as reported in the *Jackson Daily News* of that same date. [92] The *Jackson Daily News*, an afternoon paper, was able to run the story, as the sighting occurred around four to six in the morning. Officer Steve Sasser of the Taylorsville Police Department had only been on the job for eight months when he got the shock of his life. Sasser was quite shook up by what he saw that morning, which is typical when something brings your world view of reality to its knees. If my family had made the sighting, it probably would have never made the news. We would have said, "Hey there's 'The Light.'" When you are not prepared to see something that is not supposed to exist, it can shake you to your core.

A direct quote from a shaken Sasser was, "I think people ought to straighten up. The object is the Lord's work." Sasser's account goes this way. A call came into the Taylorsville Police Department sometime after 4:00 a.m. and was taken by Charles McKinley. It was from a tow truck driver that had been pulling a vehicle to Laurel. The driver was upset about something he saw out at the Leaf River Bridge near the intersection of Highways 531 and 28. He had called for the police to come out to investigate.

Sasser was dispatched to the area about 1 1/2 miles outside of town to have a look-see. The truck driver's name was not available at the time of the *Jackson Daily News* article, and may never have been reported. When Sasser arrived at the scene, the truck driver pointed to an object in the sky. Sasser reported the object to be about 100 feet off the ground and off to the left of the highway. How he estimated this was not given. The truck driver stated that when he first saw the object it was directly over the highway at an estimated height of 50 feet. Sasser and the truck driver stared at the object for a period of time, but not more than a minute or two, and then the object began to move toward them. Sasser's description stated that the object was oblong with a whitish glow. The object could not be seen well enough to distinguish any windows. He could see a glow of light and a little bit of the outline, which appeared oblong to him. Again, according to Sasser, the object made no noise and could move quickly up and down.

At that point Sasser got back into his patrol car and called in that the object was getting pretty close. Sasser said, "I remember that incident at Pascagoula (Pascagoula abduction 1973) and I didn't want that. It scared me." They called the police chief, Tony Blakney, who came out and attempted to take some pictures, which did not turn out. The camera or type of film was not mentioned, nor do we know what "did not turn out" meant. Was the image just a white smear or a burred smudge? We do not know. Sasser indicated that at that point, the object moved away and to a position higher in the sky, and stayed there for about an hour and 20 minutes.

They went back to town to police headquarters and the object was high enough in the sky to be seen from town. Several townspeople saw it as well. Shortly after that it was gone.

There is little doubt that this was a sighting of "The Light." The inability to see the object clearly due to the bright white light on the bottom, and the idea that the object was oblong, fit with seeing the light from directly underneath where Buford and Clark said the lighted portion appeared more oblong beneath a mostly round outline. Taylorsville is hardly more than 20 miles on a direct line from Polkville, but a little further by the twists and turns of the highways.

Of course, this sighting would have been a no-brainer for Klass. All the commotion was due to Venus or Saturn or Jupiter, or some other celestial body! Never mind that the object made several movements and maneuvers and was in all probability quite a bit larger than Venus. As in so many UFO sightings, the witnesses did not make a straightforward determination of the apparent size of the object or light. This can be done by holding a dime at arm's length visually next to the line of sight of the object. (Friedman recommends an aspirin tablet for the same purpose.) Then a witness can say half the size of a dime at arm's length or whatever is appropriate to measure. My brother used the end of his thumb at arm's length. The idea is to get an apparent size that can assist in attempting to identify the object. Most of the time, the object will be four or five times larger or greater than Venus, and that would rule out the ridiculous contention of a planetary sighting for an object that maneuvered around as well as hovered motionless for an hour and 20 minutes, any celestial object changes its position in the sky very noticeably over that timeframe. Something parked in the same position in the sky that long cannot be Venus, but there is no end to very implausible contentions made by debunkers. If the witnesses had also seen the object depart at high speed, then there would be very little debunking room at all. Since that was not the case, there is maybe some debunking room, but not much. Something that can hover off and on in the same area for most of two hours, maneuver around to change altitude, make some short quick moves up and down, and all the while remain silent cannot be a balloon of any type, a dirigible (as loud as a lawn mower), or a helicopter. A helicopter or a dirigible, and possibly a Harrier jet, are the only aircraft than can currently maneuver somewhat in the way described, but with very noticeable sound effects. The reactions of Sasser and the truck driver are telling in that they knew they saw something that was not supposed to exist.

The five cases presented above are the most famous UFO cases from the State of Mississippi. Three of them occurred in just a span of less than five years at a time during which the sightings from the Polkville area were also at their peak. While researching the Pascagoula abduction case, I have recently found another UFO sighting that is puzzling because it was not on the list of famous UFO cases in the State of Mississippi. It was reported nationally and perhaps worldwide at the time. It is the Petal, MS case of Saturday, October 6, 1973, just five days before the Pascagoula abduction case. [93] My wife and

I lived in Hattiesburg at that time, as I was attending the University of Southern Mississippi as a graduate student in the chemistry PhD program. I do not recall hearing about the Petal case, even though Petal is only a few miles from Hattiesburg just across the Leaf River Bridge near where Bowie Creek joins the river. The case was carried by the Associated Press nationwide. I was deep into my dissertation studies performing experiments, analyzing the data, and trying to understand the results. I haphazardly watched the local or national news on TV and did not consistently read the local paper, so I missed a case that began not more than six to eight miles away.

1973 **Petal** (24 miles straight line distance south of Laurel). October 6, 1973. The Petal, Mississippi case involved police constable Charlie Delk, who ended up chasing a UFO for 30 miles in his patrol car after some frightened Petal residents had called in a report about a strange bright light in the night sky. The most complete account of the Petal case that I found was in *UFO Contact at Pascagoula* by Hickson and Mendez where Mendez interviewed Delk in the book. On several successive nights, different residents of Forrest County had called about a strange light in the sky. The dispatcher called Delk and he had gone out to investigate, but found nothing on those prior occasions. He had begun to think that the residents had seen a bright star or something, and their minds had played tricks on them. On this night, a Saturday, Delk was loathe to go out to investigate. His favorite TV show "Columbo" was on. He basically ignored the first call, but sometime later another call came in about a strange lighted object. The three or four calls over several nights reporting a strange light in the sky is supporting evidence for an unusual phenomenon of some sort being present and consistently seen.

The TV show was just finishing up so, inspired by the show, Delk headed out, but did not find anything at first. Just before concluding that he was on a pointless call again, a bright light flying low and going slowly passed over his patrol car. Delk took notice and at that point followed it toward Petal High School. He states, "When I got there, I spotted it over Petal High School." Delk said it looked like an old-time wind-up top with yellow lights all the way around. (Notice the similarity to the description of the object involved in the later UFO case near Flora in 1977). Delk followed the object to the Jones-Forrest county line at which point it slowed down over a field and some additional lights came out of the side of the object. Delk described them as resembling blow torches.

He continued to follow the object, which was going 25 to 30 mph, as at that speed the patrol car tracked along with the object. Delk said he followed the object because its slow speed and nearness to the ground made him think it was getting ready to land, and if it did he wanted to see what came out. (I checked the weather on that date and on the 7th. I could not readily find the Hattiesburg data. Weather archives were found for Jackson, McComb, and Biloxi on those dates. Hattiesburg lies some 90 miles southeast of Jackson, about 70 miles east of McComb and 70 miles north of the Gulf Coast. The state was dominated by somewhat of a high pressure system during that time. Looking at the three weather stations, the maximum sustained wind speed was 12 mph, and average wind speed was 6 to 8 mph). This is important to show that it was not a windblown object of any type. Notice that the object stayed close enough to the roads and highways that Delk could follow it for such a distance. If fact, he later said that it led him in a big circle. This is highly improbable for a windblown object that will at some point usually be blown cross country making it impossible to follow by motor car. It was very much as if the object flew near the roads and highways to make it easier for Delk to follow.

Eventually, Delk found himself fairly close and directly underneath the object, at which point the patrol car died and would not restart for 15 minutes. Delk had several moments of anguish being in a desolate spot with his patrol car and the radio out. He checked under the hood and saw nothing obviously wrong. Suddenly the lights came on (he had not turned them off when the car died) and it started right up as if nothing was wrong. Sheriff's deputies had tried to contact him during the blackout time and were concerned because he lost radio contact while in pursuit. The object had not continued on at 30 mph for 15 minutes, for if it had it would have been some seven miles away when the patrol car restarted. If the object were wind-blown, it could have drifted about a mile and a half away in that time. Instead, the object was still within sight just a short distance away. He continued to chase the object through the Jones County town of Ovett and past a swamp when in one account he says the object did a "double flip" and was gone. In another account, he said it had just disappeared. He waited and listened for about 20 minutes, trying to hear some sort of sound, but heard nothing. He had lost the object in the upper part of the Desoto National Forest in an area that is today also designated as the Chickasawhay State Wildlife Management Area. There are two reasonable explanations of how the object just disappeared. One is that it made a sudden move so fast that if Delk blinked or looked away for an instant at the same time it would appear to just disappear. The other explanation is that the object's lights went out. Delk had been able to see and follow the object due to its lighting. Unless lit up by moonlight or a streetlight, such an object without any lighting would be hard to see and would seem to just disappear. Delk had nothing to gain and everything to lose by telling his story. The constable is an elected official. Many potential voters may look askance at such a story. He genuinely was curious about the object, and followed it for as far as he could and reported his experience.

If we plot the 1970s' cases near Jackson on a map, it shows a relationship I was not expecting. Looking at such a map, we see that Madison and Polkville almost fall along a straight line drawn from Flora to Taylorsville. In addition, the line follows a northwest-southeast track through Smith County, as seen in so many family sightings. Most of the sightings made near our farm were of an object traveling along a northwest-southeast track and vice versa. What does this mean? Is there any significance to the Smith County NW-SE track? I will explore that in the following chapters.

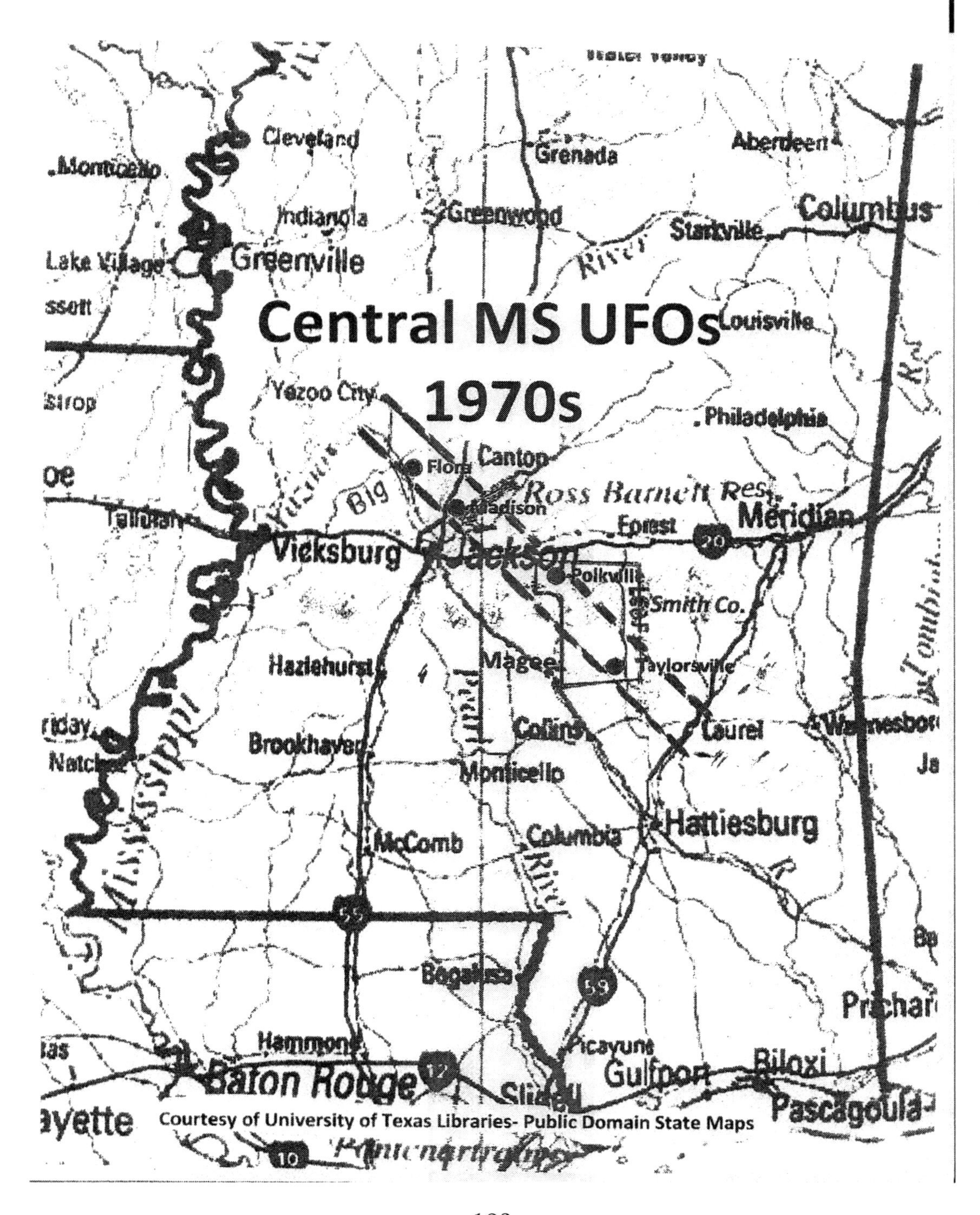
Central MS UFOs
1970s
Cleveland
Grenada
Aberdeen
Greenwood
Greenville
Indianola
Starkville
Columbus
Louisville
Yazoo City
Philadelphia
Flora
Canton
Madison
Ross Barnett Res.
Vicksburg
Jackson
Forest
Meridian
Polkville
Smith Co.
Hazlehurst
Magee
Taylorsville
Brookhaven
Collins
Laurel
Monticello
McComb
Columbia
Hattiesburg
Bogalusa
Picayune
Gulfport
Biloxi
Baton Rouge
Pascagoula
Courtesy of University of Texas Libraries- Public Domain State Maps

Chapter 19

The National Forest Connection

"The clearest way into the Universe is through a forest wilderness."

—*John Muir, 1838-1910, Scotsman & American Environmentalist whose influence helped start the conservation movement.*

When I first looked at maps of the State of Mississippi, I did not pay much attention to the locations of the national forests. In truth, I did not even think about them in any context. One day I plotted the locations of the most famous sightings in the state, added the Project Blue Book unknowns, and added the Madison, Polkville, and Laurel locations. That is when I saw an approximate straight line between Flora, Madison, Polkville, and Taylorsville. Very curious, I thought. Still looking at maps, I called up a detailed features map of the state, which happened to show the locations of the national forests. Studying this for a moment, I saw that a straight line could be drawn from the Delta National Forest northwest of Flora through the Bienville National Forest just east of Polkville to the northern section of the DeSoto national forest close to Laurel and Petal. This line just about overlays the line of sightings from Flora to Taylorsville to Petal, a NW-SE line.

When Constable Delk chased the Petal UFO, he ultimately lost it in the upper section of the Desoto National Forest, which is split into two sections in southeast Mississippi. It's as if the UFO made its way back there to give him the slip. Could there be a national forest connection to these reported sightings? Most of south Mississippi is heavily forested, not just in the national forest areas. The state is about 65% forested, and much of that is in south Mississippi. Could there be a connection between UFOs and heavily forested areas? Have I been a data junkie for so long that I like to plot everything and see relationships where none exist? Is it just a coincidence? Is this all in my graph-first-explain-later mentality with little real significance? Could a heavily forested area give visitors from somewhere else a place to hang out during their "down time," or could there be some type of phenomena associated with heavily wooded areas that we interpret as UFOs? Is it possible that the UFOs were hanging out in the national forests during the time between excursions, and maybe still are today near Flora as Frascogna stated in 2011 that the sightings were still occurring there? I presented evidence earlier that sightings are still occurring near Polkville. While the Delta National Forest is 25 miles away from Flora, there is a strip of forest some five to ten miles wide between Flora and the national forest boundary. In fact, central and south Mississippi are interlaced with cities and towns surrounded by heavily forested areas whether national forest or not.

A series of maps are presented on the following pages. The first map shows the Project Blue Book unknowns/unexplained cases for Mississippi, which occurred from 1952 to 1967. The famous RB-47 case of 1957 is not shown. It was a flyover of our most sophisticated reconnaissance aircraft of the time from the Mississippi Gulf Coast to about Jackson, and then over parts of Louisiana, Texas, and Oklahoma that was reported to be accompanied for some 700 miles by a UFO. Notice the cluster of cases along the Mississippi coast, but also notice that a NW-SE line can be drawn from Greenville to Pascagoula. The line shown is the statistical linear fit line for the locations of Project Blue Book unknowns/unexplained cases in the State of Mississippi. It stretches from about 20 miles south of Greenville and passes just south of Smith County to about 25 miles north of Pascagoula. Statistically, the greatest probability of having an unexplained UFO sighting in the State of Mississippi lay in a corridor running about 60 miles on either side of this line. Smith County lies well within this corridor. The national forest areas are not shown on this map or the next, as my first data plots were done before arriving at the national forest/heavily wooded hypothesis.

The second map includes not only the MS Project Blue Book Unknowns/unexplained cases, but also the four well-known cases from the 1970s. A second line is shown that is a best fit of the combined data, 15 data points in all. The line moved closer to Greenville and closer to Pascagoula, but overall has a similar path NW-SE.

The third map includes the national forest areas of the state and also the sightings reported for the first time in this book. The line shown is the previous best fit line for the PBB Unknowns/unexplained plus the well-known Mississippi sightings from the 1970s. The Polkville, Madison, and Laurel sightings were not included in the best fit line, but it can be visually seen that a line through these locations is essentially parallel to the best fit line. The lines on either side of the best fit line represents a 50-mile wide corridor on each side, and most of the national forest areas of central and south Mississippi lie within this corridor. Statistically, the highest probability of having an unexplained UFO sighting in the State of Mississippi lies within this corridor. If the center location of each of the national forest areas in this corridor is plotted in the same way, we get a NW-SE line that is a little less steeply sloped, but still has a definite NW-SE slope. The dotted line on the third plot shows the national forest location best fit line.

Most of my family's sightings were of UFOs traveling a NW-SE path similar to the UFO sightings best fit line, although we did not realize it at the time. This appears to be more than just a random, coincidental correlation. It appears that there is a connection of some type. The frequency of our sightings and the consistent NW-SE path suggests that we were seeing a repeating purposeful phenomenon, not something random and haphazard.

One of the points to be made in this book is that the UFO phenomena of central and south Mississippi were very consistent during the 1970s. The nature and range of the phenomena formally reported during the 1970s in Mississippi was very much consistent with the

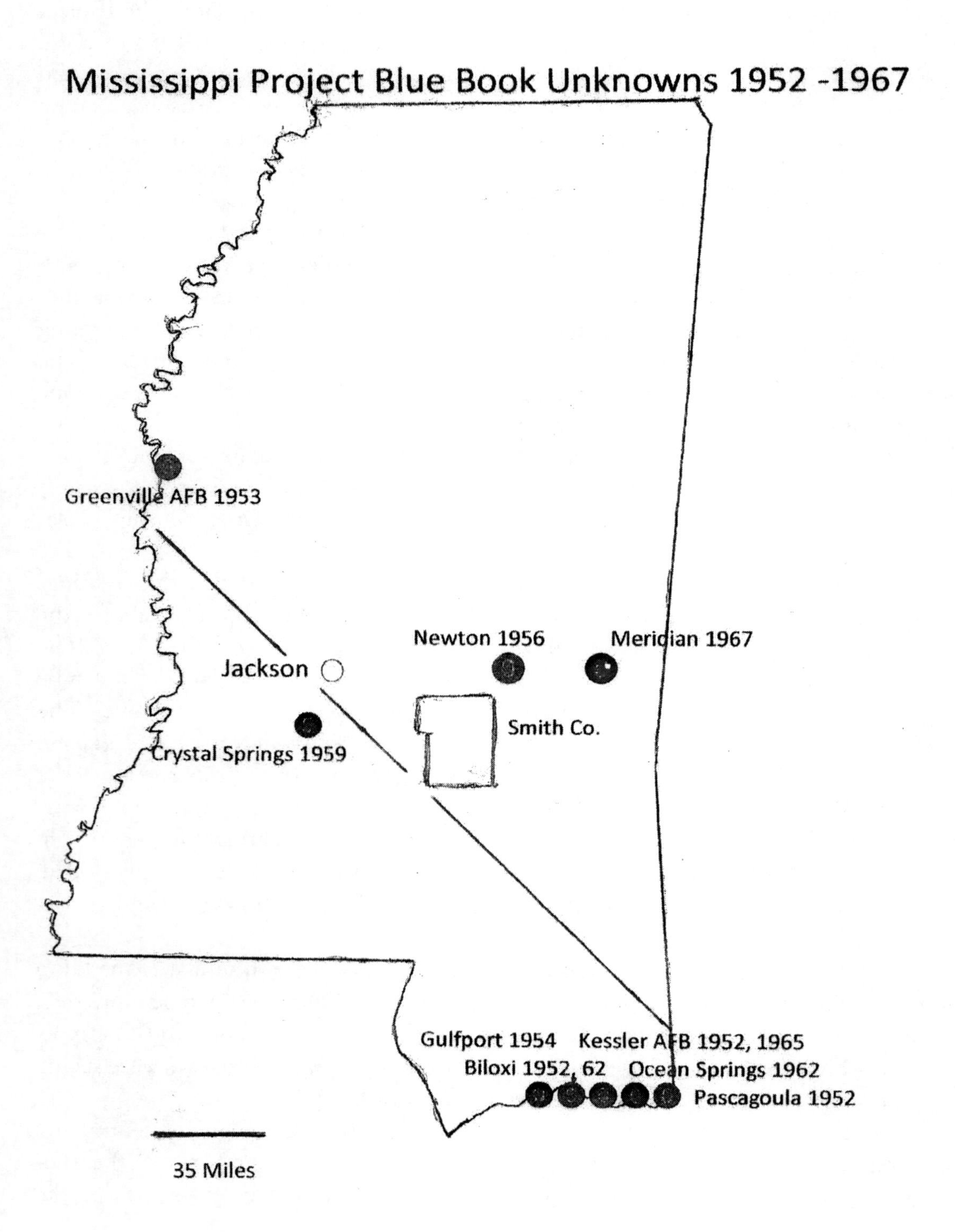
Mississippi Project Blue Book Unknowns 1952 -1967
Greenville AFB 1953
Jackson
Newton 1956
Meridian 1967
Smith Co.
Crystal Springs 1959
Gulfport 1954
Kessler AFB 1952, 1965
Biloxi 1952, 62
Ocean Springs 1962
Pascagoula 1952
35 Miles

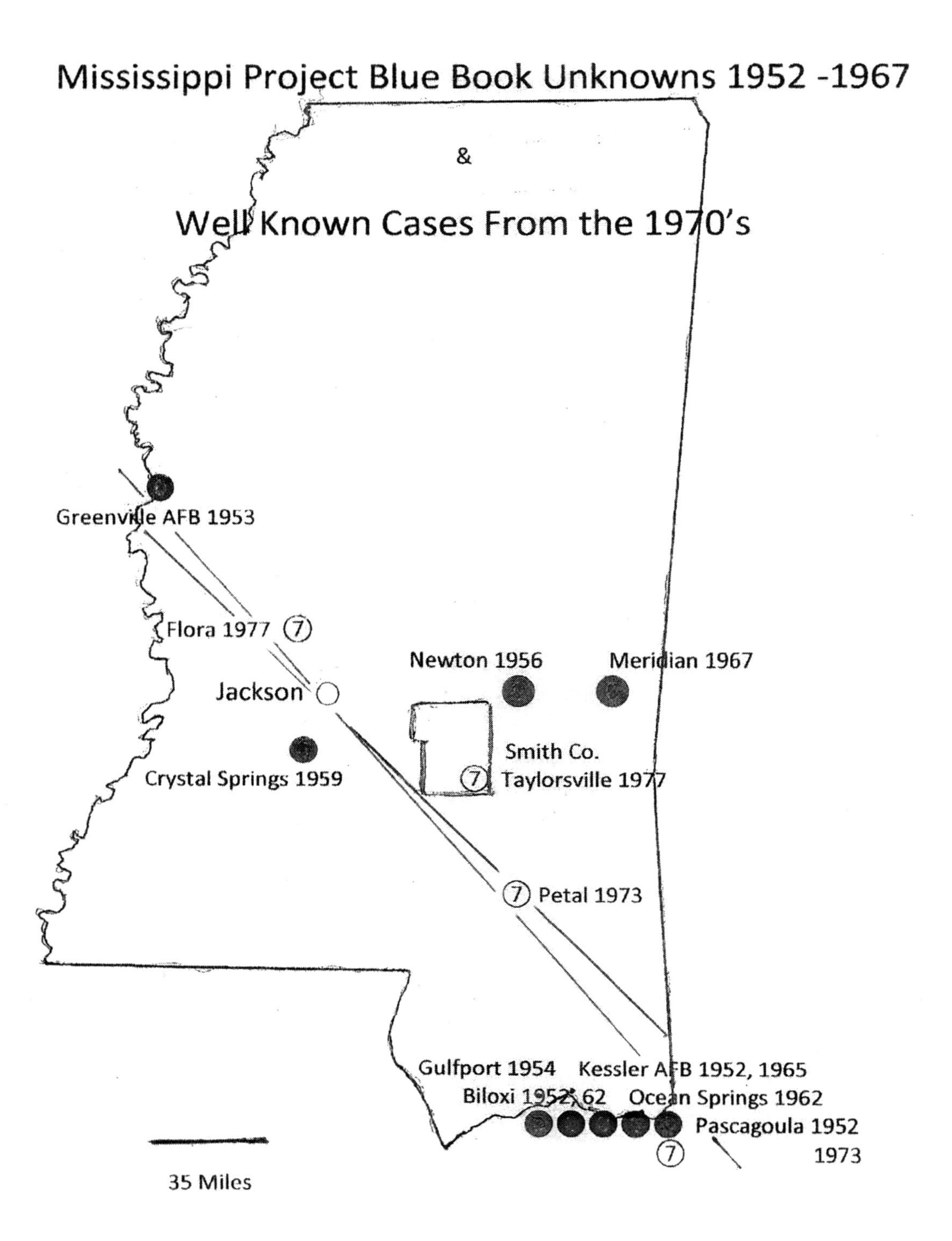
Mississippi Project Blue Book Unknowns 1952 -1967
&
Well Known Cases From the 1970's
Greenville AFB 1953
Flora 1977
Jackson
Newton 1956
Meridian 1967
Crystal Springs 1959
Smith Co.
Taylorsville 1977
Petal 1973
Gulfport 1954
Kessler AFB 1952, 1965
Biloxi 1952, 62
Ocean Springs 1962
Pascagoula 1952
1973
35 Miles

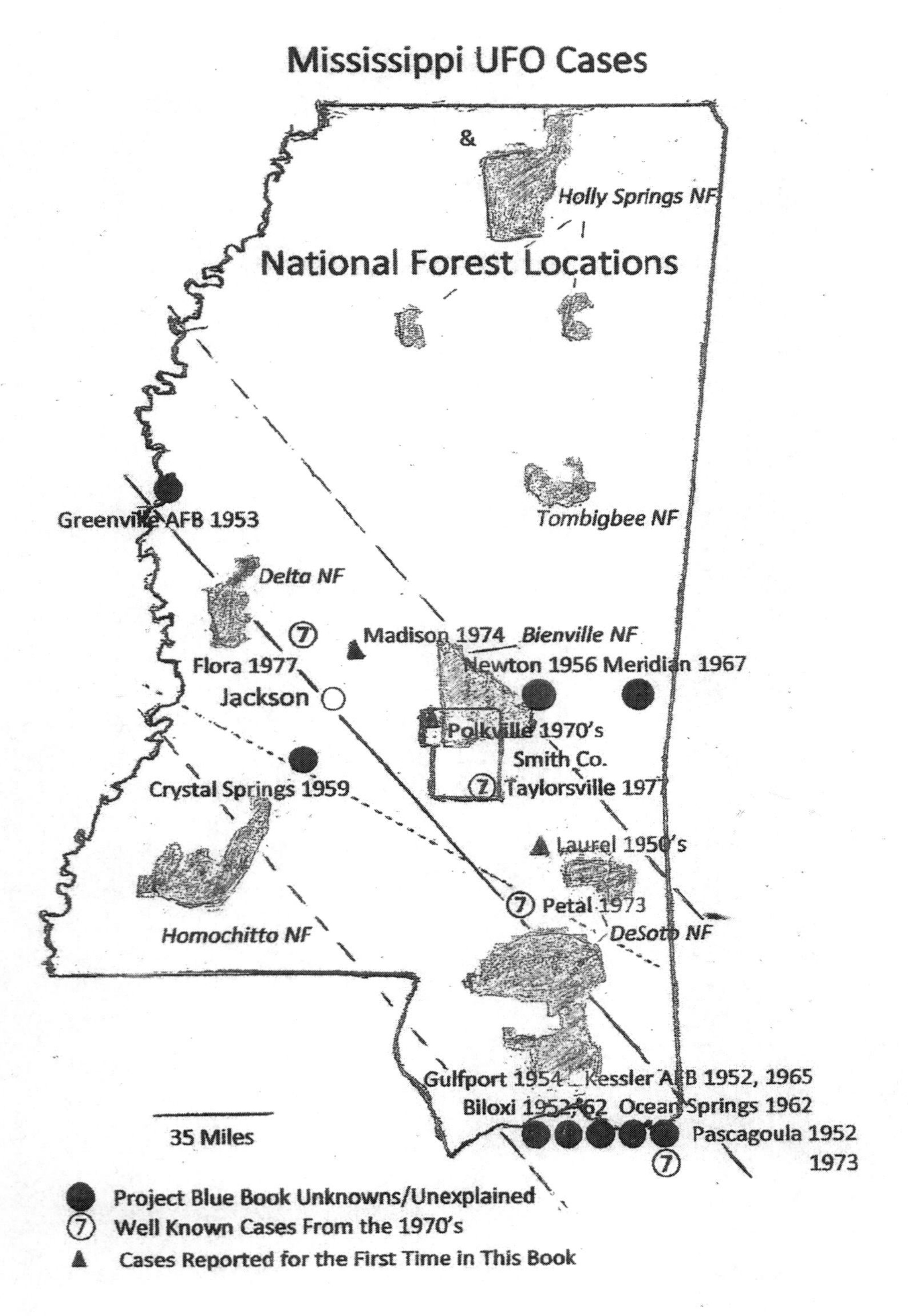
Mississippi UFO Cases
&
Holly Springs NF
National Forest Locations
Greenville AFB 1953
Tombigbee NF
Delta NF
Madison 1974
Bienville NF
Flora 1977
Newton 1956
Meridian 1967
Jackson
Polkville 1970's
Smith Co.
Crystal Springs 1959
Taylorsville 1977
Laurel 1950's
Petal 1973
Homochitto NF
DeSoto NF
Gulfport 1954
Kessler AFB 1952, 1965
Biloxi 1952, 62
Ocean Springs 1962
35 Miles
Pascagoula 1952
1973
Project Blue Book Unknowns/Unexplained
Well Known Cases From the 1970's
Cases Reported for the First Time in This Book

nature and range of sightings that my family had during this same time period. This suggests that there was an underlying related phenomenon behind most of them.

The Flora and Petal UFO flying saucers were similar in appearance to the UFO flying saucer my parents saw backlit by the moon. My parents at first thought it was just "The Light," but because it was backlit they could see the entire object even though it had the bright white light on the bottom. The Taylorsville 1977 sighting was similar to my family's November 1972 sighting, the Madison sighting, and the sightings where "The Light" was seen hovering over or near people. The Hickson family's 1974 Mother's Day sighting was initially of a bright white light that followed their car. It subsequently crossed over the highway in front of them and hovered over some pine samplings. The white light went out and they were able to see a row of horizontal lights around a rather large, hovering UFO. This is reminiscent of the hovering band of lights my family saw at the community dump.

It is clear that some type of unexplained phenomena was quite active over the region of the UFO corridor of central and south Mississippi during the 1970s to the point that the civil defense radar at Columbia, MS was lit up with unknowns for most of the decade. Together it presents a very consistent story, and I believe it is very likely that heavily wooded areas are a part of that story. I also believe that there are a number of families and communities in this Mississippi UFO corridor whose stories and experiences are similar to that of my family's. Remember the calculation earlier of the percentage of UFO sightings that are actually reported? The reporting percentage came out to be about one-half of one percent. Also remember that the oft quoted 95% of UFO cases can be explained as known natural or human-made phenomena. If the Mississippi data is consistent with this, then there could have been 220 UFO reports made to Project Blue Book from the State of Mississippi over the 18 or 19 years of PBB's existence (11 unexplained/0.05 = 220). If that were the case, then there could have been 44,000 UFO sightings in the State of Mississippi over the same time period that were never reported (220/0.005 =44,000). This suggests that there could have been as many as 2,200 sightings in the Mississippi UFO corridor up to January 1970 that were similar to my family's sightings during the 1970s (44,000 x 0.05). Four difficult to explain sightings were made public in Mississippi during the 1970s. This suggests that there could be as many as 800 difficult to explain sightings in the state during the 1970s that were never reported (4/0.005 = 800). This further suggests that there are probably dozens of families in the state with UFO experiences similar to my own family's experiences. Of the four difficult to explain sightings that were formally reported in the 1970s, three of them (Flora, Taylorsville, and Petal) involved direct sightings by law enforcement personnel, and all were not very far from a heavily wooded area or national forest. The fourth, Pascagoula, also not far from a national forest area, was reported to law enforcement about two hours after it happened. I believe that law enforcement involvement is the only reason that a record of these sightings exists today.

Is there any precedent for UFOs sightings near national forests or heavily forested areas? Yes, there is. One of the most famous abduction cases in the world was in a national forest in Arizona. The setting was in a mountainous area of the Apache-Seagraves National Forest where a logging crew had been hired to do some cutting in the area. Late in the day, as they were heading home off the mountain, seven loggers spotted a strange light hovering among the trees about 90 feet off the ground and to the side of the logging access road. They stopped the truck. One brave or foolhardy young man ventured forward to investigate and things kind of escalated from there. This, of course, is the Travis Walton case of November 1975, the case that the motion picture *Fire in the Sky* was based upon, although the movie took quite a few liberties with the actual account of the case as Hollywood tends to do. [94]

Debunkers claim that there was no abduction, and some suggest that even the UFO was a contraption rigged up by Walton and his crew foreman. The presence of a UFO hovering off the ground was confirmed by all seven present. The other crew members, including the foreman, passed lie detector tests. One crew member's test was initially inconclusive, but he later passed. This particular crewman did not pass the initial test by walking out on a section dealing with hostility toward Travis Walton and some trouble he had been in as a juvenile, and passed the other sections dealing with the UFO sighting. This crewman, Allen Dalis, had some disagreements with Walton, and was afraid he would be falsely accused of murdering him if Walton was never found. The UFO was also seen flying away from the area by Mike Rogers, the contractor and crew chief who passed a lie detector test concerning the UFO. Give it up on this point, debunkers! While one lie detector test passed by a lone individual may not mean that much, seven witnesses that passed lie detector tests means that there is something on the order of one chance in 10,000 or less that the presence of a UFO was hoaxed in some way. The minimum that can be said is that there was some type of UFO present and it flew away in a manner indicating intelligent control of the craft.

Most of the misinformation about the Walton case was supplied by Klass, which would be grounds for firing someone if they were employed on an investigation team. This is the case where Klass would have to be given the pink slip as the team skeptic if he were in my employment. Making up a baseless story later refuted by the National Forest Service about Mike Rogers, the crew foreman, shows a complete loss of objectivity on Klass's part. Klass concocted the idea that Mike Rogers came up with the UFO angle to get an extension of his contract with the National Forest Service. The idea was checked out with the National Forest Service and found to be baseless. Rogers never brought up the UFO incident during contract negotiations. Klass's completely fabricated story still circulates in skeptical and debunker circles.

The national forest connection gets stronger when you consider some other details of the Walton case of November 1975. The Walton family lived near two national forests in Arizona: Apache National Forest and Tonto National Forest. Travis Walton had never seen a

UFO before, but had, just like nearly everyone on the planet, discussed UFOs before on a few occasions. His brother Duane had seen a UFO 12 years before. This did not cause them to fixate on UFOs and become obsessed with them as debunkers claimed. Duane's sighting, being a close relative of course, leads to a mild "repeater" problem for ufologists, and is called being a "UFO buff" by debunkers, but just about anyone who has ever heard the words U-F-O before is considered a "UFO buff" by skeptics and debunkers. They continually commit the fallacy of using sweeping generalizations to try to discredit witnesses. Maybe some people are just alert and are good observers. Can no one ever be a neutral witness of events happening around them? (I did not know the fact of a prior UFO sighting in the Walton family until after I decided that I had to write an alien abduction chapter and started getting more detailed information about the case, both pro and con.) Klass's reported actions in this case got more and more unacceptable. In addition to making up all kinds of stories about Mike Rogers and the Walton family, he reportedly tried to bribe Steve Pierce, one of the logging crew, with $10,000 to say the case was a hoax. It is one person's word against another's, but Klass's behavior was over the top to put it mildly.

Compared to my own family's sightings history, the Walton family had barely had any sightings at all. The nearness to national forests gives their story a connection to my family's story. Of course, we did not have the up close and personal experience that Travis Walton reported (16 lie detector tests passed by nine people say a real UFO was there, and there is only a very slight chance that it was not). The lie detector probabilities are difficult to dismiss because of the number of people involved. The movie *Fire in the Sky* let the crew still be under suspicion of harming Walton after five of them passed lie detector tests. In reality, once they passed the tests, they were no longer under as much suspicion from law enforcement. The movie version wanted to get all they could out of keeping the crew under serious suspicion for as long as they could to keep the tension going.

Based upon this and several more cases to follow, UFOs have many times been seen near heavily wooded areas. My hypothesis is simply that there is an association between UFOs and heavily wooded areas. In a number of cases, UFOs have been reported hovering in or near national forests or heavily wooded areas, and were seen traveling to and from adjacent forests nearby. If the UFOs were flying craft with occupants, or were flying craft under remote control, the dense forests not far from cities and towns would give them a place to easily encounter humans and to quickly elude human pursuers when they wanted to. Alternatively, if UFOs near heavily wooded areas are some type of natural phenomena, then the underlying phenomena have not been adequately explained to this point in time.

One of the strongest precedents concerning UFOs and heavily wooded areas is in the Betty and Barney Hill abduction case of 1961 in the White Mountains of New Hampshire. [95] New Hampshire is a heavily forested state with more than 80% of the land area covered in forests. After the abduction incident, Betty Hill had so many more UFO sightings over the years that UFO researchers wanted her to keep silent about them for fear of hurting her credibility due to the "repeater" problem. In fact, some of the other sightings were in the presence of family, friends, or other members of the community. Some were multiple witness sightings and not due to Betty's imagination, but some sightings in later years seem to have been a bit of a stretch and appear to have been due to her active imagination. Many times in the multiple witness sightings, the UFOs would take leave by escaping into a heavily forested, inaccessible area. In fact, the subsequent confirmed sightings over the years after the main abduction event are very similar to my own family's sightings history.

More precedents are found in the Elmwood, Wisconsin cases described in Howard Blum's 1988 book *Out There*. [55] Elmwood has been designated the UFO capital of Wisconsin because of the frequency of UFO sightings back in the 1970s and 80s. Elmwood does not exactly lie next to a national forest (~80 miles away), but does lie in the edge of a generally, heavily wooded area known as the Western Coulee and Ridges Ecological Landscape. The Wisconsin Department of Natural Resources characterizes the landscape as highly eroded, driftless topography that is relatively extensively forested. In many cases there are steep slopes where the bottomland and the hilltops are sometimes cleared, but the slopes are heavily wooded. [96] About 40% of the area is heavily wooded, and the area is roughly 50 miles wide and runs for nearly 200 miles.

Some of the Elmwood UFO cases were ominous. Many of the UFOs around Elmwood were characterized as giving off an orange glow. In one account, a huge silver UFO zapped night marshal George Wheeler's patrol car on April 22, 1976. A huge silver UFO flew over the squad car, which was parked at the time. The UFO displayed an orange light beam as it stopped and hovered momentarily. As it began to rise up, a blue ray shot out and hit the squad car, knocking out the car lights, the points and plugs, and Wheeler. This sounds similar to the account of the blue light ray that struck Travis Walton. The sighting of a UFO nearby was confirmed by a second witness named Paul Fredrickson. Four houses nearby had their TV sets go off for about 10 minutes around that time. Wheeler never really recovered and died some months later after stays at two different hospitals. The stress of the case may have exacerbated already existing health problems.

A case with overtones of the Wheeler case occurred in Minnesota in 1979 near the North Dakota state line involving Deputy Sheriff Val Johnson. [97] Johnson's patrol car seemed to be hit by a UFO, or a force from a UFO, with inexplicable damage that could not be sorted out as to how it could have been done. Johnson had welder's burns to his eyes, was weak and traumatized after the incident, but survived and went on to become sheriff in a nearby county, unlike the unfortunate Wheeler case.

There were many other sightings around the Elmwood area that were less ominous in tone, and in many ways paralleled the Polkville sightings except for the orange lights. A good percentage of the town's population had had at least one sighting during this period. In

Blum's book *Out There*, some CIA agents visited the town's UFO Days celebration looking for information, but it was more of a carnival atmosphere than anything else. My question is: *Why didn't the agents stake out the area of most frequent sightings and take a look for themselves?* It could have been much more productive, or maybe the agents were there just to see what the local populace actually knew. The Elmwood sightings are not mentioned that much in the UFO literature, but I do not know why. In this case, we have a recurring phenomenon that was widely publicized in the community and became known to the most secret government agencies, and no one set up anything to study it in any systematic way. The community went so far as to build a UFO landing field nearby to invite the craft to land and contribute to the celebratory atmosphere.

Also, another area well known for UFO activity is the Gulf Breeze, Florida area where many, many people on the order of thousands have seen UFOs out over the water. [98] Some claim that the Gulf Breeze photographs of hovering UFOs were faked, but the fact is that the numerous sightings of unexplainable maneuvering and hovering lights over the water have been documented by reliable sources and videotaped. Gulf Breeze lies 30 to 40 miles from a heavily wooded area of state forest adjacent to a national forest that extends into Alabama. Coincidence? Maybe, but it fits into the hypothesis of an area with frequent UFO sightings that is not too far from a large, heavily wooded area.

Our final locale is the Uintah Basin in Utah. The 2010 book *The UTAH UFO Display* by Frank B. Salisbury, PhD, was an update of a 1974 book on the same subject. [99] The book was based upon the files of Joseph Junior Hicks. Hicks has documented some 400 sightings in the Uintah Basin from the 1960s up into 2009. There are extensive national forests and mountains to the north and west of the basin some 20-30 or so miles away from the center. The entire area has a semi-circle of national forests and mountains around it. This includes Ashley National Forest to the north, Wasatch to the NW, Uinta to the west, and Dinosaur National Monument to the east. The 2005 book *The Hunt for the Skinwalker: Science Confronts the Unexplained at a Remote Ranch in Utah* by Colm A. Kelleher, PhD, and George Knapp, describes some very strange goings on at the ranch in the Uintah Basin, and some of it is documented with cameras and recording equipment. [100] The strangest occurrence to be recorded electronically was that of an unknown force ripping the wiring loose from one camera while another camera was aimed at the first camera. The incident happened between camera frames in a fraction of a second. A number of multiple eyewitness sightings of strange creatures, strange objects and events, and unidentified lights are all discussed in the book.

Salisbury was one of the prominent ufologists in 1974 at the time of the publication of the first edition of his book. He was on the *National Enquirer* panel of judges that included some of the biggest names in ufology at the time, such as Leo Sprinkle, Jim Lorenzen, James A. Harder, Robert Creegan, and J. Allen Hynek. In 1972, the *National Enquirer* offered a $50,000 prize for conclusive proof that UFOs were of intelligent, extraterrestrial origin. Later the prize was upped to one million dollars. But after a little thought one can see that the only proof that would do would be a recovered craft or propulsion system, or something on that order, so no one came close to collecting the prize. Later, a $5,000 prize was offered for the best case of the year, and it was awarded on occasion. By 1982, the *Enquirer* quietly let the offer drop. Salisbury was also on the *Playboy* panel, which was a discussion between several ufologists and two prominent debunkers, Philip J. Klass and Ernest H. Taves. Taves and Menzel had written a book debunking UFOs in the 1950s. [101] The *Playboy* panel discussion was published in 1978.

Salisbury leaned toward the ET explorer from another planet hypothesis for UFOs around 1974, but shortly after that he began to change his mind and now leans toward a stargate or portal from another dimension as a possible explanation. The very strange occurrences at the Skinwalker ranch that suggest a portal to somewhere else, the strangeness of the Basin sightings, and his own personal beliefs have brought him to this view. Again, I focus on the hypothesis of the surrounding national forests, which are not that far from where the UFOs have been seen in the basin. The proximity of the national forests fits in with my hypothesis quite nicely, but then maybe most locales would fit in.

Let us do some "what if" speculation. Let us suppose for a moment that the UFOs are extraterrestrial in nature. What does it say about the UFO and its possible occupants if they scurry back to a heavily wooded area to hide out in between excursions out into the countryside? It says to me, if this is indeed the case, that they could be flesh and blood occupants from this world or another world and not supernatural, not from another dimension or a parallel universe or from the future, and that they cannot just pop into and out of our world at will by jumping into another dimension or parallel universe. If this hypothesis is valid, it would appear that when they travel here, they have to find a place to stay that minimizes interaction with humanity until they want to have some sort of interaction. To paraphrase Sagan and Arthur C. Clarke, beings that could travel to earth from somewhere outside our solar system will be so advanced technologically, and most likely culturally and philosophically, that their view of us would probably be much the same as we would view a group of Neanderthals, and their technology may be indistinguishable from magic to us. (Arthur C. Clarke said, "Any sufficiently advanced technology is indistinguishable from magic.") Interaction would be on their terms and not on ours. An advanced society that had been advancing technology for tens of thousands of years longer than humans, and had learned not to destroy itself, could be loath to interact with us as equals. Human civilization is in what Sagan called the juvenile stage, and there is no guarantee that we will survive to move to a more mature and sensible stage.

The national forest/heavily wooded hypothesis could explain the perception that many farmers out in the boonies report seeing UFOs. If the phenomena were some unexplained natural phenomena associated with wooded areas or isolated areas, then those out in the countryside would be in a better position to observe it. If the UFOs are occupied flying craft and their occupants are looking for a place to hide, it would be in the more remote areas away from cities and they would travel at night under cover of darkness. Another explanation for why a lot of UFO reports seem to come from less populated places is the jaded nature of people in the cities when it comes to lights in the sky.

When working in Houston, I would get to work a lot of mornings by 6:30 a.m. In the winter, my drive would be almost entirely in the dark. It was not unusual to see three or four lights in the sky of airplanes shortly after takeoff or coming in for a landing. One morning I counted six different planes; two of them with landing lights on that made them look similar to "The Light" for a while until they flew on to where I could see their blinking lights. The point is that the flying saucer my parents saw could fly around all over Houston in the nighttime darkness and not be noticed by that many people, if by anyone. Just another light in the nighttime sky hardly anyone notices. It would not be tracked by air traffic radar without a transponder, and it would not even show up on air traffic radar if it flew at only about 100 ft. altitude. If it flew at normal altitudes and its behavior was not similar to a normal aircraft, it still would not be tracked by air traffic control radar systems. Things that do not act as normal aircraft are filtered out by the flight control radar software. It would not show up as a national security target of interest flying around at 30-60 mph. Busy people in the city with a lot of other things on their minds probably would not take much notice of it.

Budd Hopkins, in his book *Missing Time*, described a UFO landing within the boundaries of a busy city that was seen by witnesses on the ground. [102] People in a busy, noisy city may not notice, as readily as people in the country, a UFO at night with a bright light that obscures the object itself as it leisurely flies by. Consider what happened in the past when unknowns or UFOs streaked through the skies at 600-1000 mph at a few thousand feet altitude near restricted air space. Interceptor jets were dispatched. But what would be the reaction if they stayed under 80 mph, flew 50 feet above the treetops, and stayed out of the restricted airspace. Not much.

If visitors from another world truly wanted to be inconspicuous, why would they sometimes hover in plain view of multiple witnesses? The need to seclude yourself away from humans at various times is understandable. The other behavior—not so much. However, Salisbury titled his book about the Uintah Basin as he did because he came to believe that the UFOs were deliberately presenting displays to certain witnesses as if trying to get their attention. The UFOs seemed to be deliberately interacting with certain people. Of course, this may or may not be the case, but it could be said that UFOs were trying to attract my family's attention, or that it was a frequently occurring phenomena that we just happened to witness on numerous occasions. As Sagan said, the behavior of advanced beings from another planet will probably be completely baffling to us. This can be broadened to say that the behavior of sufficiently advanced beings from any source will very likely be completely baffling to us.

The differences in the community reaction to multiple UFO sightings were striking to me for several of the regions involved. In Mississippi around the Smith County area, most of the witnesses hesitated to tell anyone outside of their family and close friends. In Elmwood, Wisconsin, the sightings were shared and embraced by the entire community, but I am not sure that the sightings were studied or cataloged that well. The whole thing took on a carnival atmosphere and became the basis for a town festival. In the Uintah Utah Basin, one man, Joseph Junior Hicks, became the chronicler/data logger of the sightings, but was a local chronicler. By word of mouth, he became known to most of the Basin residents as the man to report unusual sightings to. Most of the reports did not go to the major UFO investigation organizations. Three different areas had three very different community reactions to somewhat similar unexplained phenomena.

Chapter 20

Earth-Lights, Ball Lightning, and Upper Atmospheric Light Phenomena as an Alternative Explanation

"What is a scientist after all? It is a curious man looking through a keyhole, the keyhole of nature, trying to know what's going on."

—*Jacques Yves Cousteau (1910-1997), French Oceanographic Explorer*

Earth-lights came on the scene as a full-fledged theory to explain some nocturnal light UFO reports sometime in the late 1970s to early 1980s through the work of Michael Persinger and John Derr. They put forward the Tectonic Strain Theory (TST) to explain the generation of atmospheric lights in regions with considerable seismic activity. Historically luminous phenomena were often reported in the sky weeks or months before a major earthquake in many locations around the world. Recently in 2013, Persinger and Derr summarized their past work dealing with the Tectonic Strain Theory and lights seen in the sky in the *International Journal of Geoscience.* [103] The focus of the article was upon using earth-lights ("luminous shapes with unusual motions" in their terminology) to predict an impending earthquake. They point out that the exact mechanism of how the lights are generated is unknown, but there are several hypotheses for how the light phenomena could be generated. It seems to me that they have good evidence that earth-lights are real, are associated with areas of high seismic activity, and are responsible for some nocturnal light UFO reports. The light phenomena seem to be short lived on the order of 100 seconds or so, and perhaps a little longer. They claim that the phenomena can display several shapes besides just a round ball of light, and that the phenomena seem to move erratically. They indicated that lights that flew from horizon to horizon in a very stable manner were unlikely to be earth-lights. However, it seems that the light spectra, wavelengths and intensities, of these phenomena have yet to be cataloged.

In 1997, Devereux and Brookesmith greatly expanded the earth-light hypothesis so that the humble earth-light was transformed from an aimless drifter seen at a distance, into the earth-light on steroids, the super-uber earth-light. Devereux and Brookesmith took things much further in postulating that the earth generated energetic phenomena that they called earth-lights could explain all unexplained UFO reports. They postulated many properties for the earth-lights back in 1997 that would be required to explain all unexplained UFO reports and abductions. They based their ideas upon the Tectonic Strain Theory of Persinger and Derr, and the additional work of Persinger concerning the effect of magnetic fields upon the temporal lobes of humans (Persinger's God helmet). [104] Persinger seemed to claim somewhat more than what his results actually indicated, in my view, and in the view of the Swedish researchers Larsson, Larhammarb, Fredrikson, and Granqvist, who in 2005 concluded that the participants' reports correlated to their personality characteristics and to their suggestibility. [105] Persinger's God Helmet work was carried in the media for several years in the mid to late 1990s and beyond.

Now combining the hypotheses of Brookesmith and Devereux, and the work of Persinger, humans beware of the diabolical earth-light displaying some rudimentary intelligence to interact with or avoid humans, having the ability to hit humans with energy discharges similar to death rays or stun rays, having the ability to appear as a sphere, disk, or complex structure in daylight, and having such a strong electromagnetic field that a human's temporal lobes are putty in its "hands." The temporal lobes of humans are supposed to be so susceptible that any human that gets near enough will go into a trance, have a UFO encounter, and possibly progress into an alien abduction experience. I could go on, but it has already gotten beyond ludicrous. In short, they were proposing that these super-uber earth-lights could explain the entire range of UFO phenomena—everything. Most of those ideas are still unproven today and remain just a series of hypotheses without much evidence to support them. If I am wrong about this, I ask that someone please show me some studies to support these hypotheses.

Brookesmith and Devereux's book was well researched into the history of UFOs, well written, and made fascinating reading on every subject they touched upon. However, they misrepresented several famous cases by picking and choosing what details of a particular case they wanted to present and omitting some of the most important facts, because those facts did not support their contentions. This is classic behavior that is typical for skeptics, debunkers or supporters of alternate hypotheses. They make Klass look like a rank amateur, a bull in a china shop, as they are much smoother, much more erudite at debunking, while still admitting that there might be a case or two in all of ufology involving an ET type UFO, but they cannot identify which cases these might be.

Nonetheless, they do give the reader a philosophical and entertaining path down a road paved with suppositions about earth-lights. I thoroughly enjoyed their book, which shows that you do not have to agree with everything in a book to appreciate it.

I fully expected that by 2013, some 16 years after Devereux and Brookesmith's book was published, I would find some summary of the research into earth-lights telling me their typical appearance and properties, or their average light spectrum or luminosity, or some identifying characteristic or behavior, but I cannot seem to find such data or a reference to such data. Persinger and Derr's 2013 review did touch on luminosity and power output, but the type of data I was expecting to see was not there.

The reason may be that most of mainstream science is keeping a low profile concerning earth-lights, but what on earth are they waiting for? Mainstream science should be all over the earth-light situation measuring spectral power outputs and spectral characteristics, magnetic field effects, etc., etc., except that just like UFOs, earth-lights do not appear upon demand like some experiment in a test tube except in a few places on earth such as Hessdalen, Norway and at Marfa, Texas, according to earth-light proponents. Why haven't these places been studied thoroughly, and why haven't reports and useful data been generated and widely disseminated? Why can't I go outside with a spectral viewing device and say the spectral output of a light seen in the distance does or does not fall into the earth-light range? Maybe all this exists and I just haven't found it yet. Otherwise, why not? One big answer to this question may be that some skeptics claim that these places with unusual lights are just something concocted to bring in tourists. Of course they come to this conclusion without having done any investigation and with having a strong desire to dismiss any unexplained phenomena of any type. Earth-lights seem to be another type of phenomenon that is largely ignored by much of the science establishment.

Devereux and Brookesmith did not give Klass much credit for the earth-light hypothesis, as he had backed away from it by that time. Although someone superseded Klass and Jung with the idea (somewhere I recall that balls of energy of unknown origin was brought up as an explanation for UFOs as early as the mid-1950s), Klass recognized the utility of using ball-lightening and earth-lights to explain otherwise unexplainable sightings, and was an early adopter of the idea and then later he backed away from it due to criticism from all sides. Klass was mentioned a few times in their 1997 book, but they do not give him as much credit as it would seem that he deserves for being one of the early adopters of the same or very similar idea as early as 1966 (Jung had adopted the idea by 1958 and he got it from somewhere else). In all fairness, Klass deserves a major share of the credit for promoting this idea. He clearly states the idea in his 1974 book *UFOs Explained*, although he had earlier focused on debunking unexplained UFO reports by claiming they were ball lightening from electric power lines. He claimed ball lightening in his very first debunking case known as the Exeter Incident of September 3, 1965, in Exeter, New Hampshire. No one knew much about ball lightening then, but it is known now that the more than hour-long duration of the incident eliminates ball lightening as a possibility. Ball lightening near the ground usually has a lifetime measured in seconds or only a minute or so.

Klass later expanded the hypothesis to include energetic, earth-generated plasmas for sightings he could not explain as being any known phenomena. Earth-lights can certainly explain some UFO reports, but cannot explain reports involving hard surface radar targets, for instance. It cannot be overemphasized that even Klass, the greatest UFO skeptic ever to come down the pike, speculated early in his career that there were some UFO reports that could not be explained without invoking phenomena that was, and still is, without a consensus scientific explanation. Klass was invoking unexplained phenomena to explain UFOs. Even though he later quietly let this drop, as he took flak from all sides about it, this came from the mouth of Mr. Philip J. Klass and it is a statement of extreme importance on this particular point. The crowd that claims UFOs are all in the mind and the extreme debunker crowd that says UFOs are nothing but misunderstood conventional phenomena did not listen very well to what Mr. Klass said early in his career and to what, I perceive, he actually believed. The point is so important that I will paraphrase Klass's hypothesis below without changing its meaning at all:

> "Some UFOs are real in the sense that they are as yet unexplained or poorly understood electrical or earth generated plasma phenomena."
>
> —*Philip J. Klass, Paraphrased*

Could some or all of my family's sightings have been earth-lights? Some possibly could be, but the nature of the light appeared to the naked eye to be very nearly the same for most of the sightings. The sightings with the lighted bottom and the dark upper structure cannot be due to orbs of plasma. In their book, Devereux and Brookesmith furnished no daylight photographs of plasmas, but claimed that plasmas look metallic in daylight. In my experience, balls of energy do not look like solid objects no matter how much proponents claim otherwise. They make a lot a claims for earth-lights, but, in reality, earth-lights are ill-defined and mysterious, and if earth-lights behave as if they have some rudimentary intelligence, as Devereux claims, then anything that behaves that way is not a ball of energy, but is something else altogether. Most of their fantastic claims for what happens to people who come into close contact with earth-lights are ill supported by any actual data. Members of my family who were directly underneath the objects should have gone into trances or been zapped with a beam of energy, or imagined an alien abduction right there on the spot if what we were seeing were earth-lights, according to Devereux, Brookesmith, and Persinger.

To be fair, we must thoroughly consider the earth-light explanation for the lights alone that were seen flying by, and we must consider the phenomenon of ball-lightning and the related higher altitude light phenomena proposed by Tore Wessel-Berg in 2004. [106] Although I do not think the lights my family saw some 100 or a few hundred feet or so above the ground constitute higher altitudes, the one sighting in the summer of 1970 was definitely at higher altitudes, but it occurred on a clear night devoid of the clouds necessary to generate ball-lightning or higher altitude light phenomena (higher altitude ball-lightning). A check of the weather records showed that in our location after the first week of June1970, there were very few clouds or rain for nearly three weeks.

Ball-Lightning. I believe that Wessel-Berg nailed the explanation of atmospheric ball lightning in 2003 in his proposed theory appearing in *Elsevier, Physica D*, volume 182, but apparently no one else seems to think so. [107] No mention was given in Wikipedia with references as late as 2010, even though just about every other idea anyone ever came up with to explain ball-lightning was presented. [108] In reality, there are probably two or three separate phenomena grouped under ball-lightning. I believe that I saw the real ball-lightning as I was looking across my backyard near midnight on the night of March 1, 2005. Wessel-Berg's theory explains very well what I saw where there was no actual lightning to be seen, and no thunder in the distance either. What I saw on March 1, 2005, near midnight, looked to be an enclosed discharge of some type with standing light rays enclosed within a blue-green ball the size of a beach ball. Ball-lightning? Yes, according to Wessel-Berg's explanation.

Clearly, it was not due to the reactive silicon hypothesis proposed by some scientists, as this can only occur after a lightning strike hits the ground to vaporize elemental silicon. It seems that the silicon vapor explanation is, at best, describing a secondary effect and is not ball-lightning. On the night of March 1, 2005, I observed a blue-green transparent ball that came drifting down from the sky with no thunder or lightning to be heard or seen. The glowing balls of energy that have been seen in a few instances traveling along electric power lines, actually appearing to be in contact with the power lines, is probably not the same phenomena as atmospheric ball-lightning. I believe these power-line phenomena usually only exist in contact with the power line and do not usually float away from it. In any event, it is very doubtful that true ball-lightning can come from power lines.

In Wessel-Berg's theory (hypothesis really), ball-lightning is caused by an electrical potential between a cloud and the ground, but instead of a lightning strike, a closed spherical DC electrical circuit forms that travels through air in a manner similar to a wave traveling through water. The ionized air molecules do not travel with the ball. They are briefly part of the ball as the energy travels through them. The wind would have little effect on the "ball," and since the air molecules do not have time to heat up, the "ball" tends to fall rather than rise up as a heated ball of gas would. I have to disclose that I do not have the expertise to properly evaluate the equations Wessel-Berg presented, but by reading his narrative I could visualize the phenomena that I witnessed. I am surprised that I cannot find any discussion among experts able to evaluate his mathematical treatment as to its validity. So far I have found neither pro nor con about his ideas.

In Wessel-Berg's 2004 article in the *Journal of Scientific Exploration*, Volume 18, No. 3, he extends the theory to upper atmospheric light phenomena in the attempt to explain atmospheric UFOs. He says that cloud to cloud electrical potential could give rise to a similar phenomenon as ball-lightning, but it would be higher up in the atmosphere. Wessel-Berg proposed that due to the much greater electrical potentials involved in the cloud to cloud upper atmosphere type of ball-lightning, the size of the closed DC circuit can be much larger, and he presents equations that say the shape does not have to be a sphere.

Illustration of Ball Lightning Seen March 1, 2005

Blue-Green Transparent Outer Sphere About Beach Ball Size

White Inner 'Spokes' Almost Too Bright Too Look At

Dropped Steadily to the Ground From ~ 40 Feet in 6-8 Seconds

Left No Mark on the Ground

Seen From 90 Feet Away

Property Line
march 1, 2005

Wessel-Berg gives an explanation for a UFO sighting reported in Sturrock's 1999 book *The UFO Enigma* where a bright cigar-shaped object approached a helicopter at something on the order of 600 knots, decelerated, and began to match the speed of the helicopter. [109] It hovered in front and above the helicopter for a few minutes, and then took off rapidly and disappeared beyond the horizon. During the encounter the object was giving off strong red, white and green lights. A magnetic compass aboard the helicopter behaved erratically during the encounter indicting the presence of strong magnetic fields.

To explain this sighting as due to a large, football shaped, ball-lightning incident, there have to be clouds of opposite charge and correct electrical potential located at the correct position relative to the helicopter's approach. The "large ball DC closed circuit" had to have formed, and had to be residing in an unstable position within the electric field set up between the clouds. After considering this explanation for some time, it seems to me that the ball residing at the unstable position is a sticking point in the explanation. Things of an electrical nature tend to form instantly at a stability point and do not tend to form far away from it. The "ball" had to have been moving to a stable position as the helicopter approached close enough to see the "ball." In Wessel-Berg's hypothesis, as the helicopter approaches it distorts the local electric field drawing the "ball" more rapidly toward it. As the "ball" nears the helicopter, the distortion in the electric field draws the "ball" along with the helicopter. As the helicopter begins to leave the electric field generated by the clouds, the "ball" is immediately attracted to the stable position of the electric fields between the clouds. The helicopter was headed north. The "ball" approached from the east and tracked with the helicopter for a while before leaving rapidly to the west in the direction of its stability point.

As UFO explanations go, this is not that bad, not that bad at all, except that the stability point and helicopter approach had to be spatially just right for the explanation to work. Of course, if the helicopter was equipped with radar, then one could tell if you were dealing with a ball of ionized plasma or a solid object, but the helicopter appears not to have had radar. For the upper atmosphere ball lightning explanation to be valid, Wessel-Berg's theory has to be independently verified to be correct, and the clouds and the "ball" had to have been in exactly the correct locations as the helicopter approached. If the "ball" was at its stable location when the helicopter approached, then it may or may not have been drawn toward the helicopter, but, if it did approach, it would have left the helicopter to return to its original location as the helicopter flew on. We do not know if the clouds required to produce the upper atmosphere ball-lightning were there or not, but this type of explanation can explain the excessive G-forces a physical object would experience in the above maneuvers, at least 5 or 10 G for several seconds in stopping and taking off. There are no G-forces involved with the ball lightning explanation.

The explanation can account for the strong magnetic field effects as well. In this explanation, the large ball lightning would not originate near the ground, or even within 1,000 feet of the ground. If the potential between the clouds decreased over time, the energy source for the large ball would be diminished and the plasmas ball would shrink in size. As cloud to cloud electrical potential decreased, cloud to ground electrical potential could come into play and this upper atmosphere ball-lightning could morph into ordinary ball-lightning and begin to drift down to the ground. Ordinary ball-lightning has been seen to move erratically and quite slowly near the ground on the order of 5-10 mph before it ultimately fizzles out or drops to the ground and is dissipated.

How does all of this relate to earth-lights? I do not really know. We see that ball-lightning ultimately returns to ground and is dissipated. How then can a ball of supposed plasma energy leave the earth to go up into the air? Maybe it is a different type of energy that only becomes visible once it enters the air. Maybe the energy ionizes the air and creates a plasma that then travels or moves in conjunction with the earth's magnetic field or a strong electrical field of a storm cloud. Perhaps, as in Wessel-Berg's ball-lightning explanation, the earth-light is an energy wave that passes through the air like a wave through water, ionizing the air as it goes along. The plasma ball would eventually lose enough energy to fizzle out. How long can such a plasma ball exist? In Devereux and Brookesmith's 1997 book, a time of 12 minutes was reported for an earth-light actually seen leaving the ground until the time it went out. They reported other lights lasting an hour or more, but the source of the light was unknown. The light was first observed in the sky and was not seen exiting the ground, so its source was unknown. Devereux also saw what he attributed to earth-lights flying overhead at Hessdalen, Norway. He had portable radar equipment, but only reported one case of an earth-light giving a radar return. The earth-light gave a radar return on every second sweep and nothing on the intervening sweeps. It appears as if he were seeing some sort of radar artifact.

Most of our sightings were of "The Light" seen without seeing any structured object associated with it, but as seen with Charles Hickson's Mother's Day sighting, a quite large upper craft can be hidden by a bright bottom light, especially if the closest approach is a hundred yards away or more. The fact of the matter was that the light alone, and the light associated with structured objects, appeared to be virtually identical to one another. Both appeared to be a nearly fluorescent white light with a hint of faint blueness on occasion. This tended to make us think that all the lights we were seeing came from the same source, and it is a natural tendency of the observer to assign all seemingly similar phenomena to the same source.

I find it difficult to believe that a glob of undefined energy spewed forth from the earth by tectonic activity can fly in a straight line at constant altitude from horizon to horizon at 25-50 mph as "The Light" has been observed to do. My impression of such a glob of energy is that it would seem to aimlessly drift around without purpose. I must, however, admit a great prejudice against aimless drifting globs of energy being able to fly at constant altitude and speed for miles. Supporters of earth-lights say the earth-light can be accelerated by fluctuations in the earth's magnetic field, such as occur near an active fault line along tectonic plates. This is sort of like a magnetically lifted and driven train. One problem, though, is that the rapidly flipping field polarities that allow a mag-lift train to move are not going to be occurring along a natural fault line. Most everything I can find seems to say naturally occurring plasmas or other globs of energy are going to behave erratically, so I would have to suspend that perception of reality to consider the possibility of earth-lights.

I was dumbfounded to see the possible influence of Brookesmith and Devereux in a 2006 document released by the UK Ministry of Defense (MOD) through the UK FOIA. [110] The declassified document concerning Project Condign discusses the UFOs known as "Black Triangles," and attributes them to formations of electrical plasmas in the atmosphere caused by the disintegration of meteors. This latest hypothesis is that earth-lights or atmospheric plasmas can be generated by the energy of meteors breaking up in the atmosphere. So now the earth-light enthusiasts have an "explanation" for how an earth-light or atmospheric plasma or plasma-generated "Black Triangle" can be generated just about anywhere on the planet. Klass must be smiling down approvingly or looking up approvingly, or not, depending upon your afterlife beliefs. That a light in the atmosphere could be produced by meteors breaking up, I will buy. That a classic "Black Triangle" could be generated this way ... no way in the world!

For some skeptics, geo-magnetic generation of earth-lights has lost out to the meteor explanation, which can put the earth-light randomly at any location on earth to explain any UFO report. However, random atmospheric-generated earth-lights should be just that random, and there should be no locale on earth where frequently occurring meteor-generated earth-lights are regularly seen. It would logically follow that there could not be any regularly occurring UFO activity seen flying over the exact same fields anywhere on earth. Thus, the meteor generated earth-light idea fails the simplest logical tests for being the sole explanation of earth-lights or UFOs. If meteor-generated earth-lights are even possible, they could possibly explain a few nocturnal lights in the night sky. By logic alone we can dismiss the meteor-generated earth-lights as having little to do with my family's sightings. Our sightings were shown to be non-random, and to be occurring over and over again within a very small geographic area.

The idea that "Black Triangles" could be due to meteors breaking up in the atmosphere generating highly structured plasmas seems absurd. Black Triangles give strong radar returns, indicating them to be hard-surfaced objects. Standard physics would indicate that a ball of energy or plasma should absorb the radar signal or be transparent to it unless longer wavelength radar is used, according to the UK MOD. In the case of a longer wavelength radar beam, a weak radar return may be seen. So it is very difficult to see how a meteor plasma earth-light could give a strong radar return signal resembling a hard surfaced aircraft's radar return.

Black triangles also respond as if under intelligent control when fighter jets approach. Plasma formations have not even been shown to exist in the atmosphere for any length of time. All that plasma energy should rearrange into one great big ball of plasma, and it should dissipate fairly quickly as in a lightning strike. The MOD explanation for "Black Triangles" is beyond absurd. It is laughable. I think that someone at the UK Ministry of Defense must have had their head stuck in a plasma formation for quite a while to come up with such nonsense. Here are two facts that show this hypothesis to be complete nonsense:

1. Black triangles, while sparsely seen or described in the first 40+ years of the UFO era, burst on the scene suddenly in 1989 in the Belgium wave and have been seen many times around the world since then by credible witnesses. Where was this "natural" phenomenon for the last 10,000 years of human existence? Why was it not reported frequently in the first 40+ years of modern ufology when attention was drawn to every possible explanation for UFO reports? The answer is that it is not a natural phenomenon and was not out there en masse then.
2. A sharp-edged triangle shape would be virtually impossible to create from plasma in the atmosphere by natural forces. And to create this shape with bottom lights at each corner and a center light over and over again in the atmosphere all other the world in exactly the same way is blatantly absurd.

The earth-light hypotheses, geo and meteor, seems to be one reason Stanton Friedman excludes most nocturnal lights from his UFO investigations, although he does not say so specifically.

Another contention that goes along with the earth-light plasma idea is that anyone that gets close enough to one of these black triangles will start to hallucinate. I do not know how close "close enough" is, but the magnetic field strength should fall off as a function of the square of the distance or the cube of the distance depending upon one's orientation to the magnetic field center or magnetic pole. [111] Most witnesses are seldom as close as 500 feet to these craft. If being within 50 feet would cause your frontal lobes to melt down, then at 500 feet you would only get 1% of the field strength at most. At 1,000 feet, you would get approximately 0.25% of the field strength. Anyway, no two people are going to have the same hallucination at the same time, or rather the odds are at least 625 to 1 against it (25/1 x 25/1, odds of two people both being fantasy prone enough to incorporate each other's hallucination). [112]

Let us go into this magnetic field–hallucination connection just a little deeper and show what nonsense this is once and for all. A super strong magnetic field strength experiment is performed upon some 20 million people a year in the US without narrative adventure hallucinations as side effects. It is called getting an MRI scan. The strength of the MRI field is from 15,000 Gauss to 30,000 Gauss. [113] If the head is scanned, that is the field strength received by the patient's head.

Ironically, the MRI head scans are used to look for brain lesions or tumors that can cause recurring hallucinations. The field strength of up to 30,000 Gauss is strong enough to cause instantaneous or momentary hallucinations if the patient moves their head during the MRI scan. The magnetic field is strong enough in this case to interfere with the ions involved in nerve transmission according to the website *www.vias.org/physics/example_4_7_01.html* containing Benjamin Crowell's *Lectures on Physics.* [114] These types of hallucinations involve seeing a flash of light or hearing a sound such as bacon frying coincident with the moving of the patient's head. Narrative adventure hallucinations are not induced by these tremendously strong magnetic fields. If the lower body is scanned, a patient's head will experience a much weaker field, but I am guessing at least 30 Gauss, which is a far greater magnetic field strength than anyone can receive from any type of atmospheric plasma, and here's why.

Let us use the field strength of a sunspot on the surface of the sun, 1500 Gauss (Tesla units can also be used, one Tesla [T] is equal to 10,000 Gauss). Let us say that an atmospheric plasma can have the same magnetic field strength as a sunspot, even though that would be a stretch, and the real strength may be perhaps 150 Gauss. Let us also say that at a one foot distance from the edge of the atmospheric plasma the field strength is 1500 Gauss, then for an observer 100 feet away, the maximum strength of the field would only be 0.15 Gauss, which is less than the magnetic field strength of the earth (0.25-0.65 Gauss). For an observer considerably more than 100 feet away, the magnetic field strength from the atmospheric plasma would become vanishingly small. If a 30,000 Gauss head scan with an MRI can't produce narrative adventure hallucinations, it is blatantly absurd to claim that the magnetic field from an atmospheric plasma can.

My wife had an MRI scan for a kidney stone a few years ago. I asked her if she had any strange thoughts or experiences while being scanned and she looked at me puzzled and said that she felt or thought nothing unusual, but did hear a repeating pinging sound (the normal MRI coil pulses) even though she was wearing earphones playing music.

Enough of this "Black Triangle," super-uber, earth-light, super-uber nonsense. Let us give them the benefit of the doubt and say that whoever came up with it probably did so in an honest attempt to explain the observations. They are, however, pushing an untenable pet hypothesis. To test this yourself all you need to do is take two 50 Gauss refrigerator magnets and press one on each of the temples of your head and see if you start hallucinating. You are experiencing a far greater magnetic field than any earth-light can produce, even at only 50 feet away.

The super-uber, earth-light hypothesis just does not have much basis in fact. Note that I use the word hypothesis and not theory. In science, theories have tons of evidence that support them, and over time a hypothesis can become a theory because of a body of evidence that has been built up. A hypothesis is any idea that can be brought forward to explain something regardless of a lack of supporting evidence. In common usage, the word theory is used when the word that should be used is hypothesis.

Why wouldn't the earth-light plasmas affect everyone's brain the same way, if, and that is a tremendously big if, they were able to do so in some way? The individual differences in humans are such that there will be a wide range of responses to any stimulus, and no two people are likely to respond the same way, if at all. In Brookesmith and Devereux's book, they report researchers near earth-lights a few meters away (10 feet? 20 feet?) from large (six-foot diameter) earth-lights emerging from the ground, and none of these researchers seem to have been affected at all. Maybe they were just looking at the ordinary earth-lights not the super-uber type (said with your best Arnold Schwarzenegger accent). This is the multiple person test, and earth-lights failed it just as they fail any rational analysis of the possible magnetic field strengths that could be involved.

Finally, ball lightening and earth-lights cannot possess intelligence. The observer might infer intelligence because of some type of apparent behavior, but as soon as a light displays enough apparent intelligent behavior that it is clearly either intelligent itself or controlled by some type of intelligence, the observer knows it is neither ball lightening nor earth-light. Could there be intelligent beings, say as intelligent as an insect or a bird, that appear to be balls of light? Intelligence and problem solving clearly are not limited to primates (crows have been seen to manipulate a mechanical apparatus faster than chimpanzees to get at food) [115] Could there be intelligent energy beings? If physics is at a loss to explain earth-lights, could biology explain them? Could they be living organisms capable of very bright bioluminescence? Capable of appearing to be intelligent by maneuvering around and avoiding capture? Capable of flying indefinitely? Perhaps, or perhaps not. This type of life form was proposed by Trevor J. Constable, who called the unknown life forms "Critters" or "Heat Critters," and proposed them as an explanation for UFOs. [116] The idea of energy beings has also been the subject of science fiction and several Star Trek episodes that I recall, but maybe the idea of energy beings is just purely science fiction, or maybe evoking energy beings is the same as evoking ghosts. Much study would be needed to understand such phenomena and there is probably little funding available to support it.

However, the Marfa Lights near Marfa, Texas come to mind when talking about an intelligent maneuvering light. [117] A number of attempts have been made to surround a Marfa light, but without success. One such attempt was shown on national TV back in the 1990s on a show similar to *Unsolved Mysteries*, or one of the TV news magazines. It took quite an effort to get one of the lights surrounded from, I believe, three different directions. As investigators walked steadily in toward the light from about 300 feet away, the light went out when they got, say, within 200-150 feet of it, and it came back on some 200 to 300 feet away from the area surrounded by the investigators. This type of behavior and other seemingly intelligent behaviors have been reported many times. Clearly there is more going on there than an aimlessly drifting glob of energy, but what? I do not consider the authentic Marfa Lights to be earth-lights, as too much apparent intelligence has been displayed. Some skeptics say the

Marfa Lights are just cars in the distance seen directly or reflecting off the mountains, and certainly today there are car lights to contend with, but the authentic Marfa Lights were there long before cars and their headlights. The Marfa Lights seem to be another mystery to ponder altogether.

Chapter 21

Let the Hypothesis Testing Begin
Forest Coverage, Earth-lights, Population, Military Bases

"The great tragedy of science - the slaying of a beautiful hypothesis by an ugly fact."

—*Thomas Huxley, English Biologist (1825-1895)*

I do suspect that there is some common connection between earth-lights and ball-lightning, as they both seem to be some form of intense energy dissipation traveling through the air. The TST theory proposes that earth-lights will be generated in areas of high seismic activity, and that increased earth-light generation will precede earthquake activity by weeks or months. This provides us with a simple way to compare the hypotheses of UFOs associated with forested areas to the hypothesis that the UFOs are earth-lights associated with seismic activity. Let us consider the areas of UFO activity identified in a previous chapter: Polkville, MS; Snowflake, AZ; the Uintah Basin in Utah; Elmwood, WS; Gulf Breeze, FL, and the entire state of New Hampshire.

A US geological survey map was found showing all the fault lines under the continental US. [118] Right away it can be seen that there are no active faults listed for Wisconsin or New Hampshire, so earth-lights from geomagnetic tectonic strain activity seem to be a remote possibility in either of those locales. Actual seismic data from the USGS shows that New Hampshire is somewhat more active than Wisconsin, but still ranks 31st out of the 50 states. Wisconsin is dead last in seismic activity at 50th. However, Snowflake, AZ is surrounded by possibly active fault lines in about a 100-mile diameter semi-circle. Snowflake itself is probably about 20 miles away from the nearest fault line, and Arizona ranks 13th in seismic activity among the 50 states, so Snowflake is definitely in the geo-active list. The Uintah Basin in Utah is very similar to the Snowflake, AZ area, as it is surrounded by a semi-circle of faults to the west and is itself some 30 or 40 miles away from some active faults. These faults seem to be the source of somewhat frequent, low-level magnitude one or two earthquakes some 40 or so miles from the center of the basin. The State of Utah ranks ninth in seismic activity, so that places another area is in the geo-active column.

Now on to what came as a major surprise to me: The bottom 40% of the State of Mississippi is undercut by numerous faults and so is all of the State of Louisiana and most of eastern Texas, but these are old faults that are classified as Class B by the US Geological Survey. [119] The Class B designation means that it is unsure as to whether or not the fault could produce surface movement. These faults may be associated with oil and natural gas production in Mississippi, Louisiana, and East Texas. About 75-80% of the oil and gas fields in Mississippi are in the counties sitting above these faults. Some geologists think the Class B faults are older than the quaternary period (older than 1.6 million years).

They seem to be inactive, and one source thought that they are decoupled from the surface so that there is some question as to whether or not they could produce an earthquake or movement on the surface. The biggest surprise was the northern edge of this underground formation. It falls on a NW-SE line beginning north of Flora and Yazoo City, MS, and about 20 miles south of Greenville, and runs to about 25-30 miles south of Meridian. The line runs a little north of Smith County and the family farm. So, the family farm is maybe 15 miles south of the boundary between different underground formations, and the northern edge of the underground formation runs along a NW-SE line similar to our UFO sightings. This came as quite a surprise. There is almost an exact match of the Mississippi UFO corridor with the most heavily wooded areas of the state, and with the underground Class B geologic faults. Could this be more than mere coincidence? There is another possibility never mentioned by the earth-light proponents. Could the geologic faults be attracting UFO activity by generating detectable energy differentials of some type? With the entire southern portion of the state underlain with Class B faults, where does Mississippi rank in seismic activity? The answer is 38th. Not very active. Our final location, Gulf Breeze, Florida, ranks 45th in seismic activity, giving a final score of six to two for heavily forested areas versus areas of high seismic activity.

For the six areas of UFO activity that I was generally aware of there is a much better fit for the heavily wooded hypothesis than the geo earth-light hypothesis, but the choice of geologic areas was limited to my immediate knowledge at the time of writing this section. This comparison would have to be made for a wider range of UFO hot spots to be valid. However, it does show that four of the six UFO areas, Wisconsin, New Hampshire, extreme northwest Florida, and south Mississippi do not have a strong seismic driving force for geo earth-light generation. Therefore, the geo earth-light hypothesis seems to fail for those areas, and if it fails for those areas, then it cannot be a general universal explanation for the UFO phenomena, but it can explain short-lived, erratic nocturnal lights.

As for the Class B faults, I have to believe that a boundary between two dissimilar underground formations is a place where some sort of energy phenomenon could possibly be generated at the surface. I have to recognize such a possibility because the geologic difference is indisputable. However, most of our sightings did not match up with earth-light behavior—a hovering spotlight that lasted a good 15 minutes or more, a light flying back and forth between two other lights for 20 minutes or more, a lighted object no more than 100 feet in the air hovering over a group of people, a lighted object that appeared to be a classic flying saucer when backlit by the moon, a flying light that hovered and retraced its path, and more. Objects with large, dark, top structures with lights attached to the bottom of them cannot be

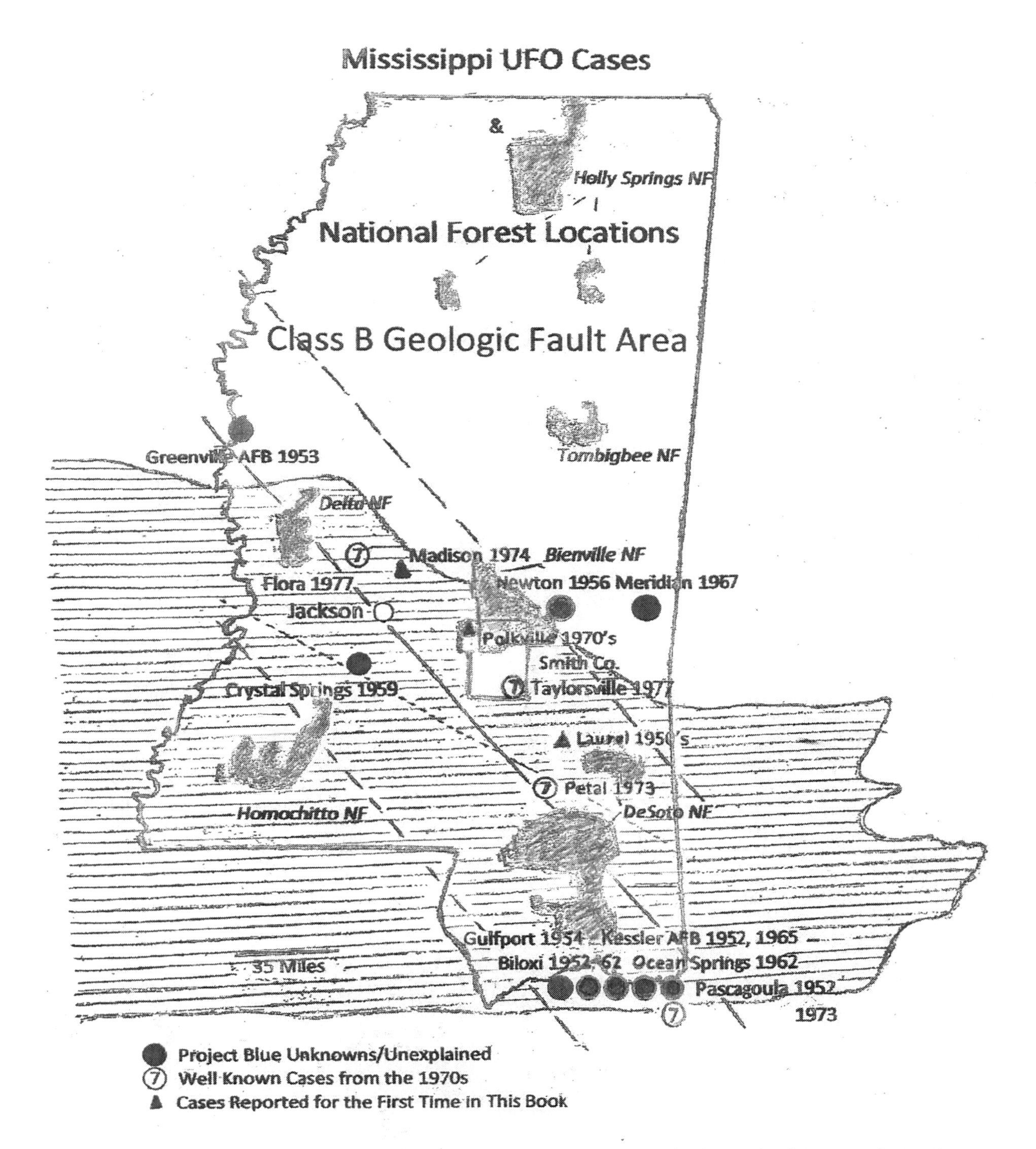

Source: US Geological Survey (USGS)

earth-lights. Remember the similarity to Charles Hickson's Mother's Day sighting with five or six witnesses, where the bright bottom light went out to reveal a rather large craft with horizontal lights in a band around the middle? Only the bottom light could be seen at a distance when it was on. The lights/objects we witnessed seemed so purposeful, so stable, and flying the same courses repeatedly. Earth-lights generated by tectonic stresses would have to be much more erratic, much more random in behavior and flight path. The probability that such natural forces could produce lights flying nearly the same path over the far edge of the fields north of our house repeatedly for over a decade is virtually zero. Remember that south Mississippi civil defense radar was one of the most active radar sites in the world for hits upon unknown targets during the 1970s. This is completely at odds with earth-light activity, since earth-lights cannot give strong radar returns.

There is another very strong argument against the case for an earth-light explanation. This argument is that these bright white lights were not seen in the decades leading up to the 1970s in the Polkville area, although Buford Jones recalled his parents seeing very similar lights slowly crossing their fields at night in the Laurel, MS area some 50 miles away in the mid-1950s. During the 1930s up through the 1960s, there were many more people outside on foot at night than during the 1970s, and they were outside for longer periods of time. Nearly every family in the area had second cars and window air conditioning units by the 1970s. Many more fox hunters and raccoon hunters were out at night for long periods of time in the 1950s and 60s, compared to the 70s. Not many people were walking two to five miles at night to get from one place to another by 1970. Besides the spotlight from the sky incident in the 1930s related by my father's family, there were no bright white lights reported traversing from horizon to horizon by anyone in my father's family or my mother's family up until about 1970. In our area, anything in the sky at night before the 1970s had the potential to be seen by many more people than after 1970. The decade of the 1960s was a transition period in so many ways from the prior ways of life.

The further back before 1970, the greater the probability that something in the sky would be seen because of many more people being out at night, but not that much was seen. Then suddenly around 1970, we have lights and lighted flying objects all over the place for more than a decade. It seems too much of a coincidence to say that we had UFOs and earth-lights flying around willy-nilly during the 1970s when both were mostly absent before that time.

To put this in perspective, the website www.larryhatch.net/LAMSAR.html states that the busiest one-week period of UFO reports during any period in history for Mississippi, Arkansas, and Louisiana was from October 10-17, 1973. [120] The Pascagoula abduction incident occurred on October 11, 1973. The January 17, 1974 issue of *Rolling Stone* magazine, which had the Pascagoula abduction story by Joe Eszterhas inside, had on the cover: "When the UFOs Fell on Dixie." Literally, there were sightings all over Mississippi and the Deep South during the 1970s, and 1973 in particular. In addition, the UFO flap in the 1970s was actually worldwide. The southeastern deep south of the United States is one of the least active seismic regions in the country. Earth-lights do not appear to be a reasonable explanation for the tremendous surge in UFO activity during the 1970s for such a seismically inactive region.

However, I must admit that the geologic data intrigued me. Finding out about the Class B faults gave me a new appreciation for what is underground. While still believing that the national forests are an important part of the overall picture, I began looking at magnetic anomaly data. From the magnetic anomaly data for North America, it is clear that all the UFO hot spot areas discussed above have local, permanent, magnetic anomalies or variations in the local magnetic field. This is mainly attributed to the existence of different mineral deposits underground. The US Geological Survey website is a repository of this magnetic data: [121] *http://irpsrvgis00.utep.edu/repositorywebsite/Default.aspx*. One has to enter the range of coordinates for latitude and longitude to get the results in table form that you have to graph yourself. In doing this, I found that from near Flora, MS to a little southeast of Polkville there is nearly a constant gradient of about 6.0 nana-Tesla, nT, per mile from a -250 nT to a +20 nT over 45 miles. Polkville itself is almost a neutral polarity near zero nT. This is a tiny magnetic gradient over this distance as 1 Gauss is equal to 100,000 nT, and a strong refrigerator magnetic can be 50-100 gauss. Could this tiny gradient drive an earth-light? Could it attract a UFO of a different type? Could the earth's magnetic gradients be an energy source for UFOs of some type? Can the Class B faults cause these magnetic gradients to fluctuate? The US Geological Survey data also included gravitational anomalies. Some believe that UFOs manipulate gravity to hover and for propulsion. I did not explore the gravitational anomalies, as I was focused upon the magnetic field claims of earth-light proponents. Could the real attraction be gravitation anomalies?

In summary, it seems that geo-magnet earth-lights fails for areas without a certain level of seismic activity and cannot be a general explanation for UFOs, but the case seems to be on a sound basis that they are responsible for some reports of short-lived nocturnal lights. However, earth-lights cannot have a strong enough magnetic field to cause anyone to hallucinate and imagine some sort of narrative adventure. Standard physics has a difficult time trying to explain the exact mechanism that produces ball-lightning or earth-lights, but some things are clear: there has to be matter at the center of a ball of light energy to maintain the ball shape, and that matter has to be the source of the light generation (plasma: ionized atoms and molecules too energetic to recombine). The energy could be traveling through the air like a wave through water, as Wessel-Berg proposes, but this plausible explanation for ball-lightning is not generally accepted as yet to my knowledge, and perhaps has not even been acknowledged by the powers that be. Meteor generated earth-lights were shown to be unable to explain frequently occurring, highly localized, repeating light phenomena. Light phenomena seemed to be mostly absent near my family's farm before the 1970s with one or two exceptions, which greatly weakens the argument for geo-magnetic earth-lights as an explanation for the explosion of lights in the 1970s.

In the final analysis, evoking ball-lightning or earth-lights as an explanation is still proposing something not fully accepted or understood by mainstream science and perhaps, even today, it is proposing something that is rejected nearly as strongly as UFOs are.

Looking deeper into what else might correlate with the Mississippi UFO corridor, it dawned on me one day that I should look at the human population density of the state to see how well that correlates. Sure enough, the UFO corridor seemed to encompass about two-thirds of the state's population. The population map shown is for the year 2000. From 1970, the state's population grew from about 2.22 million to 2.88 million. One of the areas with the fastest growth was Madison County just north of Jackson, which by 2010 had more than tripled its 1970 population. The place where our deer hunters had their UFO encounter may be a development of some sort by now. A map from 1970 would look similar, except that some of the higher population areas would be a little smaller.

While the UFO corridor does encompass much of the state's population, there is another area of the state with a high population density. This is the northernmost one-fourth of the state along the Mississippi-Tennessee state line. There is a large population area south of Memphis and mostly all along the state line, except for the national forest area. Out of 11 Project Blue Book unexplained cases for the state, none came from this area. Even in the 1950s, the population would have been four or five hundred thousand people or more.

Let us suppose that this northern one-fourth section of the state has about one-fourth of the population of the state and has had since 1950. If UFOs are just a completely random phenomena, we would expect that there would have been three Project Blue Book unexplained cases from the area. Likewise, if UFOs are just a product of the human mind, we would randomly expect the same three unexplained cases, but there were none. Coming to the 1970s where there were four exceptional cases in the state that became well-known. For randomly occurring events one of them should have been reported in the northern quadrant, but none were. From 1950 to 1980, we have 15 exceptional UFO cases from the UFO corridor and none from the northern quadrant where we would have expected to have about four. The UFO phenomena in the state do not appear to be randomly distributed with the human population. The correlation with national forests looks a little weaker as well, since the northern quadrant has sufficient national forests and human population, but no highly unusual sightings on record. There are two possibilities. One is that the reports from the northern quadrant of the state were just as unusual, but never received as much attention for some reason, or that there is a regional UFO phenomenon occurring in the southern half of the state similar to the area around Elmwood, Wisconsin and the Uintah Basin in Utah, and other areas.

What else besides forested area, quaternary geologic faults, and population could be correlated with this UFO corridor? Of course, why didn't I think of it sooner? Military bases. Nine of the ten or eleven Project Blue Book unexplained cases for the state were sightings near military air bases. Seven were not that far from Kessler Air Base on the Gulf Coast, one was near the Greenville, MS Air Force base (closed in 1965, now a regional airport), and one was not too far from the Meridian Naval Air Station. However, even this correlation breaks down in two ways. Three of the four exceptional cases in the 1970s were not near military air bases and there are two military air bases, one near Memphis and one near Columbus, MS, near the northern population quadrant with no exceptional sightings making it to the Project Blue Book unexplained list or gaining very much attention. The correlation with military air bases looks pretty good in the 1950s and 60s, but not so good later.

Of course we are talking about the exceptional UFO cases that standout in some way and are difficult to dismiss with the usual explanations. I have not examined all the Project Blue Book case reports that came in from the state as I write this, but I would expect that there are plenty of UFO reports from the northern quadrant. It seems very peculiar that none of these made it into the exceptional category and made the "Unknowns" list. It is equally as peculiar that the exceptional cases that do exist are in or very near the NW-SE UFO corridor of central and south Mississippi. After all of this, I have to conclude that the location of the UFO corridor correlates better with the location of the quaternary faults that underlie the southern half of the State of Mississippi than anything that I have considered so far. The Mississippi UFO corridor has it all: much of the state's population, forest coverage, geologic faults (but little seismic activity), and some airbases. The questions still left to be answered about the geologic faults are: Do the quaternary Class B faults have some role in all of this? Do they attract the UFOs, or is it all just a coincidence of location? If there is a quaternary fault connection, what role do the heavily wooded areas and national forests play? The famous sightings from the 1970s and several of our sightings were of flying craft that appeared to be flying saucers, not just nebulous lights. This seems to suggest that the area attracts flying saucer-type UFOs, and if these UFOs wanted to get out of sight in a hurry there are the national forests and heavily wooded areas nearby. On the other hand if these UFOs were natural phenomena generated by the quaternary Class B faults, perhaps combined with other natural processes, then Mother Earth sure knows how to play one heck of a trick upon humanity.

The Mississippi UFO Corridor and The Population Density

Nat. Forest
Nat. F.
Nat. F.
Nat. F.
Nat. F.
Nat. F.
Smith Co.
Nat. F.
Nat. F.
Nat. F.

35 Miles

- ⓪ Project Blue Unknowns/Unexplained
- ⑦ Well Known Cases from the 1970s
- ▲ Cases Reported for the First Time in This Book

Source: US Census Bureau - 2000 Mississippi Profile

CHAPTER 22

National UFO Hotspots and Seismic Activity

"I cannot give any scientist of any age better advice than this: the intensity of the conviction that a hypothesis is true has no bearing on whether it is true or not."

—*Sir Peter B. Medawar (1915-1987), Nobel Prize Winning British Biologist*

After looking just at the State of Mississippi and seeing an apparent association of UFOs and national forests, and an association of UFOs with the quaternary faults underlying the state, I decided to look at this entire Class B fault system that underlies parts of Texas, all of Louisiana, part of Mississippi, a little of Alabama, and a tiny bit of extreme northwest Florida. I was looking at the relationship of this area of Class B geologic faults with Project Blue Book Unknowns. I later branched out to look closer at several other states as well. Finally, I decided to look at the entire US from the geo fault/seismic activity and forested area standpoint. I stumbled upon a website dealing with the Yakima Indian Reservation UFOs in Washington State that had a link to the website of *Popular Mechanics* magazine dealing with an article published in the magazine in 2009. [122, 123] The subject of the article was the top UFO hotspots in the US from 1947 through 2005. The data came from CUFOS (the Center for UFO Studies, HQ Chicago, IL founded by Dr. Hynek) and gave the top five urban counties and top five rural counties with the most UFO reports over the time period. The Uintah Basin, Utah sightings at 400 would place that location near the top five overall in any category, but those sightings apparently were not reported to national UFO investigation organizations such as MUFON (Mutual UFO Network, HQ Cincinnati, OH) or NUFORC (National UFO Reporting Center, HQ Davenport, WA). The CUFOS data actually added five more areas of frequent UFO sightings to my list:

Los Angeles County to San Diego County in California
King County (Seattle) to the Yakima Indian Reservation (near the top secret Hanover Nuclear Area) in Washington State

Saguache County, Colorado
Cook County (Chicago), Illinois
Westmoreland County (Kecksburg, 30 miles from Pittsburg), Pennsylvania

The CUFOS list included Maricopa County, Arizona, near Phoenix, but I already had the Walton case from Snowflake, AZ on the list. Maricopa County is on the southern side of the same mountains and national forests involved in the Walton case, which was on the northern side. Santa Rosa County, Florida, was also on the CUFOS list, but that is the same as Gulf Breeze, FL. The final entry on the CUFOS list was Rockingham County, New Hampshire where the Exeter UFO incident occurred in 1965.

I already had New Hampshire on my list of UFO hot spots, even though Project Blue Book only had three cases from NH on its unknown/unexplained list from 1947-1969. Blue Book did not list the Hill abduction case as unexplained, but everything cited by Blue Book to dismiss the case for just a simple UFO sighting has been proven to be incorrect by any objective assessment. There was not a temperature inversion to explain two sets of radar picking up an unknown target on the night in question, as Blue Book tried to claim, only to be shot down later by the National Weather Service. Skeptics have to resort to contending the Hills saw the planet Jupiter that night and that started the entire episode; a quite comical assumption. The same UFO was also reported by another set of witnesses that night. But the reason I listed New Hampshire was that Betty Hill had later UFO sightings on several separate occasions in the presence of witnesses that confirmed the sightings. The odds against this happening are astronomical, unless the area is a hot spot for UFO activity. I know that Betty went a little too far later on and skeptics claimed she was calling every streetlight a UFO, but the sightings that were confirmed by other witnesses cannot be easily dismissed.

The best way to see the relationship of the UFO hot spots to heavily forested areas is to look at a map showing forest cover with the hot spot locations labeled. The map also has marked St. Clair County, IL, which is just across the river from St. Louis, MO. The most famous black triangle case in the US occurred there in January, 2000, involving a number of local law enforcement officers in neighboring jurisdictions. The case was reported on the Discovery Channel in November and December of the same year. The area is not that far away from the Piedmont area of Missouri, studied by scientist Hartley Rutledge, which had a number of strange lights and UFO sightings similar to the Uintah Basin and Yakima sightings as discussed in Greg Long's 1990 book *Examining the Earthlight Theory: The Yakima UFO Microcosm*. [124] Long found some of the same weaknesses in the Tectonic Strain Theory for the myriad of sightings on the Yakima Indian Reservation in the 1970s. Night watchers at several fire lookout towers on the reservation noted a huge spike in unexplained lights in the nighttime sky throughout the 1970s. The sightings dropped down considerably by the end of the 1970s, and even more by the mid-1980s. Some of the sightings involved much more than nocturnal lights in the sky, seemly too much more to be just erratic earth-lights.

There are 11 hot spot areas on the map. I consider Los Angeles and San Diego Counties to be one area; King County, WA to Yakima to be one area; Maricopa County, AZ and Snowflake to be one area; and, after some deliberation, I consider the Mississippi UFO corridor and Santa

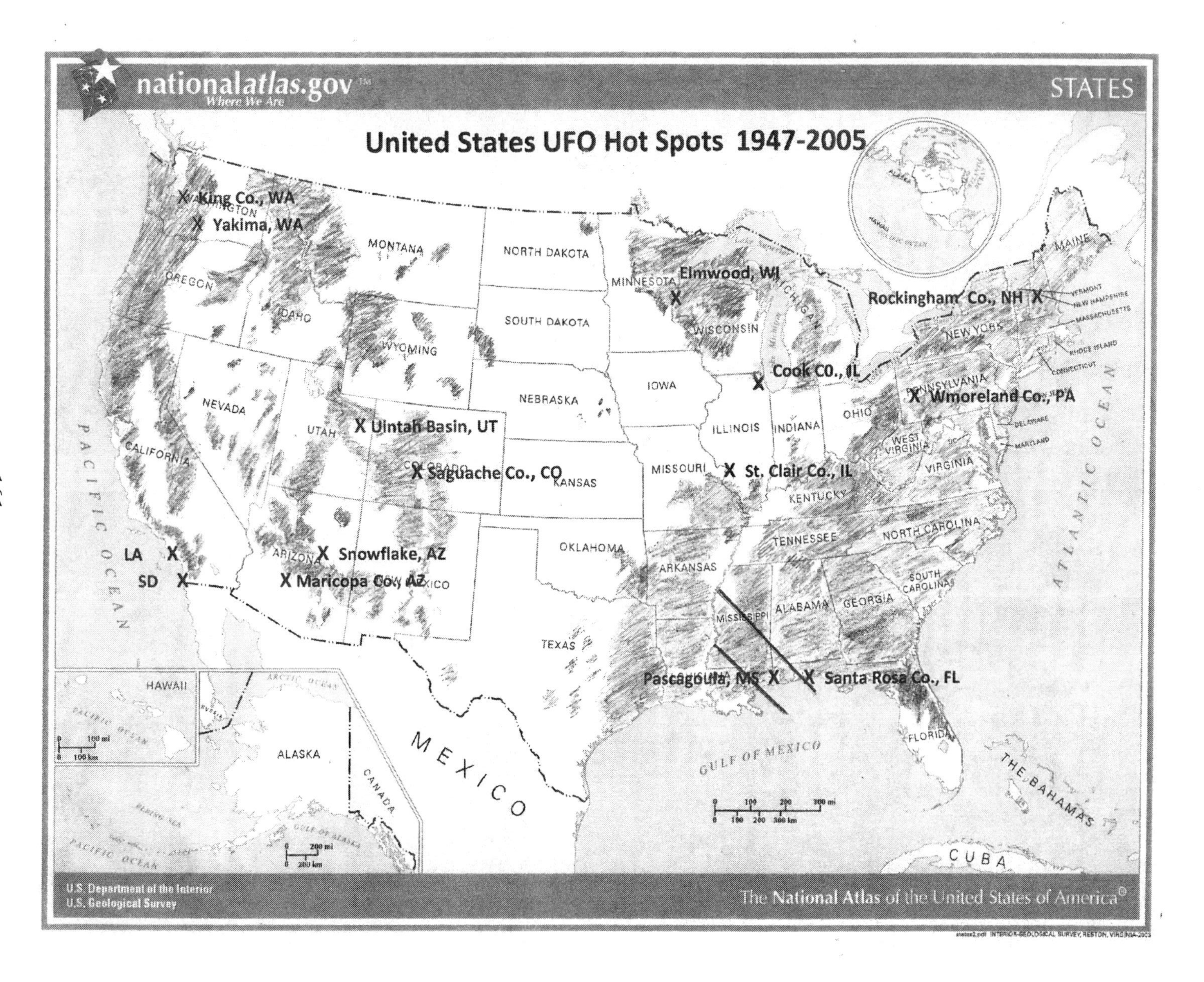
nationalatlas.gov
Where We Are
STATES
United States UFO Hot Spots 1947-2005
X King Co., WA
X Yakima, WA
X Uintah Basin, UT
X Saguache Co., CO
X Snowflake, AZ
X Maricopa Co., AZ
LA X
SD X
Elmwood, WI
X
X Cook CO., IL
X St. Clair Co., IL
Rockingham Co., NH X
X Wmoreland Co., PA
Pascagoula, MS X
X Santa Rosa Co., FL
MONTANA
NORTH DAKOTA
SOUTH DAKOTA
NEBRASKA
KANSAS
OKLAHOMA
TEXAS
MINNESOTA
WISCONSIN
IOWA
MISSOURI
ILLINOIS
INDIANA
OHIO
KENTUCKY
TENNESSEE
ARKANSAS
MISSISSIPPI
ALABAMA
GEORGIA
SOUTH CAROLINA
NORTH CAROLINA
VIRGINIA
WEST VIRGINIA
PENNSYLVANIA
NEW YORK
MAINE
VERMONT
NEW HAMPSHIRE
MASSACHUSETTS
RHODE ISLAND
CONNECTICUT
DELAWARE
MARYLAND
FLORIDA
OREGON
IDAHO
WYOMING
NEVADA
UTAH
CALIFORNIA
ARIZONA
PACIFIC OCEAN
ATLANTIC OCEAN
GULF OF MEXICO
MEXICO
CANADA
CUBA
THE BAHAMAS
HAWAII
ALASKA
U.S. Department of the Interior
U.S. Geological Survey
The National Atlas of the United States of America®

Rosa County, FL (Gulf Breeze) to be one area. It is possible that the Mississippi UFO corridor should be widened slightly to include the Gulf Breeze area, or maybe the MS UFO corridor is just an overflow area adjacent to the Gulf Breeze main event.

Of the 11 hot spots, only one is not in, or fairly near, a heavily forested area or national forest, and that one is Cook County, IL. However, Cook County sits next to Lake Michigan, which offers an 80-mile wide and hundreds of miles long area to hide in. Witnesses the world over have described UFOs entering and leaving large bodies of water.

Four of the areas (Elmwood, WI; Cook County, IL; Westmoreland County, PA; and Rockingham County, NH) do not have any type of geologic fault present. Four of the hot spots (Los Angeles-San Diego; King County-Yakima, WA; Missouri Piedmont-St. Clair County, IL; and the Uintah Basin, UT) sit on or near major, class A geologic faults and three of the areas (Saguache County, CO; South MS – Santa Rosa County, FL; and Maricopa County – Snowflake, AZ) sit over class B or lesser class A faults. The overall score is forest coverage 10 and geo faults 7, but that was based upon the physical presence of any type of geologic fault. I went further and made a table of the UFO hotspot ranks versus the seismic activity ranks. This seems to be the fairest way to compare the data.

UFO Hotspots and Seismic Activity US States

Location	UFO Hotspot Rank CUFOS Data	Seismic Activity Rank USGS 1973 - 2003
Southern CA Coast LA-SD	1	2
King County – Yakima, WA	2	5
Cook County, Illinois (Chicago)	3	18
Maricopa County, Arizona (Phoenix)	4	13
Westmoreland County, PA	5	32
Uintah Basin, Utah	6	9
Saguache County, Colorado	7	14

Santa Rosa C., FL (South MS & Gulf Breeze)	8	45
Rockingham County, NH (Exeter)	9	31
Piedmont Area Missouri (& St. Clair, IL)	10	16
Elmwood, Wisconsin	11	50

The data gives a linear correlation coefficient of 0.697.

The correlation coefficient of nearly 0.7 shows some association of UFO hotspots with areas of high seismic activity, but there is also an association of some UFO hotspots with areas of low seismic activity, as clearly seen in the table. Ten of eleven UFO hotspots were near a heavily forested area. The correlation coefficient for that is over 0.9. One could reasonable conclude that the association of forest coverage and UFO hot spots can be said to be very strong, and the geo faults/seismic activity association can be said to be moderate. The data strongly points to geo fault/seismic activity generated earthlights not being a universal explanation for UFOs, but does suggest that geo earthlights can be an explanation for some UFOs, such as erratic, short-lived nocturnal lights.

This UFO hot spots data deals with the total number of reports coming from a given area. What if one looked at not just the total number of reports, but reports that remained unexplained after investigation? I eventually decided to study the entire list of Project Blue Book unexplained cases from the NICAP website, 564 in all from 1947-69, to see what stories these could tell. After two days and nights of focused effort, I managed to enter the data into a spreadsheet on my computer. While doing this, it gradually dawned on me that I needed to look at a wide range of explanations that have been put forward over the years concerning UFOs, not just the implications of forest coverage and seismic activity. The next few chapters describe my evaluations of the data for all 50 states from the Project Blue Book Unknowns list comprising the years 1947-1969, and doing what is called data mining and hypothesis testing.

Chapter 23

Project Blue Book Unexplained Cases

"The only relevant test of the validity of a hypothesis is comparison of its predictions with experience."

—*Milton Friedman, (1912-2006) American economist, statistician*

This chapter is about the database of Project Blue Book Unknown/Unexplained cases and hypothesis testing. What can we learn from the 564 cases in the NICAP database? It turns out that we can learn a huge amount. The first thing that we learn is that ufologists dispute the number of unexplained cases as being far too low, and they do so with some valid statistical justification. Project Blue Book closed its doors at the end of January, 1970, and said there were 701 unexplained cases out of about 13,000 total cases on file (about 5.4% unexplained). However, only 564 unexplained cases could be found in their released information.

The number of truly unexplained cases has been put as high as 1500 or higher. The Battelle Institute studied about 3,200 cases given to them by Project Blue Book staff (some of these cases never made it into the Project Blue Book official case files) up through 1952 and found an unexplained rate of about 20% and another 9% with insufficient information to make a determination. That gives about 600 or so unexplained cases through 1952 with several hundred more as possibly unexplained. Yet Project Blue Book files show that 3,200 cases takes one nearly to the end of 1954, and PBB files only show 293 unexplained cases through 1954, a discrepancy of hundreds of unexplained cases. It seems that the Battelle engineers were "played" by the Air Force in that they were given official cases and letters from the public that never made it past Project Blue Book criteria for inclusion as an official case. They were also not given other basic information that the Air Force had such as balloon launch times and locations, weather, military test flights times and locations, and other data. It is not clear exactly what information the Air Force gave them. The declassified Air Force files that came to light later did not have some of the pertinent information for a number of cases.

Whatever the imperfections of the Project Blue Book dataset are, I believe that it contains some useful information for hypothesis testing. It is, after all, 564 cases that have been vetted in one way or the other. All of the usual explanations of UFOs have been weeded out from mirages to weather balloons to misidentified aircraft. We know that the cases from 1947 up to 1953 were reviewed more thoroughly than any of the others by Blue Book staff under Ruppelt, and their stated goal was to be as objective as possible. However, Ruppelt described an incident that occurred nearly directly over Project Blue Book Headquarters at Wright-Patterson AFB in Dayton, Ohio that calls into question this objectivity. From the incident one can see that if there was any possible way to distort a sighting to put it in the known or explained category, then that is where it went.

Consider the case in 1952 of a radar-visual sighting by base/PBB staff personnel. A bright light was seen hovering in the middle of the day fairly high up in the sky. It was seen on radar that indicated a location consistent with the visual sighting from the ground. A jet interceptor was sent out and the pilot had a visual sighting in the air of what he described later as a large, metallic object. The object was not a balloon of any type, as it did not drift with the wind. However, the intermittent presence of ice clouds and the planet Venus in the same general area of sky led to the sighting being put in the known category. How, you ask? The committee reviewing the case met right away, but did not interview the radar operator or the pilot, and did not go outside to judge the visibility of Venus for themselves, which they could have done the next day. Ice clouds moved in and out during the sighting and finally stayed the rest of the day. A staff member who doubted the official explanation checked on the visibility of Venus, which was in a similar position in the sky the next day and would have a similar visibility as the day before. While Venus can sometimes be seen during the day as a faint spec when at its maximum brightness if you know where to look, the staff member found that nothing could be seen with the naked eye. The committee attributed the sighting of a visible object some 10 or 20 times brighter than is possible for the planet Venus, as being due to the planet Venus. The radar signals were attributed to the ice clouds, which can sometimes give radar returns. The radar operator had data showing that the radar returns could not be due to ice clouds. The pilot was never interviewed by the committee. So much for objectivity! We know that at some point around 1953 or so, the Blue Book staff changed their investigation/evaluation methods to be even less inclined to label a sighting as due to an unknown cause, but there is a dataset to be examined no matter how imperfectly it was determined.

As I entered the data into the spreadsheet, I saw that the cases were not just from the US, but from all over the world. There were eight or ten from Canada, one or two from Mexico, some from Europe, a number from Japan, the Pacific Ocean, the Atlantic Ocean, and elsewhere. About 90 cases were from other countries, leaving 474 cases from the US. These foreign cases were mostly US airbase and military cases, with a few exceptions. Some 75 cases in all were sightings from air force bases with a few from naval air stations.

Some very intelligent people have asked me if this could mean that an air base sighting was of some secret operation going on at the base at the time. My reply was there is no way in the world for that to be the case. Think about it. Project Blue Book staff would have an extremely easy way to explain the sighting if that were the case. The case would have never made it into the unexplained file. Typically, it was highly trained and skilled base personnel involved in secret operations that reported the sightings. These are some of the highest quality sightings on record. There are several instances of an oval or saucer-shaped object flying by at high speed while the base personnel were observing a test flight or balloon launch. On some occasions of secret skyhook balloon launches, the UFO came in and seemed to examine the balloon by hovering and circling before taking off at high speed. In his book, Ruppelt told about making the assertion that maybe they were seeing their own balloons to the balloon project personnel at a General Mills research site in Minnesota (site of development of the enormous polyethylene skyhook balloons) where a number of UFO incidents had been reported. He was nearly thrown out on his ear into a blizzard. There was literally no possible way for the research personnel to mistake one of their balloons for a UFO. Most of the time, they were observing the balloon when the UFO came on the scene. Some of the balloon researchers had seen UFOs so frequently that they came to mostly ignore their presence.

When the locations of the other Project Blue Book unexplained sightings were examined, nearly half of them were within twenty to thirty miles of a military air base, again raising the question of the sighting being of some activity going on at the base, and again the answer is that the first reaction of Blue Book investigators was to look at base operations. If there was any similarity to something going on at an air base, the sighting would be attributed to that and listed as explained. PBB investigators were not outsiders looking in as many civilian UFO investigators were. PBB personnel had access to know when classified projects had been underway and the overall nature of the project (aircraft, rockets, balloons, etc.) without knowing the rest of the classified information or true objectives of the project. The presence of military air bases became high on my list of variables to check on further in the dataset.

Perhaps the most startling thing in the dataset was the number of unexplained sightings making the list in 1952. The number of reported sightings skyrocketed in 1952, along with the number of cases listed as unexplained. The 1952 unexplained sightings numbered 209 out of 1283 reported cases for the year (16.3% unexplained). This one year, 1952, had 37% of all unexplained cases in the entire dataset from mid-1947 through 1969, a total of 22.5 years. No wonder the CIA was so concerned about the public panicking over the subject of UFOs and called together the Robertson panel in January of 1953 to try to defuse the subject.

Even though there were three other years with elevated reported sightings above the 570 yearly average during this time period, no other year came anywhere close to the number of unexplained cases. Spikes in reported sightings occurred in 1957 with 1023, 1965 with 920, 1966 with 1110, and 1967 with 924, yet the number making the unknowns or unexplained list averaged about 19 a year in those years. With 1952 as a guide, these years would have averaged over 150 unexplained cases a year. We already know, as discussed earlier in this book, that Project Blue Book under Ruppelt, who took over in late 1951, examined the prior cases and the cases into 1953 as thoroughly and objectively as possible within their staff limitations, with lots of resources and specialists of every stripe available for consultation. The 1952 cases were some of the most thoroughly investigated UFO reports by Project Blue Book.

After the Robertson Panel findings in early 1953 (closely controlled by the CIA), we can tell exactly when the investigation/evaluation methods of Project Blue Book changed drastically. The first chart below shows the number of Blue Book 'Unknowns' by year plotted against the total number of reports for the year. From this, one can see that the time of change was by the end of 1954, or early 1955. The number of unknowns per year correlated very closely to the total sightings per year from 1947-54. The statistical correlation coefficient for those years turned out to be 0.95 (1.0 would indicate a perfect linear correlation). The number of unknowns per year after 1954 dropped drastically, and did not necessarily show a corresponding uptick when the reported sightings for a given year ticked up (correlation coefficient after 1954 was 0.66). Excluding the distortions for 1952, the unknowns dropped from an average of 38 per year in 1953-54 to 15 per year thereafter. The only year after 1954 to show more than 20 unexplained cases was 1966 with 29 unexplained out of 1110 reported sightings.

The second chart shows the percent of unexplained cases per year from 1947 through 1969. Before 1955, the unexplained cases per year averaged over 10 percent. After 1955, the average was less than three percent.

It could be argued that maybe the phenomenon behind the UFO reports changed in 1955 instead of the Project Blue Book methods, or that they both changed. We already knew that Project Blue Book reduced resources and personnel, and changed their attitude toward investigations at some point in the 1950s. I believe that the data simply shows the timing of those changes. It could be argued that the sightings that made the unknowns list after 1954 were the best of the best. PBB staff would have to judiciously use their resources. They would limit their cases for more in-depth investigation to only the cases with the most credible witnesses, and among those, perhaps only the cases with the most supporting evidence. I do not know their exact methods, but with limited resources, only a limited number of cases could be investigated well enough to be put on the unknowns list. Many cases could be dismissed due to questions of credibility of the witnesses or special military maneuvers were being carried out, or a balloon launch or recovery was nearby, or a planet was particularly bright in the same general area of the sky at the time of the sighting, or there was a strong temperature inversion present. Any of these things could possibly place a case in the explained file with no further investigation.

What we have is a tale of two datasets with one set of investigation/evaluation protocols for 1947-54 and another set of protocols from 1955-1969. I have split the data into these two time periods of 7.5 years and 15 years, and restricted the sightings to the current 50 US states (Alaska and Hawaii became states in 1959). That gives a dataset of 474 US unknowns with 293 or 61.8 percent occurring in the time period of mid-1947-1954, and 181 or 38.2 percent occurring in the time period 1955-1969. This great reduction in unknown cases after 1955, while sightings remained high, and the examination of the case records, has led a number of

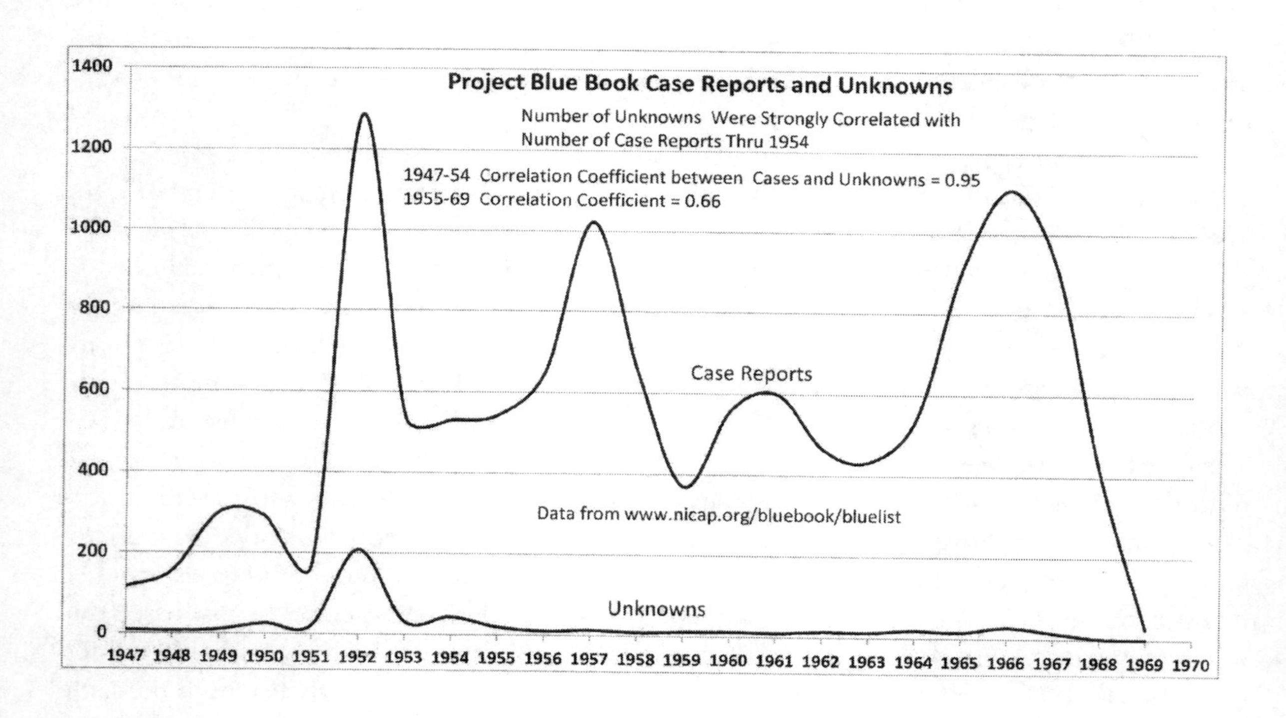
Project Blue Book Case Reports and Unknowns
Number of Unknowns Were Strongly Correlated with
Number of Case Reports Thru 1954
1947-54 Correlation Coefficient between Cases and Unknowns = 0.95
1955-69 Correlation Coefficient = 0.66
Case Reports
Data from www.nicap.org/bluebook/bluelist
Unknowns
1400
1200
1000
800
600
400
200
0
1947 1948 1949 1950 1951 1952 1953 1954 1955 1956 1957 1958 1959 1960 1961 1962 1963 1964 1965 1966 1967 1968 1969 1970

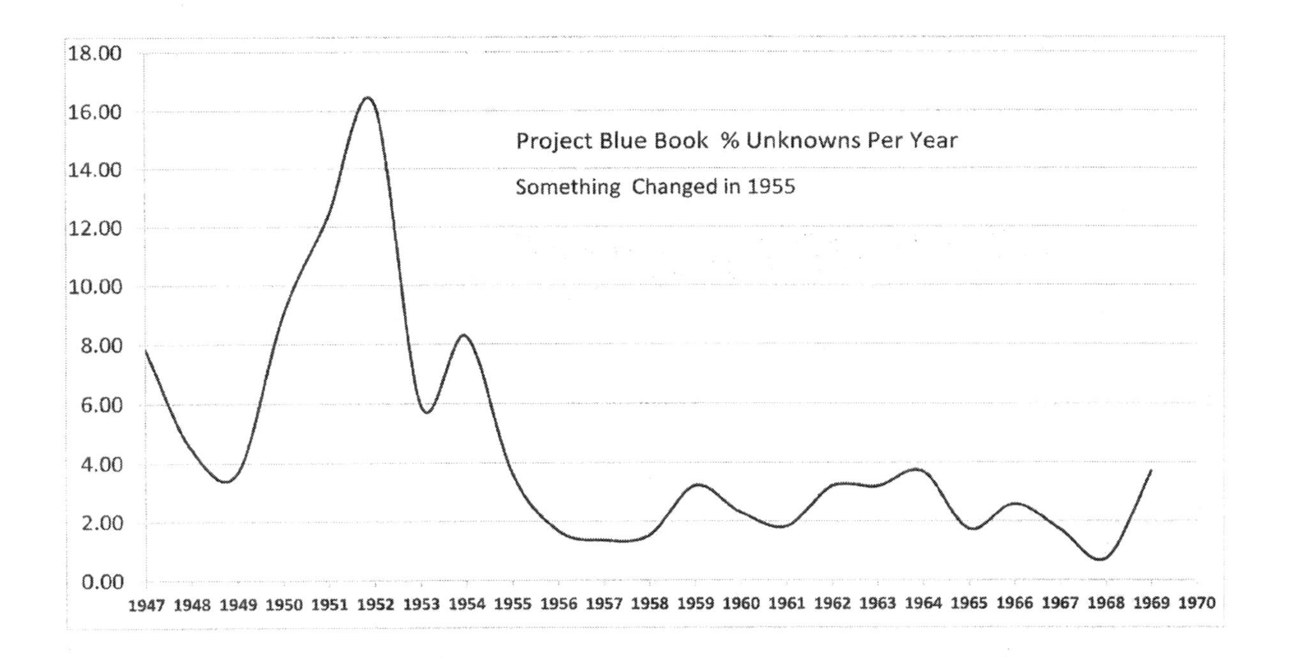

Project Blue Book % Unknowns Per Year
Something Changed in 1955
18.00
16.00
14.00
12.00
10.00
8.00
6.00
4.00
2.00
0.00
1947 1948 1949 1950 1951 1952 1953 1954 1955 1956 1957 1958 1959 1960 1961 1962 1963 1964 1965 1966 1967 1968 1969 1970

ufologist to say that the number of Project Blue Book unknowns is really more like 1,500 rather than 564. My analysis of the data certainly supports this contention, if the same investigation/evaluation protocols had been in place after 1954 as before. If the investigation/evaluation protocols had been truly objective throughout 1947-1969, I would expect the unexplained cases to number around 3,000. My evaluations of the data will be valid as long as the vast majority of the reports from 1947-1954 were all handled in the same way, and the vast majority of reports from 1955-1969 were all handled in a similar, but obviously different way from earlier.

I created one master spreadsheet of the data, and created several copies. In the copies, I performed various sorting of the data. I sorted one spreadsheet of the sightings by state and date before breaking the sightings into the two time periods of 1947-1954 and 1955-1969. Some states had many unexplained sightings while others had only a few. Once I spilt the sightings into the two time periods, I discovered that some states with a large number of unexplained sightings overall had far fewer in the later time period than would be expected. It also became clear that there was a west to east shift in unexplained sightings over time.

The first question that came to mind was whether or not the west to east shift was related to a change in Project Blue Book investigation/evaluation methods. For purposes of comparison, I have used the Mississippi River as the east-west dividing line. There are 26 states east of the Mississippi and 24 west of it. For the purpose of answering the question of whether the shift was due to PBB methods or not, I plotted the percent of cases east or west of the Mississippi River by year. Looking at the chart one can see that an average in the early years would put about 55 percent of the sightings in the west, but one can also see that in any given year the preponderance of unknowns shifted back and forth from west to east seven or eight times until the year 1961. After 1961, the unknowns in the east averaged about70 percent of the yearly total year after year. I will get to the explanation of this in due course, as it turned out to be a result of the interworking of Project Blue Book that changed over time.

Ruppelt had said July and the months immediately around it were the most active for UFO reports in the US. These are the summer months in the northern hemisphere when most people are outside doing things, observing nature, etc. This trend was indeed seen in the Project Blue Book Unknowns dataset. June, July, August, and September, with the peak in July, turn out to be the months with the most Unknowns in the 1947-1954 timeframe, just as Ruppelt had indicated. However, things are not as clear for the data from 1955-1969. The months June, July, August, and September are only slightly elevated. The Unknowns show about the same level of activity from February through November with sharp drop offs in December and January only. I do not have a good explanation for the difference between the two time periods. It could be due to the change in PBB investigation/evaluation procedures, a change in the phenomena itself, a change in people's behavior, or something else.

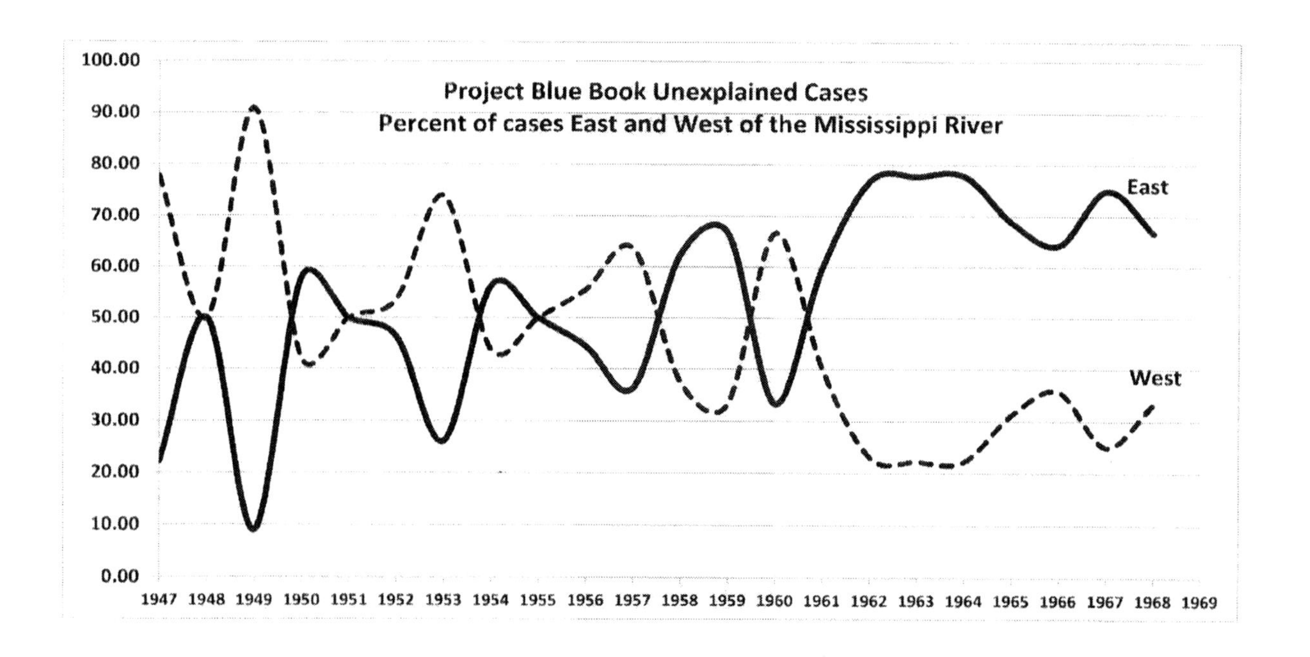
Project Blue Book Unexplained Cases
Percent of cases East and West of the Mississippi River
East
West
100.00
90.00
80.00
70.00
60.00
50.00
40.00
30.00
20.00
10.00
0.00
1947 1948 1949 1950 1951 1952 1953 1954 1955 1956 1957 1958 1959 1960 1961 1962 1963 1964 1965 1966 1967 1968 1969

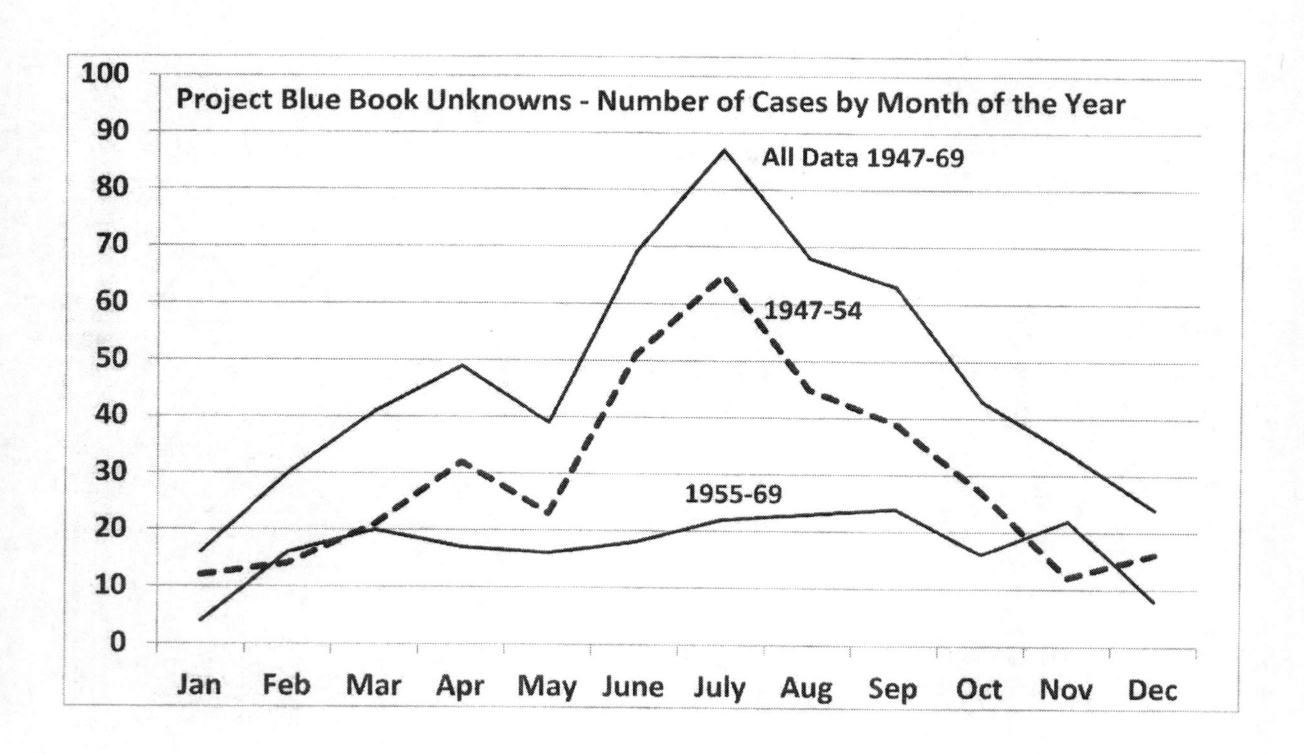
Project Blue Book Unknowns - Number of Cases by Month of the Year
All Data 1947-69
1947-54
1955-69
100
90
80
70
60
50
40
30
20
10
0
Jan
Feb
Mar
Apr
May
June
July
Aug
Sep
Oct
Nov
Dec

Now let's get on with the basis for hypothesis testing. What do I mean by hypothesis testing? Well the data set has 474 data points of UFO unknowns in the US with date and location of the sighting. We have already discussed why the data should be divided into the unknowns before 1955 and the unknowns after 1955. Some states have a large number of unknowns and some have only a few. Some states had a huge number of unknowns before 1955 and very few unknowns recorded after 1955. Some states showed a consistent level of unknowns throughout both time periods. One has to ask oneself what variables will affect the number of unknown reports and what variables does the data correlate with. Can I derive an equation using these variables that predicts fairly well how many unexplained sightings a given state had? Let us examine the main UFO hypotheses and see what variable or variables should correlate with UFO reports according to a particular hypothesis.

The purely psychological hypotheses say that UFOs are all in the mind. It follows that the population of the state alone should correlate well with the number of UFO reports. The greater the population, then the greater the number of "nut cases," and the greater the number of UFO reports. However, population should correlate with just about every UFO hypothesis one can think of, because people are making the reports. The more people—the more eyes are looking out to observe something. So population would be expected to correlate with every UFO hypothesis, but if the psychological hypothesis is in fact true, and all the others are false, then no other variables should really correlate at all.

Remember that as I was entering the data into the spreadsheet, I saw that an association with military airbases will correlate with the unknowns at some level before performing any mathematically rigorous correlation tests. Looking a little closer, I found that three-fourths of the military airbase Unknowns occurred in the 1947-1954 timeframe, with only one-fourth occurring from 1955-1969. I also pointed out that these are some of the highest quality sightings on record, because highly trained aeronautical specialists observed UFOs performing with capabilities far beyond anything we had then or have now. Therefore, it follows that the purely psychological hypothesis is absurd and should have never been proposed. It was nonsense when first formulated and it is nonsense today, but it is the number one hypothesis of mainstream science. The number two hypothesis of mainstream science is that the witness is mistaken. The highly trained specialist sightings, radar-visual sightings, multiple witnesses from different locations, and more, indicate that the mistaken witness hypothesis does not hold up for these Project Blue Book Unknowns. If the PBB investigators could have found some way to play the mistaken witness card, they would have, and they did, as they eliminated many cases because they judged the witness or witnesses not to be credible or capable. Anything not covered by the first two hypotheses of mainstream science is judged to be a hoax. The vetting of witnesses and other scrutiny meant that PBB investigators eliminated all suspicious cases from consideration. Mainstream science seems to basically ignore the PBB

Unknowns because this dataset contradicts all of the science establishment's hypotheses concerning UFOs. It should not exist, and would not exist, if their hypotheses were correct. Also, because UFOs are a taboo subject, most scientists do not want to dirty their hands by examining the PBB Unknowns or by even acknowledging them.

So now I have identified two variables that should correlate in some way to the dataset, population, and military air bases. What else? Well, we have the earth-lights hypothesis, geomagnetic and meteor. The meteor generated earth-lights seem at first glance to be a difficult hypothesis to test. They could be generated anywhere at any time. However, they fail to explain any repeating phenomena, and there are lots of repeating phenomena in the data. The same airbase or top secret government laboratory has been visited up to three or four times in the dataset over a short period of time. This is highly improbable for something completely random. The meteor generated earth-lights can, it turns out, be easily eliminated from consideration, because it should be totally random. Because of this random nature, the probabilities are that there will be few second occurrences, and certainly no third occurrences at any one place throughout the 22.5 years of PBB's existence.

The geomagnetic variety of earth-lights should correlate to geologic faulting and seismic activity. A look at the US Geological Survey map of the geologic faulting of the US mainland shows that the western US has by far the most widespread geologic faulting. There are a number of Class A faults capable of causing surface movements. When Class B faulting is included, the faulting area swings down into east Texas, through all of Louisiana, and most of south Mississippi. Another area of serious Class A faulting occurs in the center of the country near New Madrid, Missouri, with another area of Class B faulting in southern Illinois and Indiana. There are a few other small, isolated faulting areas in the eastern US, but the bulk of the serious faults are in the far western states. I have assigned the seismic activity risks to each state based on the USGS's seven levels of risk for the purpose of seeing how well the locations of the unknowns in the dataset correlate to the serious faulting areas. If the geomagnetic earth-lights have any validity at all, then there should be more unknowns from the higher risk/higher activity faulting areas than from the low risk/low activity areas. The areas of higher tectonic stresses should generate more reports, and more reports that end up classified as unknowns or unexplained cases.

So now we have three variables to test: population, military air bases, geologic faulting. What's next? What if UFOs are real objects from some unknown source flying randomly overhead? What would correlate with that? Population correlates, of course, because someone has to witness the over flight, but what about the land area of the state? The bigger a state is in land area, the greater the probability that a random flyover will occur. For two states with great differences in land area, the bigger state has the greater probability of a random flyover. Of course, population density comes into play here. Population density and physical size of the state are two more variables to check for correlation with the dataset. That brings us to

Earthquake Hazards Program

By State and Region

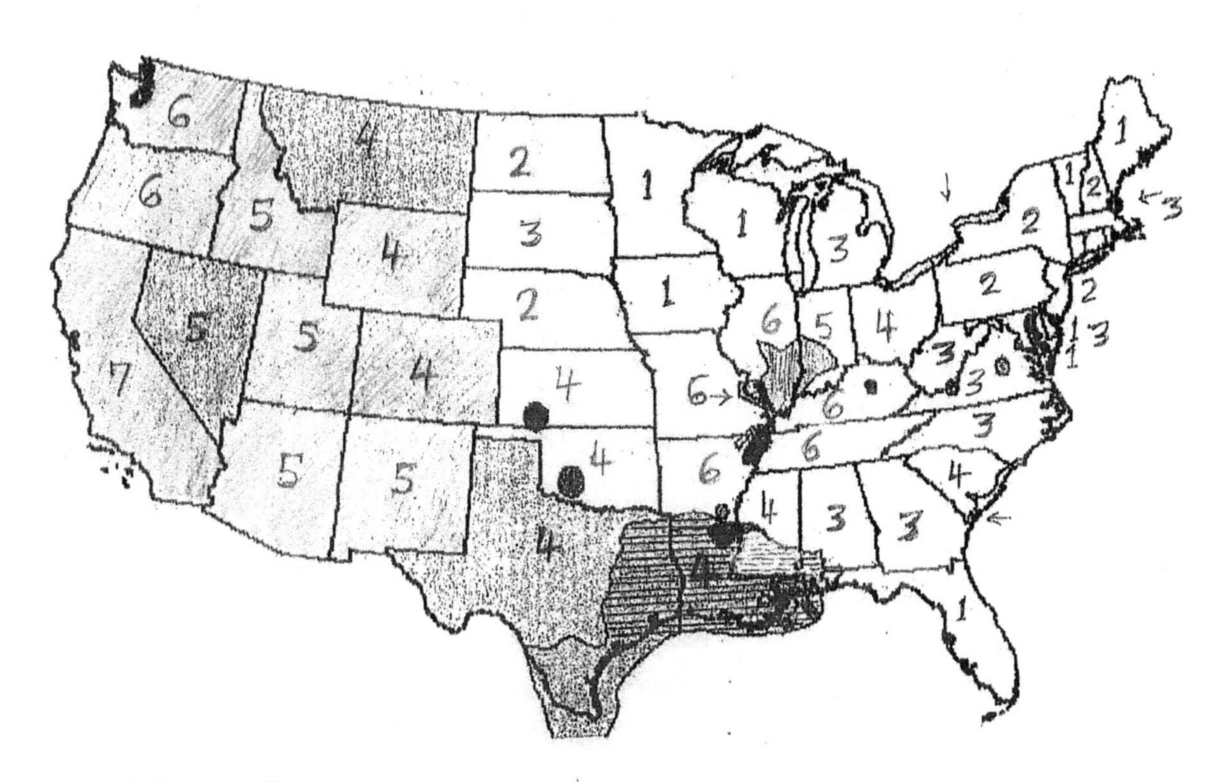

Island of Hawaii - 7
Alaska - 7

When you use this data, please provide proper acknowledgment.

five variables to check for correlation: population, population density, military air bases, geologic faulting, and the land area of a state.

Suppose UFOs are not randomly flying overhead—suppose that they are controlled by, and/or piloted by, some unspecified group such as ETs or otherwise. What would correlate with that? My hypothesis of a convenient hiding place, such as forest, cover could correlate with such a scenario. Visiting military airbases to snoop around could correlate also. Also, what about the amount of mountainous terrain as a hiding place? Also, consider USOs, Unidentified Submerged Objects. The History Channel UFO series had a few programs discussing USOs, which are basically UFOs underwater. Witnesses have seen UFOs coming from beneath the surface of the ocean or a large lake and flying away. One of the best documented cases was the Shag Harbour incident in Canada in October of 1967, which was discussed on one of the History Channel programs. What would correlate with USOs? How about the amount of coastline due to ocean or Great Lakes that a state has? If the USO flies inland, it has got to cross a coastline. Finally, what about skyhook balloons (a US Naval research program)? It is known that these have led to UFO reports.

I modeled the skyhook balloon hypothesis by considering a radial distribution from the launch site. The further away from the launch site, the lower the probability of seeing the balloon. Think about a 250-mile radius around the launch site that gives the balloon a nearly 200,000 square mile area to land in. Then consider a 500-mile radius that would give nearly an 800,000 square mile area, or four times larger. In the following states where there is evidence of recurring skyhook balloon launches: New Mexico (headquarters of the skyhook program), Arizona, California, Colorado, Montana, North Dakota, Missouri, Georgia, and Minnesota (General Mills basic polyethylene balloon research site of the 1940-1960s), I assigned a numerical value of 10. In all neighboring states, I assigned a value of five and in states further away I assigned a value of three. The model predicts a center of sightings near the launch site and diminishing sightings at greater and greater distances. The complete list of variables that I came up with for hypothesis testing is given below, but I am sure there are others that I have not considered yet. The idea is to see how well each one of these individual variables correlates with the number of unknown reports grouped by state in the dataset. How well does any one of these variables explain the differences in the number of PBB unknowns from state to state?

<u>Variables for Hypothesis Testing</u>

Population in 1950 & 1960	(State's Total Population)
Population Density 1960	(Entire State Pop/Land area in square miles)
Military Air Bases	(Number of air bases in the state 1950-1970)
Land Area	(Square Miles)
Total Forested Area	(Fraction Forested x Land Area)

Total Mountainous Area	(Fraction Mountainous x Land Area)
Geo Faulting/Seismic Activity	(Faulting/Activity Map Assignments)
Miles of Coastline	(Ocean or Great Lakes)
Skyhook Balloon Sites	(Radial Distribution from Launch Sites)

Most of the values for the variables for each state were readily available in standard references, but the forested area was the most difficult to find for the 1950s and 1960s. I visited each state's forestry website and looked for information. Sometimes the site would say something about the current forest cover and how it related to the past forest coverage, and in a few cases the statement was made that the forest coverage has changed little since 1960. It seems that the greatest change in forest coverage occurred when a state's lands were first settled, as the best land for farming was cleared of timber in short order. Later, great waves of exhaustive logging occurred in many areas of the country in the late 1800s and early 1900s. With some ebb and flow, the nation's forests seem to be mostly recovering from this extensive logging ever since.

Chapter 24

Correlations of the Variables with the PBB Unknowns from 1947-1954

"The goal is to transform data into information, and information into insight."

—*Carly Fiorina, Executive Hewlett-Packard Co., 1999-2005*

I was concerned for a while about whether or not the nation's forests and the population were too closely linked. At first glance, population maps and forest coverage maps look remarkably similar. This would mean that the effects of forest cover could not be separated from the effects of population. I examined the data more closely and came to realize that there were states with high to moderate population with extensive forests (Eastern US and far West), states with high to moderate population and not much forest cover (Midwest), states with low population and extensive forests (upper New England and Southern US), and states with low population without much forest (Great Plains). It's not perfect, but there is a reasonable coverage of each sector of possibility. The map below gives the number of Project Blue Book Unknowns for each state during the 1947-1954 time period along with the US forest coverage from about the year 2000. I did not find a US forest coverage map for the 1950 to 1960 timeframe. I do not believe that it would be that different, but maybe some will have an issue with this.

It can be readily seen that forest coverage was not the deciding factor in determining where PBB Unknowns occurred. Some heavily forested states did not have many Unknowns, and some less forested states had quite a few. Forest coverage will, at best, be a secondary factor. Population has to be more important for spotting something in the sky, but I did want to test a number of hypotheses to see how things sorted out.

There was a book published in 2011 with the title "Aliens in the Forest" by Noe Torres and Ruben Uriarte about the encounter of Donald Shrum in the foothills of the Sierra Nevada Mountains near Cisco Grove, CA in the Tahoe National Forest of Northeastern California in the fall of 1964. [125] I found out about the book in mid-2012, well after I had formulated the hypothesis concerning forest cover and UFOs. There was a Project Blue Book Unknown north of this area in 1963, and one south of the area in 1964. Both Unknowns were sighted in the Sierra Nevada range and forests, and both sightings involved more than one object flying overhead. On further investigation, the sighting in September of 1963 lends supporting evidence of Shrum's account, which occurred about one year later. Shrum described a larger elongated or cigar-shaped object from which emerged a smaller, rounded, silvery module that landed on the ground. The 1963 witness was E. A. Grant, a fire lookout trainer for Forest Service personnel for 37 years. Grant described a smaller flying object intersecting a larger elongated object over the forests of the Sierra Nevada range and disappearing as if it had attached itself or entered the larger object. As far as I know, no one else has noted the similarity in the two reports. Something very similar was seen flying in the area one year before Shrum's ordeal. He had absolutely no possible way to know about the earlier report until sometime after 1975, and this lends support to his case. Project Blue Book investigators labeled his case psychological, apparently just because of the abduction nature of the case.

I ventured on thinking that population would be the number one correlating variable for each set of data. However, I was immediately surprised to find the correlations that follow for the Project Blue Book Unknowns/Unexplained cases from mid-1947-1954. The statistical quantity r squared is also listed, because r squared times 100 gives the percentage of the variation among states that can be accounted for by the variable being tested. It must be cautioned that correlation does not prove cause and effect, but things that are related by cause and effect will show correlation. After finding a correlation, one must examine the situation to see if a reasonable cause and effect case can be made for the correlation. The variables are given in order of strongest correlation to the weakest. The most surprising thing about the data is that population came in fourth instead of first. The correlation with military air fields and bases came as no surprise, as I had seen the association of air fields and bases as I entered the data into the spreadsheet before doing any statistical correlations. About 52 of the total 293 unexplained sightings for this time period were reported from air bases and air fields. The surprise came in military air facilities giving the number one correlation. A closer look at the data for these years shows that many more unexplained sightings were made in the vicinity of air bases and air fields; nearly 50 percent in all, so the correlation is completely valid, but there is more to the story that I will get to just a little later.

Project Blue Book Unknowns mid-1947-1954
(State to State Variation - Correlation With Variables)

Variable	r	r^2	% Explained	Cross Correl r
No. Military Air Facilities	0.783	0.613	61.3%	
Forest Cover in Sq. Miles	0.700	0.490	49.0	0.540
Land Area in Sq. Miles	0.654	0.428	42.8	0.475
Population, State, 1950	0.486	0.236	23.6	0.341
Mountains in Sq. Miles	0.376	0.141	14.1	0.274
Miles of Coastline.	0.349	0.122	12.2	0.495
Geologic Faulting	0.278	0.077	7.7	0.226
Skyhook Balloon Facilities	0.274	0.075	7.5	0.110
Pop. Density, Pop/Sq. Miles	0.039	0.002	0.2	

The data spreadsheet is given in Appendix I.

Linear correlation coefficient r level of significance for 50 sample points.

r = 0.273, 95% confidence that the correlation is real

r = 0.354, 99% confidence that the correlation is real

A 95% confidence interval is commonly used.

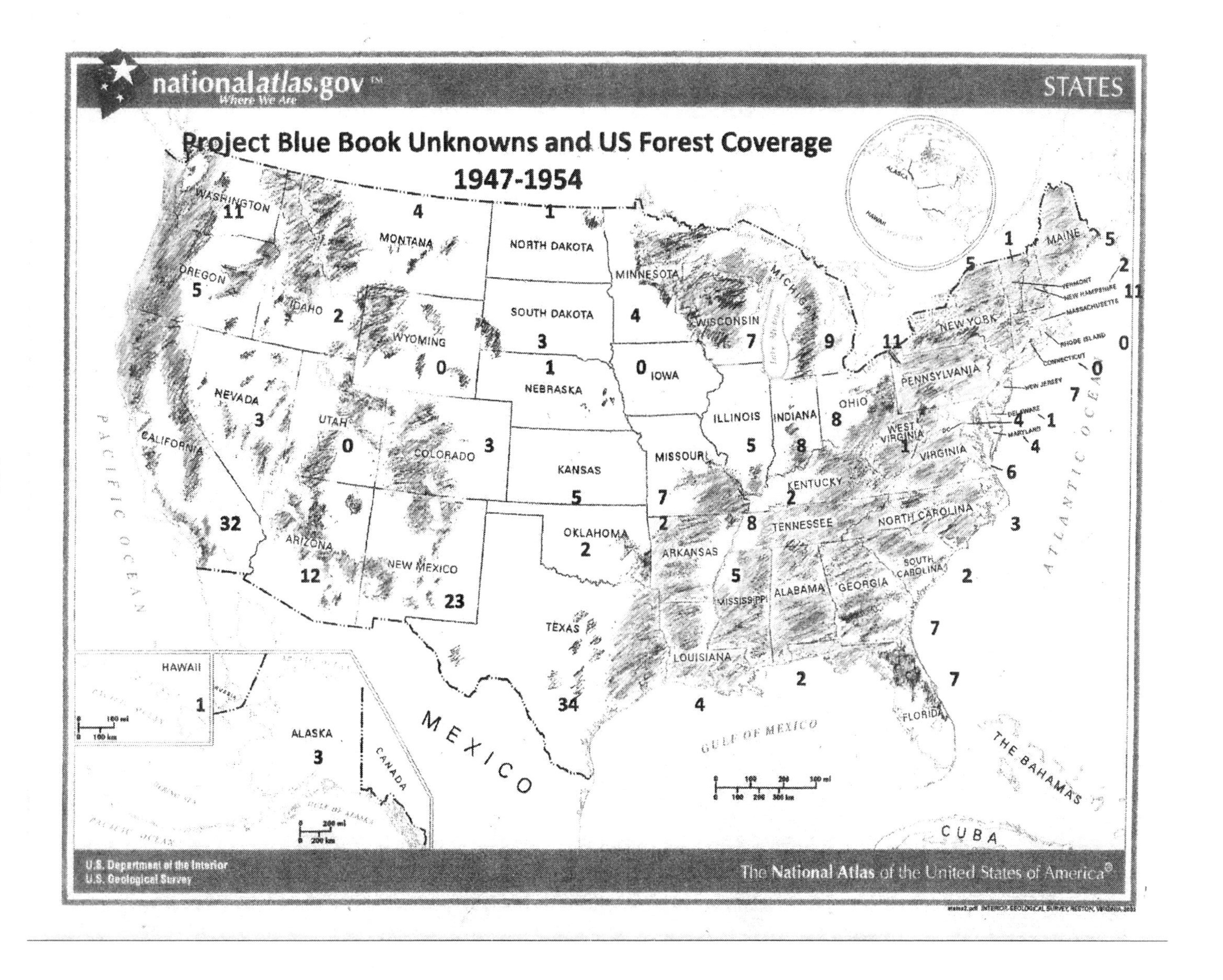
nationalatlas.gov
Where We Are
STATES
Project Blue Book Unknowns and US Forest Coverage
1947-1954
WASHINGTON 11
OREGON 5
CALIFORNIA 32
IDAHO 2
NEVADA 3
MONTANA 4
WYOMING 0
UTAH 0
ARIZONA 12
COLORADO 3
NEW MEXICO 23
NORTH DAKOTA 1
SOUTH DAKOTA 3
NEBRASKA 1
KANSAS 5
OKLAHOMA 2
TEXAS 34
MINNESOTA 4
IOWA 0
MISSOURI 7
ARKANSAS 2
LOUISIANA 4
WISCONSIN 7
ILLINOIS 5
MICHIGAN 9
INDIANA 8
OHIO 8
KENTUCKY 2
TENNESSEE 8
MISSISSIPPI 5
ALABAMA 2
GEORGIA 7
FLORIDA 7
SOUTH CAROLINA 2
NORTH CAROLINA 3
VIRGINIA 6
WEST VIRGINIA 1
PENNSYLVANIA 11
NEW YORK 5
MAINE 5
ALASKA 3
HAWAII 1
PACIFIC OCEAN
ATLANTIC OCEAN
GULF OF MEXICO
MEXICO
CANADA
CUBA
THE BAHAMAS
U.S. Department of the Interior
U.S. Geological Survey
The National Atlas of the United States of America®

Another surprise was that population density gave no correlation at all. However, if a state's total population correlated with the unexplained sightings within a state, and the state's total land area correlated, then the state's population density, which is total population divided by the land area, would not be expected to correlate. The population density at the exact sighting location may well correlate in some way, but tracking down that detail for 474 sightings was not done.

Which hypotheses can best explain the 1947-1954 correlation results? The situation is complicated by the fact that the variables are not completely independent of each other. Population, forested area, land area, and the amount of coastline all correlate above the statistically significant level with the number of military airbases in a state. The last column in the table is labeled Cross Correl r and represents the expected correlation coefficient of the variable when airbases have a correlation coefficient of 0.783. Most of the correlation coefficients, r, are above the cross correlation r except for the miles of coastline and population density, which shows no correlation. Alaska is usually left out of the correlations because of its huge size, small population, and vast unpopulated areas. It is quite different from the lower 48 states in this regard. The interpretation is complicated, but then it usually is in real-world data analysis. However, we are left with air bases, forested area, land area, population, mountain area, geologic faulting, and skyhook balloon launches that scored above their expected cross correlation and above the cutoff for statistical significance with a 95% confidence level of an r value of 0.273 for 50 data points.

The data clearly indicates that something real and unexplained was flying around visiting or flying near military air fields, and that possibly these UFOs needed a place to lay low for a while in between areas such as a forest or mountain range. We can discard the purely psychology explanations, PSH and others, because nothing but population would show any correlation with these hypotheses. The part of the geomagnetic earth-light hypothesis calling for effects upon the human mind was tossed out earlier on the basis of magnetic field strength arguments. The simpler version of generating erratic lights in the sky just made the cut-off value of the correlation coefficient, r, at the 95% confidence level of 0.273, which means the correlation is weak. The wide scope geomagnetic hypothesis is on such a poor footing from its dependence upon impossibly weak magnetic fields to cause havoc with a person's temporal lobes that it is not a reasonable explanation of the data. The weak correlation of the geo-earth-light hypothesis leads me to conclude that it can be tossed out as a general explanation of PBB Unknowns.

We can also toss out the skyhook balloon hypothesis as too weak to be a general explanation for UFOs. It barely scored above the $r = 0.273$ cutoff for statistical significance, and it is an idea from the early 1950s that Ruppelt fairly well demolished in his 1956 book. The hypothesis is that since the UFO phenomena and the skyhook balloon program both began in 1947, and both peaked in 1952, the skyhook balloon program was responsible for the UFO craze. Some later years of the Blue Book era had nearly as many UFO reports as 1952, but with far fewer classified as unexplained cases. In addition, the year 1973 topped 1952 in UFO activity, but there was no official government agency that accepted UFO reports after 1969. Ruppelt talked about having access to secret balloon launch activities and checking on those to see if they could explain a sighting. Every sighting that was listed as unknown in Blue Book files had been checked to see if the sighting could be explained by a balloon launch or the balloon's path of either a weather balloon or a skyhook balloon. Yet that did not keep B. D. Gildenberg from speculating in 2004 in *Skeptical Inquirer* magazine that skyhook balloon activity could explain most of the UFO phenomena. [126] Gildenberg suggested this with reference to the excess number of sightings in New Mexico that we will get to shortly.

The skyhook balloons that Gildenberg described were enormous in size and capable of carrying an instrument payload weighing tons. The balloons were also capable of making maneuvers of sorts that no one outside the classified project would know about, and this has been confirmed from other sources. Helium or hydrogen could be let out by radio signal so that the balloon could descend to about 3,000 feet and release the payload to parachute down to the ground to be recovered by helicopters, or be snatched directly out of the air by special aircraft. Gildenberg pointed out that these activities generated UFO reports; very accurate UFO reports. Witnesses described the exact behavior of the balloons and related activities on a number of occasions. This was one of few skeptical articles that showed how accurate eyewitnesses could be for describing something completely unknown and strange to them. Usually skeptical articles talk about how people are very poor at witnessing unusual phenomena. Skeptics, in general, seem to gravitate to whatever depictions of eyewitnesses suits their debunking purposes at the time. Also, Gildenberg was part of the skyhook balloon program for 30-something years, so he must have known about the balloon launch crews and balloon tracking crews that reported UFOs checking out the big balloons. Like so many skeptics, he played to his audience's lack of knowledge on the subject of UFOs, but his contention was valid that skyhook balloons have a high potential to generate UFO reports. He cited three or four cases where skyhook balloons generated UFO reports that went unexplained, and parlayed that into saying that skyhook balloons were one of the main things that helped define the UFO era. We shall see whether that statement can stand up to further scrutiny or not. Certainly, in the general correlation with PBB Unknowns from 1947 through 1954, skyhook balloons ended up as a deflated explanation according to my radial distance model.

However, Gildenberg has presented a testable hypothesis, and I do not mean testable only by my radial distance model, although I think that the model is legitimate. The hypothesis can be tested further by examining each of the 23 PBB unexplained cases for New Mexico from 1947-1954 to see which, if any, can be explained by a skyhook balloon or weather balloon. According to Ruppelt's book, very few of them can be. This is something that I will look into just a little later, but first let us look at typical statistical scatter plots to get a feel for this sort of data handling.

The first data chart below shows a typical statistical scatter plot of PBB unknowns versus the number of air bases and air fields in a given state. It shows the scatter in the data and a best linear fit line calculated from the data. Scatter plots gives an overview of the data and what it means to have a correlation coefficient of 0.783. If the correlation coefficient were 1.0, all the data points would lie near the straight line.

The second chart below shows an equation that attempts to describe the number of UFO PBB Unknowns per state for the 1947-1954 time period. The heavy smoothed line was calculated from this equation using the r squared coefficients from military airbases and from forest cover, and gives a prediction of the number of PBB unknowns from that state during this time period. Together they give a moderately better fit of the data than either one does alone. Combined, they can explain about 66% of the variation in the number of unknowns from state to state. In this kind of data fitting equation there will be an equal number of points above and below the calculated line.

One can see why population did not correlate that well for this dataset. On the chart are listed the top 11 states in number of unknowns. The top six states have unknowns that exceed by a good margin those predicted by population alone. Seven states in all exceed the number solely predicted by population (Texas, California, New Mexico, Arizona, Washington, Massachusetts, and Tennessee). Three states are well below their population predicted values (Pennsylvania, Michigan, and Ohio) and only one state (Indiana) out of the top 11 had the number of unknowns predicted by population.

In any data of this type there usually will be outliers. The worst outliers in this plot are New Mexico, with excess unknowns above the predicted number, and Alaska, with far fewer unknowns than predicted. Alaska has a huge amount of forests in all, and a small population clustered in a few places. Alaska will be an outlier on anything predicting the number of sightings will go up with forest cover. The real question is why does New Mexico have 23 PBB unknowns when the predicted number is five from the best fit equation using correlations for air bases and forest cover together, and the number predicted based on population alone is two? If we look at it with a little wider view, we see that all the far west states from Washington down through Oregon to California to Arizona to New Mexico, and to Texas have unknowns in excess of that predicted by population. The unknowns during this period were dominated by the far western states and the desert southwest states, and most of these states had an inordinate number of air bases or air fields, except for New Mexico, which still had eight. But, New Mexico also had a very special air base, the nuclear airbase near Roswell (named Walker AFB until 1967), and a top-secret government nuclear laboratory at Los Alamos. However, when the unknowns for New Mexico are plotted on a map of the state, it is clear that many of the excess Unknowns were clustered near Albuquerque. There were

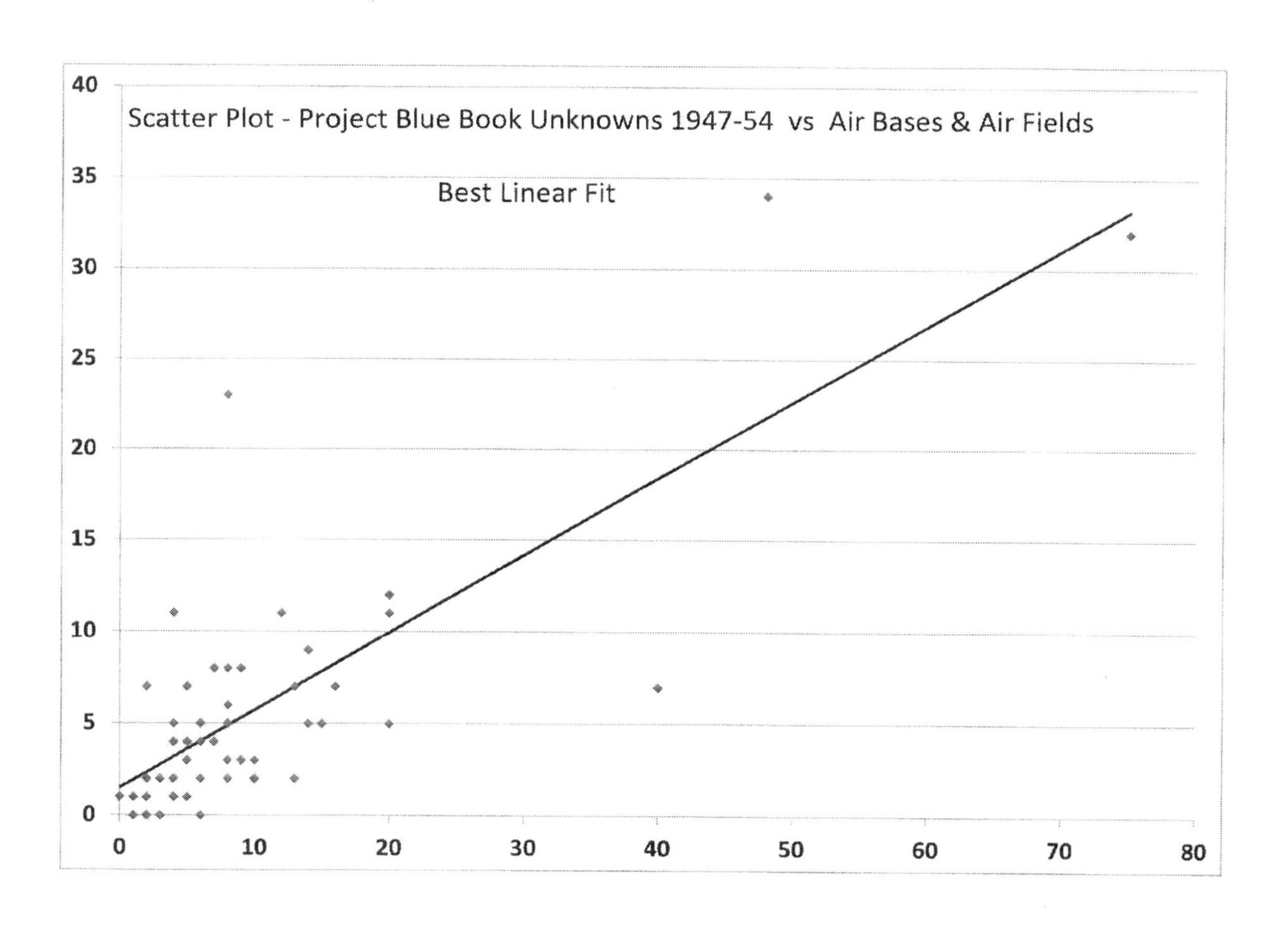
Scatter Plot - Project Blue Book Unknowns 1947-54 vs Air Bases & Air Fields
Best Linear Fit
0
5
10
15
20
25
30
35
40
0
10
20
30
40
50
60
70
80

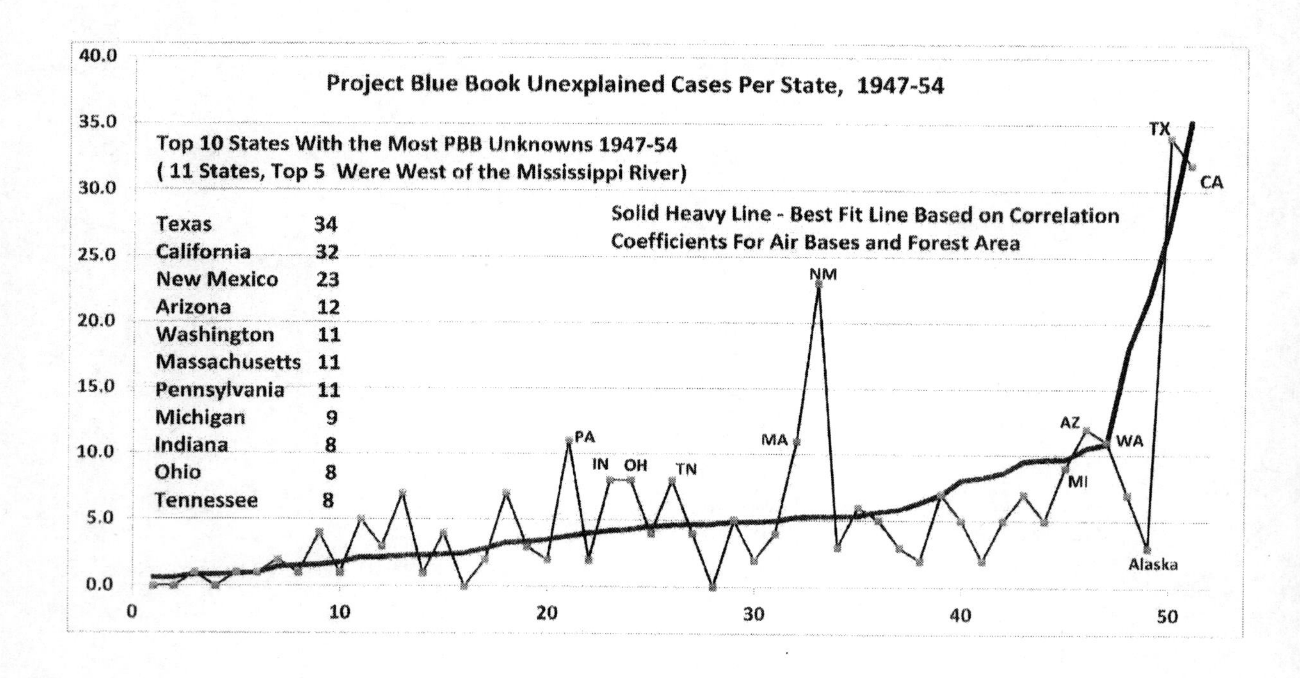
Project Blue Book Unexplained Cases Per State, 1947-54
Top 10 States With the Most PBB Unknowns 1947-54
(11 States, Top 5 Were West of the Mississippi River)
Texas 34
California 32
New Mexico 23
Arizona 12
Washington 11
Massachusetts 11
Pennsylvania 11
Michigan 9
Indiana 8
Ohio 8
Tennessee 8
Solid Heavy Line - Best Fit Line Based on Correlation
Coefficients For Air Bases and Forest Area
TX
CA
NM
MA
PA
IN
OH
TN
AZ
WA
MI
Alaska
40.0
35.0
30.0
25.0
20.0
15.0
10.0
5.0
0.0
0
10
20
30
40
50

three at Los Alamos and several along the line of air bases in southern New Mexico where the Roswell AFB was, but nine Unknowns were clustered around Albuquerque alone, which has Kirtland AFB just seven or eight miles from the city center. Other cities during this period, especially around 1952, with an inordinate number of Unknowns were San Francisco, CA, San Antonio, TX, and Washington, DC.

I was eventually able to find information about the 23 New Mexico Unknowns by going to various websites, but during the course of doing that I found the NICAP website with Project Blue Book monthly status reports for 1952 that had been declassified, and I began to read the reports. [127] My eyes were opened by the contents. I found some of the New Mexico reports with initial comments, but the Blue Book monthly status reports also give great insight into the operation of Project Blue Book during these early years.

First of all, it became clear that Project Blue Book was mainly considering military reports during this time period. Civilian reports were deemed so nebulous as to be worthless in most cases. The period of July through October of 1952 had 880 military reports and 800 reports from the general public. The vast majority of the reports that eventually were not able to be explained were from military sources and their contractors, scientists, engineers, and technicians of all sorts. Reports that did not have the details of azimuth (angle of deviation from due north), angle above the horizon, precise description of the object's appearance, behavior, location, duration of the sighting and the exact time and date were chunked out. Most single witness cases of any type were chunked unless the witness was unquestionably reliable. (I wrote earlier about the Florida scoutmaster case with impossible to hoax physical evidence that was tossed out because the scoutmaster had a questionable past and was known for telling tall tales and exaggerations). A report with a lone civilian witness with no technical training would not be looked at a second time during this period. All reports of silent silvery objects in the sky below 60° to the horizon were chunked out or explained as normal aircraft (PBB monthly reports state that experience has shown that nearly all regular aircraft observed at 60° or above can be heard by the observer as long as the background noise is low). In general, the higher the angle in the sky, the closer a normal aircraft is to the observer.

Untrained civilians would have no way of knowing how to present their case to the Project Blue Book staff, and the staff really did not want the public input (later in 1952 or early 1953, a questionnaire was developed for the public to report a case). The staff wanted all the information above and more, if possible, from highly trained military personnel with security clearances (background checked and mental stability checked). The staff handling all of these reports amounted to just eight or ten individuals. There was no realistic way to deal with the huge volume of reports. Judgments had to be made. Extreme judgments and screening measures had to be implemented. Reports from reliable sources containing the above information could be compared to balloon tracking data, aircraft flight data, weather data, temperature inversion data, and astronomical charts. Weather balloons in those days were launched at 3:00

New Mexico and West Texas

Project Blue Book Unknowns 1947-54

Proximity to Air Fields and Air Bases

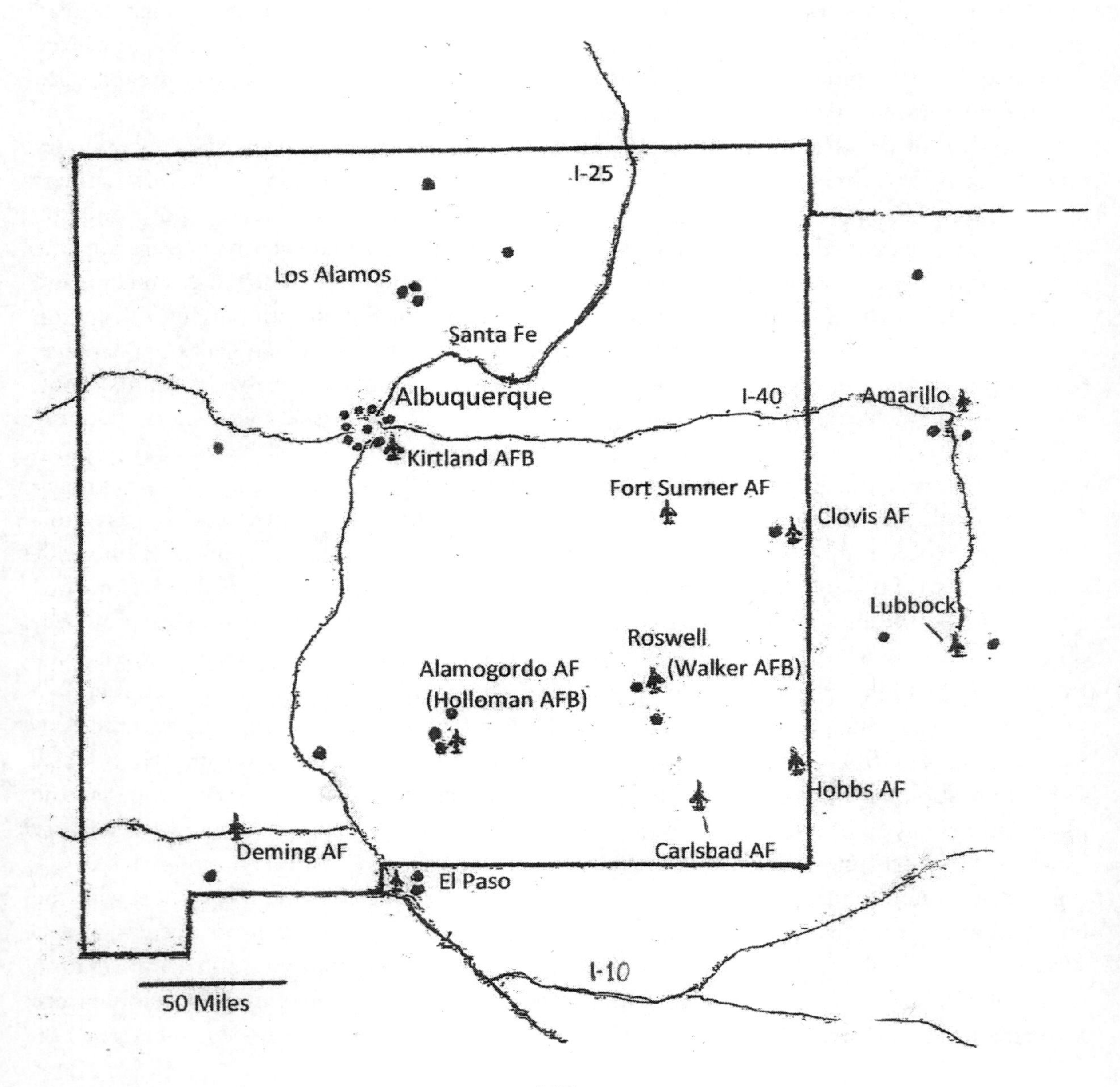

a.m., 9:00 a.m., 3:00 p.m., and 9:00 p.m. (today it's at noon and midnight). Any report within 30 minutes of a nearby balloon launch was designated as explained by the balloon (most weather balloons rise up and burst within an hour and a half or so of being launched). Any report made with a known aircraft in the air in the general area at the time was explained as due to that aircraft. Any report that placed the UFO in the area of sky where a bright planet or star was located was explained as a sighting of that planet or star, even if there was radar data to confirm the sighting and multiple witnesses attested that the object moved. The vast majority of civilian reports of the time were never considered.

Suddenly it dawned upon me, the data correlated with military air bases because that is where Project Blue Book considered the most reliable observers to be. Project Blue Book had preselected the correlation by mostly not considering other sources. We do not really know what the general phenomena correlated with unless all the sightings reported by the public could have actually been investigated. Logic indicates that if 880 military cases generated 150 cases that ended up as Unknowns, then 800 cases from the general public would probably generate 50 to 100 additional Unknowns if the cases had been reported in the proper format. The difference was that the public, in general, did not know how to observe and report a case in the format for Project Blue Book to evaluate it. These letters from the public are probably not preserved in the National Archives with the rest of the Blue Book files. It would be interesting to look at those cases if available, but it would be a tremendous amount of work.

So what about the New Mexico Unknown cases of the 1947-1954 time period? I eventually found at least a brief description for all 23 reports. Was I able to confirm or deny Gildenberg's hypothesis that skyhook balloons can explain most of the UFO phenomena? It turned out that four of the Unknowns involved weather balloon/skyhook balloon launch or tracking crews who were watching their balloon or balloons when a UFO or UFOs came by. This is just the opposite of Gildenberg's hypothesis. Most times the UFO or UFOs circled around the balloons as if getting a closer look before zooming away. Most of the other sightings involved military or military contractors or personnel with security clearances. One report was of a flying wing or delta wing type of craft with various lights and other details discussed in a previous chapter. This unknown did not appear to have registered on area radars, making the case similar in some ways to the Lubbock Lights case of 1951, and Phoenix Lights case of 1997. The flying wing UFO also sounds similar to some of the sightings in the Hudson Valley (New York and Connecticut) in1983 into the 1990s. [128] In only one case, the Mount Taylor, NM Unknown of March 27, 1953, Blue Book staff had initially marked the case as possibly due to a balloon, but that explanation did not hold up. The Mount Taylor Unknown, a bright orange sphere, was observed by and chased by the pilot of an F-86 jet fighter flying at 700 mph. The object flew at 900 mph and executed three fast rolls; so much for the balloon idea.

I expanded my search for Unknowns that could possibly be explained as balloons by reviewing Don Berliner's synopsis of all 560+ Project Blue Book Unexplained cases compiled for the Fund for UFO Research [86C] (a few had no case number). This is the same set of cases as on the NICAP website, but with brief descriptions added. I was looking for Unknowns whose appearance and behavior could possibly be due to a skyhook balloon or a weather balloon. Along the way, I found about 11 cases where balloon observers were watching their balloon with a theodolite (telescope that gives azimuth and angle above horizon) when a UFO came along. There were several other weather watcher reports that did not specify whether or not they were watching a weather balloon when they spotted a UFO. I eventually found three Unknown cases that might possibly be explained by the descending and rapidly ascending skyhook maneuver, and three cases where a spherical object was dragging something underneath it similar to a weather balloon with its attached radiosonde instrument package. I will concede three more cases that might possibly be construed to be a balloon of some sort. That is it; three to five cases, or about one percent of the 564 Project Blue Book Unknowns that could possibly be explained by skyhook balloon behavior. That hardly defines the UFO phenomena. Talk about making an overstatement of the situation.

I can only conclude that Gildenberg's hypothesis is without merit, and Gildenberg appears to be guilty of purposely trying to mislead the public, since as a skyhook balloon insider he should have known all about the contradictory information that did not support his hypothesis. He did not mention any of it. The editors of *Skeptical Inquirer* magazine should have known all of this as well, but did not require that it be included. There is no question that skyhook balloons did generate some UFO reports, and that they have a high potential to do so, but during the Blue Book years the vast majority of those reports did not make it onto the Unknowns list.

Further examination of the files revealed some very interesting findings. I was especially interested in radar-visual cases. Of the 28 radar-visual cases listed for all the Unknowns in Project Blue Book files, most were military cases, and 25 of them came from the 1947-1954 time period.

Of the five UFO Unknowns cases in the 560+ PBB Unknown files involving UFO occupants, three occurred in the 1947-1954 time period. One of the occupant cases in 1954 may be the first report of a Nordic human piloting a UFO. Occupant cases involving UFO occupants seen inside the craft, or seen briefly on the outside, made the Unknowns list on very rare occasions. Any abduction reports during this period were sent straight to the dumpster or labeled psychological, and were usually not even given a case number. Ruppelt referred to this in his book.

Looking further, none of the Unknown reports of this period described the black triangles later seen on a large scale for the first time in Belgium in 1989, and only a few Unknowns dealt with a triangular type of object, but there was a sensational flying triangle report in 1952 coming out of a multi-nation naval exercise in the North Atlantic called "Operation Mainbrace." Witnesses from several different countries saw a blue-green triangle flying at 1,500 mph.

Electrical interference cases, a mainstay of UFO lore, did not appear in the Unknowns from 1947-1954. Only a single instance of a radio going to static while a UFO was present was found in the Unknowns from this time period. Electrical interference does occur later in a few cases during the 1955-1969 time period.

And finally, referring to farmers (or someone out in the boonies), unknowns that identified the witness as a farmer are actually quite rare in the PBB Unknowns list with only four for the entire set of files of 560+ Unknowns. So either farmers did not make that many reports or were not identified as such, or the reports were lacking in the kind of details needed. I did find that the urban population of a state correlated with PBB Unknowns a little better than did the rural population in both time periods. Oh well, so much for determining much about the farmer out in the boonies hypothesis from the Project Blue Book Unknowns.

Admittedly, the Unknowns were the cases that survived a rather arbitrary, sometimes irrational, elimination process clearly biased toward dismissing a report, if at all possible. A farmer or anyone reporting multiple sightings such as in my family's case, would have been ignored by Blue Book staff, and I suspect that reports from rural areas were generally dismissed during the Blue Book era due to a bias against reports perceived to be from less educated observers in general, or observers less educated about aerial phenomena. Many farmers will disagree with this particular Blue Book bias, as farming is an education unto itself, but also because many farmers were, and are, well educated.

In summary, I can conclude that purely psychological hypotheses, earth-lights, and skyhook balloons were not responsible for the UFO phenomena of the 1947-1954 period that Project Blue Book was not able to explain. The other common explanations had already been eliminated by Blue Book staff (misidentified aircraft, bright planets, stars, meteors, mirages, etc.). During this period, Project Blue Book excluded most of the general public's input, and generally sought only those reports from highly qualified Air Force personnel, contractors, scientists, engineers, and technicians. This is similar to what Richard Haines, a retired NASA scientist, does today by collecting reports mainly from aviation personnel. The possibility of mental cases, mistaken witnesses, and hoaxes was tremendously reduced in the early Blue Book data. The data correlations support the idea that the phenomena were real and unexplained, and were consistent with flying craft from an unknown source that could have taken advantage of forests or mountains for refuge or cover.

Chapter 25

Correlations of Variables for Project Blue Book Unknowns 1955-69

"The pure and simple truth is rarely pure and never simple."

—*Oscar Wilde, (1854-1900), Irish Writer, Poet, Playwright*

Now let us consider the data from 1955-1969; the period in which the percentage of Unknowns relative to total reports dropped to below three percent on average, year to year. We saw from the earlier plots that something drastically changed after 1954. The explanation for this is that cases from the general public were making up most of the reports during this latter period. I believe that most of the best military reports were now being analyzed by Air Force Intelligence off the records of Project Blue Book. With the rigid rules in place for screening and evaluating reports, most of the civilian reports were lacking in the details needed for Blue Book evaluation. I believe that prior to the 1955 time period, most of these reports were discarded and not even given a case number. By 1955, the cases from the public were being included and were being marked explained by any explanation that might half-way fit the details given. Only a small percentage of reports containing sufficient information for evaluation were considered further, and some fraction of these went on to become listed as unknown. This is what I believe happened during this latter time period. It explains the drastic drop in the percentage of Unknowns.

With the inclusion of more and more reports from the public, the average quality of the reports would have dropped drastically in the 1955-1969 time period. Yet this time period does better reflect the true extent of the phenomena seen by the general public. It is clear that in the earlier1947-1954 time period the phenomena was not being seen solely in the vicinity of military air fields and bases, but reports from the hinterlands were being filed in the dumpster. After 1954, the situation changed and the variables that correlate from this latter period would have had fewer distortions had the military reports also have been included. A truer national picture could have probably be derived from an inclusive view of all types of reports, but we are left with only half the picture; a different half than from the earlier military period.

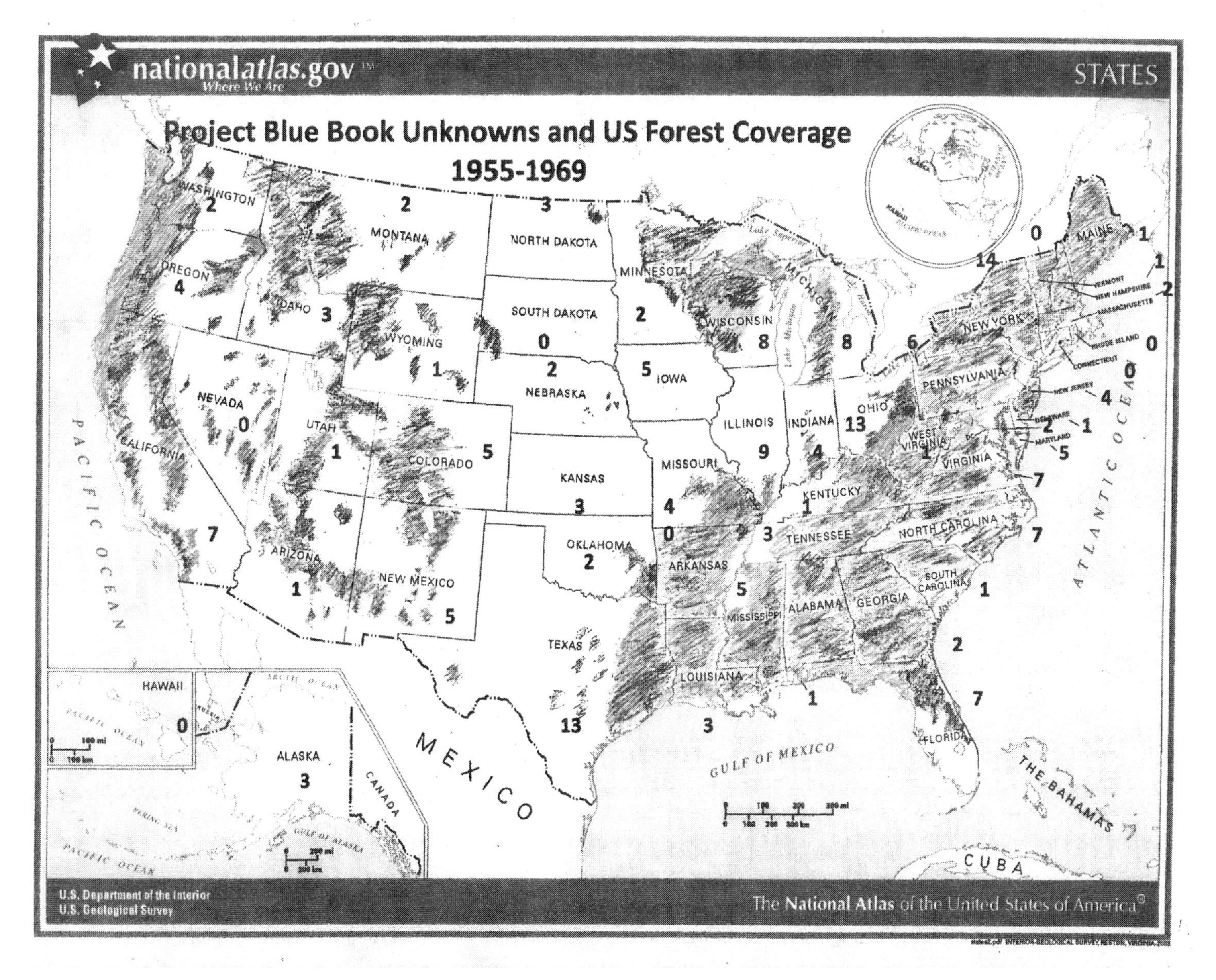
nationalatlas.gov
Where We Are
STATES
Project Blue Book Unknowns and US Forest Coverage
1955-1969
WASHINGTON 2
OREGON 4
CALIFORNIA 7
NEVADA 0
IDAHO 3
MONTANA 2
WYOMING 1
UTAH 1
ARIZONA 1
COLORADO 5
NEW MEXICO 5
NORTH DAKOTA 3
SOUTH DAKOTA 0
NEBRASKA 2
KANSAS 3
OKLAHOMA 2
TEXAS 13
MINNESOTA 2
IOWA 5
MISSOURI 4
ARKANSAS 0
LOUISIANA 3
MISSISSIPPI 5
WISCONSIN 8
MICHIGAN 8
ILLINOIS 9
INDIANA 4
OHIO 13
KENTUCKY 1
TENNESSEE 3
ALABAMA 1
GEORGIA 2
FLORIDA 7
SOUTH CAROLINA 1
NORTH CAROLINA 7
VIRGINIA 7
WEST VIRGINIA 1
PENNSYLVANIA 6
NEW YORK 14
VERMONT 0
MAINE 1
NEW HAMPSHIRE 1
MASSACHUSETTS 2
RHODE ISLAND 0
CONNECTICUT 0
NEW JERSEY 4
DELAWARE 1
DC 2
MARYLAND 5
ALASKA 3
HAWAII 0
CANADA
MEXICO
CUBA
THE BAHAMAS
PACIFIC OCEAN
ATLANTIC OCEAN
GULF OF MEXICO
U.S. Department of the Interior
U.S. Geological Survey
The National Atlas of the United States of America

Project Blue Book Unknowns mid-1955-1969
(State to State Variation - Correlation with Variables)

Variable	r	r2	% Explained	Cross Correl. r
Population, State, 1960	0.786	0.613	61.8%	———
Forest Cover in Sq. Miles	0.438	0.192	19.2	0.338
No. Military Air Facilities	0.391	0.153	15.3	0.426
Land Area in Sq. Miles	0.358	0.128	12.8	0.189
Miles of Coastline.	0.313	0.098	9.8	0.283
Skyhook Balloon Facilities	0.105	0.011	1.1	0.215
Pop. Density, Pop/Sq. Miles	0.055	0.003	0.3	———
Mountains in Sq. Miles	0.048	0.0024	0.2	0.030
Geologic Faulting	0.011	0.0001	0.01	0.065

The spreadsheet data is given in Appendix II.

Linear correlation coefficient r level of significance for 50 sample points.
r = 0.273, 95% confidence that the correlation is real
r = 0.354, 99% confidence that the correlation is real
A 95% confidence interval is commonly used.

Population is correlated stronger than any other variable during the 1955-1969 time period, as it should for all hypotheses if reports are processed from the general population. Forest cover is again correlated stronger than predicted by the cross correlation r value, indicating some real effect. Military air bases did not correlate quite as well as predicted from cross correlation with population, and is not a factor during this time period for Project Blue Book unknowns (military sightings generally were being handled at a more secret level by Air Intelligence and most were not being handled by Blue Book). Land area correlated stronger than its cross correlation r value, and appears to be a real effect. The larger a state is, the more opportunity to see something flying over. This is hugely significant in demolishing purely psychological explanations for UFO sightings. There should be little correlation of UFO sightings with land area unless something real was flying around. The miles of coastline also correlated stronger than its cross correlation r value, but by only about 10 percent more, and seems to be a weak effect. Skyhook balloon facilities, population density, mountains, and geologic faulting did not correlate at all. After 1960, nearly 70 percent of the unknowns came from east of the Mississippi River. This would diminish the impact of skyhook balloon facilities, mountains, and geologic faults. The lack of correlation with military air bases would also diminish the correlation with skyhook balloon facilities, which were located mostly at air bases.

The correlations suggest that something was flying overhead, and the greater the population and land area of a state, the greater the chance of seeing it. The correlations further suggest that forest coverage played some part in where these unknowns chose to fly. The correlations from this time period are consistent with flying craft of unknown origin flying overhead that are sometimes attracted to forested areas. Psychological explanations, earth-lights, and skyhook balloons have little support from the reports that made the Unknowns list for the 1955-1969 period.

Unknowns from New Mexico dropped off radically during this period to just five unknowns over the 15-year period. This probably reflected the switch from military reports to civilian reports, and the shift of the majority of Blue Book Unknowns from the western US to the east after 1960. However, one New Mexico report from 1964 stands out. It is one of just a total of five UFO occupant reports in the Unknowns list. PBB Unknown with case number 8766 deals with patrolman Lonnie Zamora's sighting of April 24, 1964, near Socorro, NM. Zamora reported two beings in white coveralls smaller than normal adults standing beside an egg-shaped craft that had landed. The craft had taken off upon seeing Zamora leaving landing marks and scorched vegetation. Hynek himself had investigated the case trying to poke holes in it. He could not. Zamora was well respected and held in high regard by his colleagues. Hynek had spent a year trying to trip up Zamora and could not. Debunkers Menzel and Klass took their shots at the case and had to fall back on calling the case a hoax, which is what they did when they had no way to explain a case (Menzel's accusations were comical; Klass claimed Zamora and the Mayor were trying to draw in tourists). Two days after Zamora's sighting, another New Mexico family that swore they had not heard of the Socorro sighting reported exactly the same thing. Investigators found the same landing marks and scorched vegetation.

At this point I need to mention some very interesting information concerning Menzel. While reading Project Blue Book monthly status reports from 1952, I came across a brief rebuttal of Menzel's mirage hypothesis for explaining radar-visual cases. The document was classified at the time, so the public never heard the rebuttal. It was found in a paragraph dealing with the Bellefontaine, Ohio case of August 1, 1952, and simply said that electronic (radar) and visual mirages of meteorological phenomena will not occur simultaneously in the same place. Elsewhere in the PBB status reports the statement was made that the purported explanations for UFOs that had been published at that time could not explain the vast majority of cases marked as unknown. The statement further said that it was doubtful if the explanations could explain any of the unknown cases. Hynek had gone to a conference attended by Menzel and was convinced that Menzel had not studied any of the PBB Unknowns. Menzel appeared to not have done his homework, and was also purporting something without scientific basis as an explanation for radar-visual cases. He was called out by Jung for his ridiculous debunking explanations, but was it all an act? Was he a member of MJ-12 with access to all the reports PBB had and more (Stanton Friedman's quest)? It is logical in a way, because otherwise it would have to be concluded that Menzel was very flawed in his scientific approach to UFOs. Atmospheric physicists and others must have known that Menzel was proposing something without merit, but chose to remain silent because he was attacking the validity of UFOs. One that did not remain silent was Dr. James McDonald, as mentioned earlier.

Returning to the Project Blue Book Unknowns for the 1955-1969 time period, the correlation coefficients were used to develop a predictive equation using a state's population and forest cover. Excluding Alaska, the combined variables were able to explain about 70% of the data. Population alone was able to explain about 62%. One test for a variable's significance is to see if adding it to the predictive equation improves the equation, and forest cover does. The chart below shows a comparison of the predictive equation and the actual number of Project Blue Book Unknowns from 1955-1969. New York tops the list with 14 unknowns followed by Texas and Ohio with 13 each. Eight of the top ten states lie east of the Mississippi River, representing the shift to the east after 1960. California remained in the top ten, but with its large population, it, along with Alaska (huge amount of forest cover), have the greatest negative deviations on the chart below their predicted numbers. After 1961, California had only three PBB Unknowns, and one did not have a case number.

On the opposite side of the line, Ohio had the greatest positive deviation with 13 when its predicted number of unknowns is eight. A closer look at the Ohio Unknowns shows that they were not too far off the main highways from Cincinnati to Toledo, and Cincinnati to Cleveland, the population corridors. Still, seven or eight of the Unknowns were not that far away from air bases. Four were not that far from Wright-Patterson AFB in Dayton where Project Blue Book headquarters were located. Kentucky joins Ohio's southern border, but had only one PBB Unknown during this period.

Ohio's population of nearly 10 million was a little over three times the population of Kentucky. Being right next door, Kentucky might have been expected to have perhaps four unknowns instead of just one, but such is the nature of real data. The pure and simple truth is elusive. Still, one can see that an inordinate amount of activity seemed to be going on in Project Blue Book's backyard. Could this just have been a coincidence, or could the phenomena have been targeting Ohio as it seemed to target New Mexico in the 1947-1954 time period? For the entire period of 1947 through 1969, Ohio was fourth in total Project Blue Book Unknowns with 21; Texas with 47, California with 39, and New Mexico with 28 were the top three. The majority of the western states' Unknowns occurred before 1955. If one believed in the purely psychological explanations for UFOs, then one might be tempted to conclude that most of the "nut cases" from the western states began moving east after 1954,

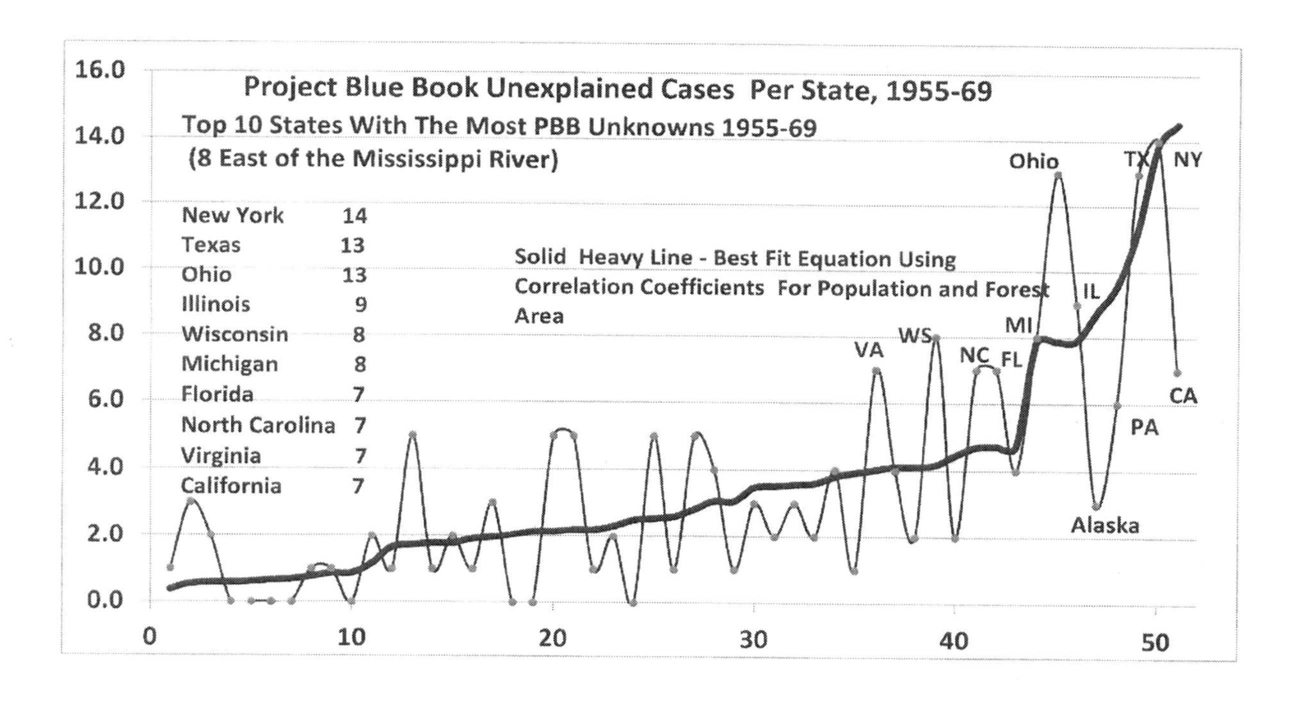

Project Blue Book Unexplained Cases Per State, 1955-69
Top 10 States With The Most PBB Unknowns 1955-69
(8 East of the Mississippi River)
New York 14
Texas 13
Ohio 13
Illinois 9
Wisconsin 8
Michigan 8
Florida 7
North Carolina 7
Virginia 7
California 7
Solid Heavy Line - Best Fit Equation Using Correlation Coefficients For Population and Forest Area
VA
WS
NC
FL
MI
Ohio
IL
Alaska
PA
TX
NY
CA
16.0
14.0
12.0
10.0
8.0
6.0
4.0
2.0
0.0
0
10
20
30
40
50

Ohio and Kentucky

Project Blue Book Unknowns 1955-69

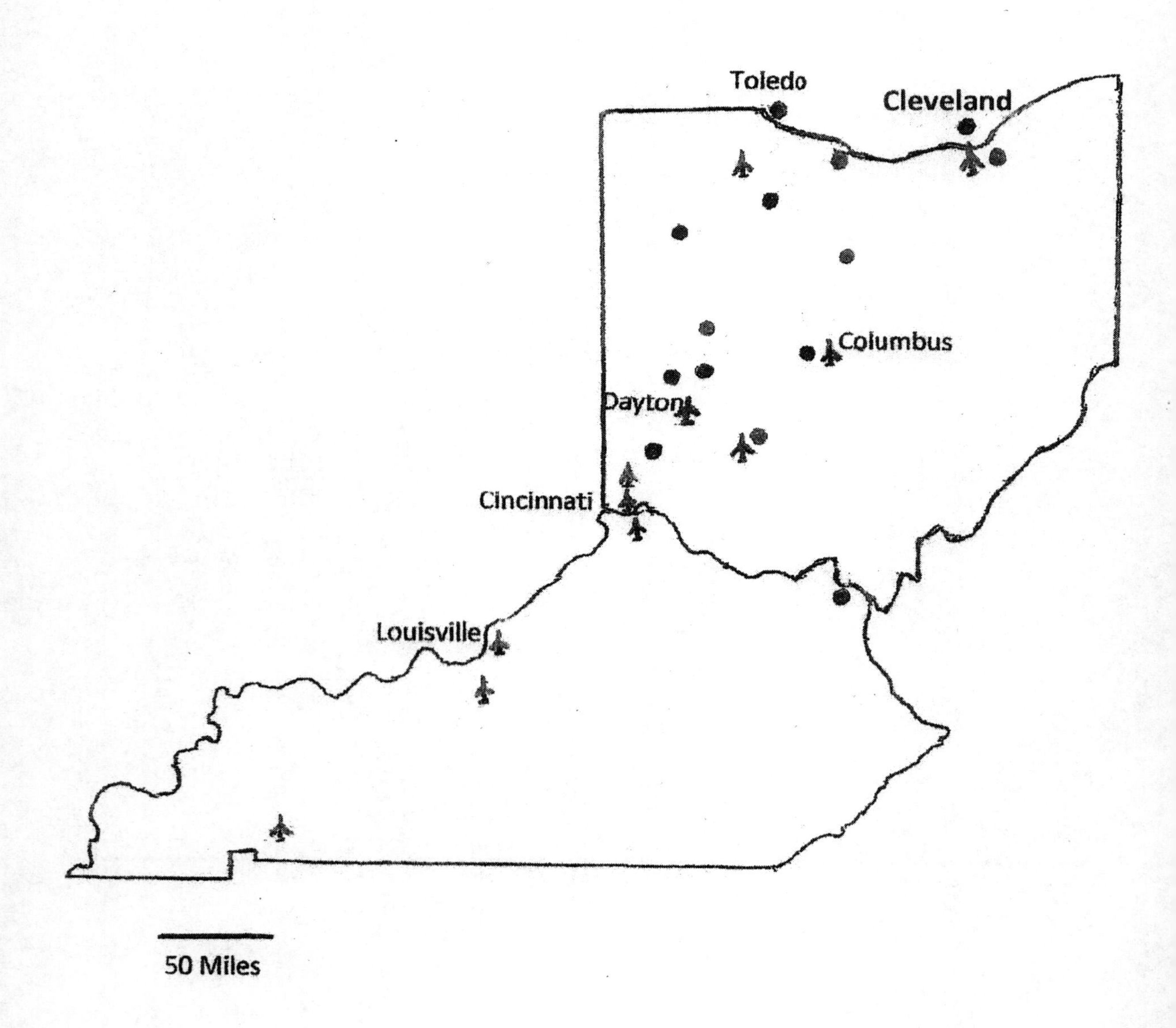

and most had completed the move by 1960. In actuality, the dramatic shift from west to east after 1960 is easily explained, and many astute readers already know the answer: population.

I confirmed the answer by looking at the type of witness involved in each case. I reviewed each case complied by Don Berliner for the Fund for UFO Research. After some consideration, I divided the witnesses into three categories: Military, Trained Civilian (scientists, engineers, technicians, airline pilots, private pilots, and law enforcement) and Untrained Civilian. In the 1947-1954 period, the cases started out about 65-70% military and decreased during the 1955-1969 period to less than 20% military. The trained civilian cases remained about 18% on average throughout both time periods, but the untrained civilian cases increased to about 70% by the end of 1968. The flip from west to east coincided with the flip from military cases to untrained civilian cases, with both crossover points near the year 1960. This confirms the shift of military cases from Blue Book to more classified status. The Air Force never really wanted to investigate civilian cases because of the difficulty of verifying the credibility of witnesses. The Air Force preferred to investigate military reports from highly trained sources and keep the information from the public. The Air Force encouraged the scientific establishment to belittle UFOs and the people that reported them, and has been very successful in creating a stigma about UFOs in the public's mind, especially after 1969.

With all that said, the flip from west to east is explained by using the Mississippi River as the east-west boundary. While about 68 percent of the land area of the continental US lies west of the Mississippi (Alaska and Hawaii not included), about 67 percent of the population lay east of the Mississippi River at that time. This corresponds exactly with the shift from military to civilian cases and the correlation of the data with population.

The contrast of the data from 1955-1969 with the earlier data is striking. Only one balloon observer case is listed with the Unknowns after 1954, although there are a few cases where the witnesses were labeled weather observers at airports. All nine Unknowns that could possibly be explained as due to skyhook or weather balloons occurred during this period. It is as if the PBB staff was not checking for balloons as diligently as during the earlier time period. Only three radar-visual Unknowns were listed for this later time period, pointing to military cases going deeper into classified territory. Seven of the eight electrical interference Unknowns occurred during the 1955-1969 time period. One large, flying triangle Unknown was reported during this time period. It was dark gray with purple and blue lights. It was reported to have a purple spotlight that played upon the ground and was estimated to be 150-200 feet on a side. The report still did not match the black triangles of 20+ years later. The idea that the black triangles could be generated by atmospheric plasmas caused by meteors disintegrating in the atmosphere is complete nonsense. Where were they during the 22.5 years reviewed by Project Blue Book? This is one of the most absurd hypotheses that I have ever heard put forward.

In summary, the correlation of PBB Unknowns during the 1955-69 time period with population is moderately strong, but the correlations with forest coverage and land area, while

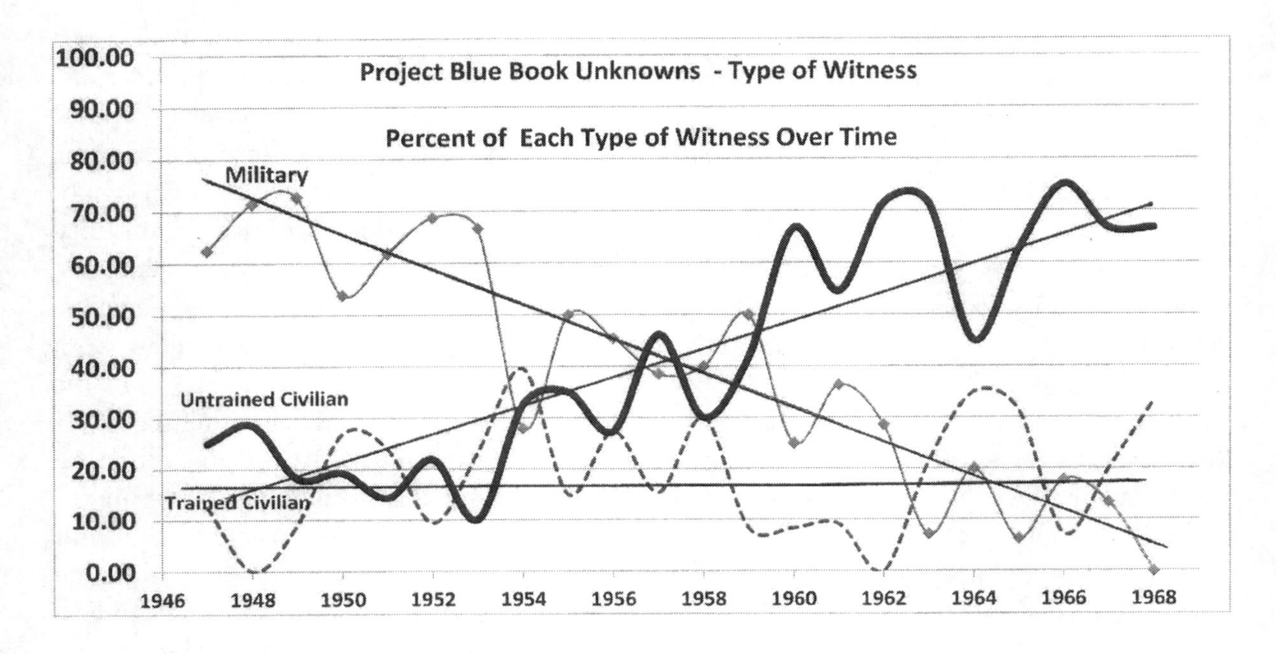

weaker, appear to be statistically significant and a part of the story. Earth-lights and skyhook balloons were seen not to correlate at all, and are judged not to be viable explanations for the UFO phenomena. Psychological explanations were eliminated earlier. The overall picture still points to something real and unknown flying around, either of this earth or from somewhere else.

In my quest to determine if the forest cover idea had merit, I found that it did, as it correlated at a statistically significant level with the Project Blue Book Unknowns during both time periods. The Mississippi UFO corridor has both population and is heavily forested. The fact that the corridor sits over the Class B geologic faults lying underneath the southern half of Mississippi may just be a coincidence. East Texas is underlain by the same fault system, but the main UFO activity was noted outside the fault area. Nationally, geologic faulting came up empty as a predictor of UFO Unknowns in Project Blue Book.

Chapter 26

A Most Disturbing Family Story (The Minefield of Alien Abductions)

"...people seeking celebrity status...little nobodies..."

—*Philip J. Klass, (Typical response about alien abductees claims)*

"They absolutely do not want their names to be made public. What they do want is to be taken seriously, and they want someone to help them understand and make sense of what they believe is happening to them."

—*C. D. B. Bryan, author of Close Encounters of the Fourth Kind about the abductees attending the 1992 MIT conference*

"It is clear that this is some sort of powerful subjective experience. But I do not know what the objective reality is. It's as if the evidence leads us in both directions."

—*Dr. John G. Miller, Emergency Room Physician, in Close Encounters of the Fourth Kind*

Shortly after I let it be known in August of 2011 that I was going to write a book about our family's UFO sightings back in the 1970s and 80s, I got quite a surprise and shock when one family member told me, "I had an alien abduction experience." I was taken aback and did not know what to say for a few seconds.

My reply was, "What? When? Did you have to be hypnotized to recall what had happened?"

Family member: "No, I did not have to be hypnotized. It was a real experience, and I recalled everything consciously. It happened sometime between 2:00 a.m. and 4:00 a.m. one night."

"Are you sure it was not a dream? Did you tell anyone the next day?" I found myself saying as I felt myself morphing into Philip J. Klass.

Family member: "It was one of the most real experiences I have ever had. What's the point in telling someone? No one will believe you."

As the transformation into Klass continued: "How do you know it was not a vivid dream?"

Family member: "I'll tell you how. I knew nothing about alien implants at that time."

Fully channeling Philip J. Klass now, I said, "You believe you had an implant? What timeframe are we talking about?"

After some discussion, we figured out that it was sometime in 1998 that the experience occurred.

Continuing on as Klass: "By that time the alien abduction cases had been well into the public eye for 25 years or more. Many implant cases had been discussed in books, on TV, and in the movies."

Family member: "But I paid little attention to it except that I knew that some people claimed to have been abducted by space aliens. I literally did not know any more than that. Only some years later did I bring myself to investigate the alien abduction cases and find out that implants had been reported before."

"You did not watch any of the TV shows of the late 1990s, especially on some cable channels, that went into the abduction cases in depth?"

Family member: "No, I did not. I just had other interests. Oh, another thing is that I have never had a single dream with the abduction setting in it. All my dreams that have a home setting are in my childhood home. I cannot think of a single dream with the abduction setting in it. Also, I interfere with digital radio signals."

"You do what?"

Family member: "I sometimes cause an interruption with the channel display and signal on digital radios." Later another family member verified that they had witnessed this radio interference first hand. The family witness had not been told about the abduction at that time. The family member just said something to the effect that interference with electronic equipment happened to them sometimes.

"You think this is due to the implant?"

Family member: "Yes, I do. It never happened before."

I, myself, cannot recall seeing a digital radio display suddenly go blank and the signal go to static with nothing else to cause it. Our family member reports having this effect on a number of different radios over the last decade. The interference is brief, as if a signal were being received or sent for that brief period of time. The frequency of occurrence was too great to be a random thing.

"Where did they put the implant?"

Family member: "In the back of my pallet, in the upper part of my mouth."

"Shouldn't this show up on dental X-rays?"

Family member: "Maybe, but I think it is too high up and too far back. It left a metallic taste in my mouth for a short while."

Suddenly, I blurted out from nowhere, "I'd like to include your abduction experience in the book. Can you describe what happened step by step."

Family member: "I will tell you about it and you can use it in your book, but I have to remain anonymous."

I agreed to anonymity, but I was really more shocked than I had let on with my family member. In fact, I was so disconcerted by the whole thing that I could not see how to move forward with the book for a month or two. I had not intended to get deeply into alien abductions, but one cannot get close to the subject of UFOs without doing so. I finally resolved to report "Family Member's" experience exactly as it was relayed to me; nothing more, nothing less.

Perhaps a little paraphrased Shakespeare is called for: ...to believe or not to believe, that is the question. I make no claims one way or the other. Here is the family member's account of their alien abduction experience. It differs a little from most of the bedroom abduction accounts, as it more resembles a physician's house call than the usual abduction account.

It happened sometime in 1998. It was late at night and everyone else in the house was asleep. We will call our family member FM for short. FM woke up to find the bed surrounded by four of the little (~4 ft. tall) gray aliens. Most of their physical appearance is hazy to FM. The clearest memory is of the eye, which was very large and completely black with no lighter areas. FM remembers a large rounded head with a very small nose and a slit mouth. Ears were not noticed as being present. FM never saw the aliens in profile. They begin communicating with FM telepathically. They communicated that they would not hurt FM and a feeling of trust was established. FM was not frightened, unlike many "experiencers" (a term taken from Bryan's 1995 book *Close Encounters)*. [129] FM, like myself, was aware of some of the family stories and of the high probability of something beyond the normal being out there in this world. However, I would have been quite frightened to experience this, real or imagined. FM cannot remember how the "Grays" got into the bedroom or how they left.

Here is where the abduction account differs considerably from most. FM never left the house. They floated FM out of the bed, out of the bedroom, and to the front entryway of the house. Here FM and the four Grays were all floating in the front entryway. This is where they performed the implant procedure, not in a flying saucer hovering overhead. They used a long thin tube to insert the implant into the upper back of the mouth or pallet. FM tasted a metallic taste and FM's tongue felt thick and heavy. At this point, the beings communicated that the feeling would soon pass and that FM would soon be okay. In spite of all this, FM liked the Grays. To FM, their bedside manner or mid-air hovering manner was better than most doctors. There were no uneasy feelings about them. At this point, FM was floated back to the bedroom and back to bed. FM cannot remember how the Grays came and went. They just appeared and disappeared. FM went back to sleep as if not much had happened. The next morning FM did not mention the encounter to anyone, and had not told more than a few people during the intervening 13 years until telling me in August of 2011, and later on giving me the full account in October of 2011. FM has not, to FM's knowledge, had any other encounters besides this one.

FM fully believes the encounter was real, and that the interrupted radio signals are confirming evidence of it.

My reflections on this account led me to another way of looking at alien abductions. The literature that I had read to that point had been filled with people who were initially terrified of the encounter. These were people who doubted their own sanity and were deeply troubled to the point of having posttraumatic stress disorders related to it. Later, after years of grappling with it and sometimes years of therapy, many usually come to view the encounters as a positive experience, opening their minds to new possibilities. Those who initially had completely closed world views that everything was known and that there were no strange phenomena out there, may be the ones most devastated by the experience. They are the ones more likely to report alien abductions and be nearly driven mad by them. FM's encounter leads me to believe that there is an entire population of "experiencers" who viewed the whole thing as positive from the beginning, and we have barely heard a peep from them, as they would be the least likely to report anything or seek psychiatric help. Besides, who would believe them anyway?

Chapter 27

Susan A. Clancy and 1,001 Witnesses (Type 1 and Type 2 Statistical Errors: Open Mind, Closed Mind, or Somewhere in Between)

"...the most grossly obvious facts can be ignored when they are unwelcome."
—*Eric Arthur Blair, pen name George Orwell (1903-1950), British Novelist, 1984, Animal Farm*

At this point, I need to introduce the concepts of type 1 and type 2 statistical errors. A type 1 error occurs when a bad or false data point is included into the body of data. The type 1 error is also called a false positive. To prevent this, one can set the criteria for the data so strictly that virtually no type 1 errors ever occur. This is where scientific journals and mainstream science itself normally tries to operate in order to keep any errors out of the scientific literature. It sounds strict and it is, but this very behavior leads to an opposing statistical concept of type 2 errors. It is similar in a way to yin-yang in Chinese philosophy. One can never be absolutely sure about a particular case, whether it is believed to be true or it is believed to be false.

Type 2 errors, also known as false negatives, occur when you set the criteria so tightly that real data gets excluded. The data is believed to be false, so it is thrown out. This is where science generally operates with the criteria for unusual phenomena set so high that real cases may be excluded, and this is the mindset where debunkers generally reside. UFO debunkers go even further in refusing to consider various forms of supporting evidence for a case. They live in the land of potential type 2 errors, but one thing is sure: They will never commit a type 1 error of accepting a false case as fact. This is commonly known as having a completely closed mind. Welcome to the land of type 2 errors of rejecting something as false when, in fact, it may be true. This is where such skeptics live, and they do not care if they are rejecting something that may be true no matter how much supporting evidence there is for it. That is the type 2 error mindset. They embrace it. They feel very secure in it. It is their comfort zone.

In the chapters above, I discussed correlation coefficients and 95 percent confidence intervals. I stated that a 95 percent confidence interval is generally used, but I did not say exactly why. The reason is that the 95 percent confidence interval gives a good balance between type 1 and type 2 errors. If I had set the criteria for the correlation coefficients at the 99.9% level, I would be very sure that a variable that correlated at that level would be real, but at that high confidence level I would exclude some variables that had a real but weaker effect.

I would be committing a type 2 error by omitting a real variable. In the same way, skeptics omit potentially real UFO cases because they will only accept a case where the UFO lands and is put on public display. They will only accept absolute proof beyond any doubt, and that puts them in the position of rejecting any real cases that might come along, if the case does not meet their 99.9999999% confidence interval criteria.

At what confidence interval was Project Blue Book operating at during most of its existence? Project Blue Book was clearly not operating at a 95% confidence level, but more like a 99.9% level. They relied on an extremely rigid set of rules that "explained" many cases that should have been marked as unknowns. The case that occurred near Project Blue Book headquarters that was discussed earlier is a prime example. A bright object was seen hovering stationary in the middle of the day, high in the sky, by numerous witnesses, and it gave a strong radar return. It was clearly an Unknown, but the visual was explained by Venus being in the same region of the sky, although Venus was at least 10 times too dim. The radar was explained as ice clouds, but the object was re-attained on radar above the ice clouds, and was witnessed by a pilot sent to intercept it as a large metallic object. Based on this and other examples already given, the Blue Book files should have contained many more Unknowns, perhaps as many as 2,000 or 3,000.

Skeptics like to reject eyewitness testimony, and they may be justified sometimes in doing so, but could all the witnesses have been wrong since 1947? When investigating a case of any type, two confirming eyewitnesses for something is good evidence that something happened much in the way the witnesses said, once the investigator is satisfied that the witnesses are reliable. Project Blue Book only considered cases involving highly credible witnesses. Every day, police all over the world take action based upon eyewitness accounts. People have been put on death row by one eyewitness on many occasions. That is going a little too far, but it happens. I say life in prison unless the evidence is there to prove the perpetrator did it beyond any doubt, and in that way very few innocent people could ever be put on death row. Most skeptics live deep in type 2 error land, rejecting all eyewitnesses to unexplained phenomena.

Susan A. Clancy also lives in the world of type 2 errors in her 2005 book *Abducted: How People Come to Believe That They Were Kidnapped by Aliens*, but it is alright for her purpose for doing the study on how people come to believe such a thing, and how it affects their lives. [130] I actually recommend her book, and do not disagree with a good portion of it. She rejects just about everything paranormal, such as UFOs, alien abductions, the Yeti, and I guess Bigfoot, too, plus ghosts, etc. A mile long track way in the snow at 18,000 feet in the Himalayas as discovered in 1951 by Sir Edmund Hillary would be easily dismissed by Clancy as non-evidence, even though hoaxing it would have been next to impossible. I take it that she does not watch "Ghost Hunters" on the SY-FY channel or a similar program on the Travel Channel or A&E a few years ago, or did not watch any of psychic John Edwards shows around the year 2000, or "Long Island Medium" currently running on TLC in 2012. There was also the "Fact or Faked" show and "Destination Truth" on the Sy-Fy channel that investigated lots of strange phenomena in the 2010-2012 time frame.

I watch the ghost chaser shows sometimes to see what kind of evidence they have gathered. The shows are supposed to be on the up and up, and I believe that they are. They routinely record electronic voice phenomena, EVPs, that are sometimes impressive, such as, "who's there" or "get out of here," and many others over the years. The EVPs sometimes seem to answer questions posed by the paranormal researcher. It is claimed by skeptics that stray radio waves are being picked up by the EVP devices. If so, why not put on convincing debunking demonstrations showing this very thing? Sometimes the ghost chasing TV programs record actual voices that are audible, along with human forms and moving shadows. Sometimes objects are thrown by some unseen force, and they have recorded chairs moving across the floor, lamps moving across a table, and similar things. What is the explanation for these things? People who live in the type 2 error world will just say it is hoaxed and not think about it anymore, but I am far from sure about these things. Joe Nickell, a debunker of paranormal phenomena, recently accused the "Ghost Hunters" of faking a lamp moving across a table by pulling the cord with their feet. [131] I saw an episode several years ago with a lamp moving across a table with no one near it. If that was faked, it had to be done some other way. If real, as it was professed to be, there is no normal scientific explanation.

When it comes to knowledge of UFOs, Ms. Clancy is extremely ignorant and probably proud of it. As almost all skeptics and debunkers do, she defaults to the Condon Report's summary page. That the summary page was contradicted by the body of the report is of little concern to her. The Condon Report was a farce set up to get the Air Force out of the business of investigating publicly generated UFO reports, 90% of which were a waste of time, so that the Air Force could spend their efforts secretly investigating high strangeness/high credibility reports of their own (Alexander 2011). Defaulting to a 1968 report whose content contradicted the summary page shows she never looked into UFOs at all. Also, it implies that no new information has come along since 1968, which is ludicrous. It is very easy for those that live in a type 2 error mindset to cite a report summary that dismisses the phenomena so that they do not have to look into it at all. It makes things much simpler that way. It is a form of self-deception and reader deception.

Ms. Clancy dismisses eyewitnesses from one to a thousand as of no consequence. One thousand eyewitnesses testifying to something will not sway her. She is very deep into type 2 error territory. I have seen few people who reside that deep into type 2 error land. Okay, let's say that one day she is walking down the road and someone comes running by and says, "Run for your life. There's a tiger loose." Well, she's certainly not going to run, as there is no confirming information for this eyewitness account. There's no proof. Just then another person comes running by yelling the same thing. Again she says it's just an eyewitness. There is no proof of such a thing. She continues on this way as eyewitness after eyewitness run by yelling the same thing, but she is a scientist. She must have absolute proof. And then suddenly she has her proof as she is attacked by the tiger. Of course, in a real situation like this, she would run away because of the possible dire consequences of not heeding the eyewitness testimony. Making such statements as "a thousand eyewitnesses cannot sway me" is pure foolishness and arrogance. It is a way to dismiss the eyewitnesses so that their information does not have to be considered, and so that her current mindset does not have to be disturbed. Eyewitnesses' statements alone are responsible for criminal convictions and sending the convicted to jail for life, but such statements cannot be considered by science? It is pure fallacy to dismiss eyewitnesses, especially when at the stage of forming hypotheses and figuring out a way to test them. Sagan made the same fallacy statement about one or two eyewitnesses, but said it oh so much more elegantly. It was still a fallacy no matter how elegantly stated, however. Sagan seemed to be saying that no eyewitness could ever be correct in their observations concerning unexplained phenomena.

One more point from Ms. Clancy's book about eyewitnesses and memory. She constantly puts forth the idea that our memories cannot be trusted. She points out that memory is constructed from several different areas of the brain, and can be malleable and made into something very different from the reality of the past event. Maybe, but the real memory is usually still there, as I will demonstrate from her own examples in the book. She gives an example of her memories from a Colorado vacation of having a very pleasant time skiing with an attractive companion, of large snowflakes drifting down, and a hot tub, etc. Only her friend reminds her that she, the friend, also was there and that the day did not turn out so well as Clancy remembered, because many little things went wrong. Clancy's memories had been turned into a glorious fantasy, and her friend's accurate recollections brought her back to reality, but how did Clancy know her friend's memories were the correct ones? Well, she obviously had the correct memories stored somewhere in her own mind where they could have been recovered by hypnosis, for instance, but were recovered by simply spurring her memory. Her example, while showing her memory was fraught with fantasy, showed that some people have incredible memories, accurately recalling minute details years later. So the reader was supposed to take from this example that no one's memory can be trusted? This sweeping generalization fallacy is made by all those wishing to dismiss eyewitness testimony. She did not dismiss the eyewitness testimony of her friend, as she realized it was correct, because the correct memories were also stored somewhere in her own brain. Things that make a sufficient impression on us can be remembered accurately for many, many years. For many things that do not make much of an impression, we forget them in a relatively short period of time, but these mundane memories can usually be recovered by, you guessed it, hypnosis or memory stimulation. Correct me if I am wrong, but it seems to me that people who fantasize and remake their memories into a glorious fantasy have got some kind of psychological problem. Yes? No?

Psychological professionals that do not believe alien abduction is even remotely possible usually could not recognize a potentially real case if it fell into their laps. She certainly does not have a clue that at least one of the cases in her book needed to be investigated further, as it involved a possible abduction of several people in a relatively remote area, but that is not of interest to her, because she knows such things are not possible. A case that comes to a psychological professional that believes alien abductions are not possible will not be investigated at all. The psychological professional will explain the case in purely psychological terms with little further investigation. The reader must understand that any case can be explained in psychological terms whether that is the correct explanation or not. This cannot be over emphasized. With people like Ms. Clancy there can be no other type of explanation for any and all alien abductions, or any unusual or unexplained phenomena.

I recommend her book *Abducted: How People Come to Believe They Were Kidnapped by Aliens* because there is a great insight in the end. As they used to say about pornographic literature, her book does have some redeeming social value. Her conclusions hold great promise for those who are struggling with the belief that they were abducted by aliens, whether real or imagined. You will have to read the book to see exactly what her findings are and what she concludes just don't fall into her sweeping generalization fallacy trap.

You will find that the skeptical literature has more opinion and a lot of dismissing of witnesses' statements, and shows very little skepticism or critical review for most anything that fits into their existing world view. Certainly it is part of our basic nature for almost everyone to have this tendency, but it is not fair play to twist, omit evidence, or downright mislead, or to not objectively investigate a case (prejudge or ignore without investigation). Of course, with Ms. Clancy, even a thousand witnesses would not sway her from her extreme position in type 2 error land. One has to be careful with what is presented in the skeptical literature as fact, because many times it is speculation or a distortion based on the omission of certain case details, and Ms. Clancy passes this along without objective review. And what of the lunatic fringe UFO crowd? They generally show less critical thought and a willingness to go deeper into type 1 error territory than seems possible, accepting almost anything no matter how dubious. Let there be no mistake about that as well.

I would recommend that Ms. Clancy study the Belgium wave of 1989 and 1990, where the black triangle UFOs first became prominent. [132] The wave has many eyewitnesses, probably well over a thousand in all (maybe a thousand and one would work to convince her), radar data, and an authenticated photograph or two. Out of about 2,000 reports, some 650 reports were investigated, and 500 remain unexplained. This is an unprecedentedly high percentage of unexplained cases for any UFO wave. And no it was not a secret American military aircraft or plasmas created by meteors breaking up in the atmosphere. We had nothing then and we have nothing now that can maneuver the way those craft did, and plasmas cannot even begin to explain what was seen. F-16s were sent to intercept one of these triangles that was slowly moving along. When the jets approached, the craft was hovering or moving very slowly, and it suddenly accelerated. It was documented to accelerate to 960 mph in three seconds. This is faster than the speed of sound, but no sonic boom was heard. This acceleration was at about a constant 15 Gs, enough to be very uncomfortable if not fatal to a human being. Perhaps thousands of witnesses in the past had talked about the sudden acceleration of a UFO, but this case had the documentation to prove it.

CHAPTER 28

Another Example of Type 2 Error Thinking (Dr. Ronald K. Siegel and the Jack Wilson Case: How Glorious Was the Glory Explanation?)

"Perception is a clash of mind and eye, the eye believing what it sees, the mind seeing what it believes."

—*Robert Brault, (Freelance Writer)*

Dr. Siegel, in the 1992 book *Fire in the Brain*, wrote about a number of cases he had dealt with in his practice as a psychological professional. [133] These were cases where he considered the patients involved to have a delusion or hallucination that caused them to believe some strange things. One was a UFO abduction case involving Jack Wilson and his son Peter. The case gives an excellent insight into the thinking of mainstream psychological professionals who believe that there is no such thing as unexplained phenomena. Dr. Siegel embraces the sighting-driven, delusional episode explanation for UFO sightings. Let me explain delusional as used in this regard. Something real is sighted apparently flying through the air that is one of the conventional things that we see, such as airplanes, helicopters, weather balloons, research balloons, ordinary party balloons, Chinese lanterns, or small hot air balloons, birds, meteors, ball lightning, earth-lights, advertising airplanes, spotlights reflecting off of clouds, sky-divers with flares, etc., and the witness or witnesses believe they are seeing a UFO. Many people seem to yearn to see something truly unusual, and may get excited enough to let themselves get carried away. That is something my family usually did not do. One or two of our sightings were understandable exceptions where something large was directly over our heads shining a light down upon us, but still we reacted appropriate to the stimulus and did not hallucinate in any way.

If our hypothetical excitable observer sees ball-lightening or earth-lights, they are seeing some very poorly understood and unusual phenomena. I would have to say that ball-lightning and earth-lights are UFOs, Unexplained Floating Orbs, and it is not really a delusion to see them and think they are something strange and unexplained, because they are. How seeing something of that nature can lead to a complete breakdown of logic and reason is a pretty far stretch of the imagination for me to understand. By just seeing something unusual flying around to lead the observer to hallucinate and create a completely imagined story, that observer would have to have already been a complete "nut case." For the psychological professional to explain alien abductions, the delusion must then progress into a whooper of a hallucination whereby the witness believes themselves to be abducted by space aliens. This can only happen to someone who is already a complete mental case. A normal person will not progress in this way. For the psychological professionals that believe this delusional induced hallucination can happen to a normal person, all I would say is that this is one of the biggest crocks of nonsense I have ever encountered. (I just want to be perfectly clear on this point).

One other enormous problem for this psychological explanation is that two people virtually never have the same hallucination or dream at the same time, no matter what the stimulus. A dream or hallucination is generated internally in each person's brain, and the likelihood of two people witnessing something and progressing into the same hallucination or dream is nil. This is why UFO cases with two or more witnesses are generally more reliable than single witness cases. Although a normal single witness can be very accurate, those that wish to discredit UFOs and alien abductions knock eyewitnesses and hypnosis much more than is warranted, just like defense attorneys knock the evidence, any evidence, which points to their client's guilt.

Getting back to our sighting-delusion-hallucination sequence, for one person to progress all the way through this sequence alone, they would have to be what most people would call "nut cases." There is one problem here for the debunkers, which is that most people reporting alien abductions test out to be as normal as the average population. Skeptics and debunkers like to find the inevitable 30 or 40% of abductees that are "nut cases" and imply that all the other abductees are as well. Usually they dress it up in nice psychological language and say that the other perfectly normal group, before and after the episode, had this one psychotic episode for some reason. If two people are involved, this one-time psychotic episode explanation is completely off the table.

Now on to a case that is the poster child of the sighting-delusion-psychosocial explanation crowd: the Jack and Peter Wilson case, as described in the 1992 book *Fire in the Brain* by Dr. Ronald K. Siegel, an associate research professor at UCLA at the time. Wilson and his son Peter had an unusual experience just two days before as they were stopped near the Arizona-New Mexico state line on their way to San Diego, California from Florida. They had been taking turns driving non-stop for some 30 hours, and were very tired. Wilson believed he'd had a genuine UFO abduction experience and that his son was involved, too. Jack Wilson had contacted Dr. Siegel because he had seen him on TV before, discussing UFOs. Thus, this is a two witness case reported from conscious memory only about two days later, all of which makes the case very interesting.

Dr. Siegel's writing is very entertaining and gripping. Somehow I get the mental image of Donald Sutherland at age 40-45 portraying Dr. Siegel, and a picture found online confirms that would be perfect casting. Dr. Siegel gives the impression that he is a boundary dweller on the boundary between a type 1 and a type 2 error mindset. In other words, he sounds reasonably open to something other than a purely psychological explanation for the case, if one can be found and the evidence is there. He even makes a promise to Jack Wilson that if he cannot find a reasonable psychological explanation for the events described, he will contact various groups to let them know he has a "*g. d. gen-u-wine UFO abduction case*" to report, in Jack Wilson's words. The misdirection in that promise is that it is usually no problem to find a psychological solution once one or two little sticky details are figured out.

Careful reading shows that Dr. Siegel really resides much further into type 2 error land than he lets on. He basically tells the reader that he thinks all UFO sighting reports and abduction reports are dreams or are purely of a mental origin before getting very far into the case. He is basically completely ignorant of the history of UFOs and Project Blue Book's early years where only the most credible and reliable witnesses' cases were considered. What he actually shows the reader is that he is completely ignorant of UFOs, and certainly has never read about UFO cases involving radar and valid photographic evidence, such as the Belgium wave only two years before the book was published. He cites Jung's psychological ramblings about UFOs, but he never tells the reader that Jung was a severe critic of Menzel's distortions and ridiculous attempts to debunk some major UFO cases in the 1950s, and that Jung said radar proves some UFOs to be real and possibility extraterrestrial. Any book that cites Jung's psychological musings about UFOs without taking the reader to Jung's later conclusions on the subject is either ignorant of the facts or is trying to be highly misleading. This shows the author or authors to be heavily into a purely psychological explanation, regardless of any evidence to the contrary. It is completely misleading to cite some of Jung's musings about a possible psychological explanation for UFOs without citing that he eventually rejected such a view and said UFOs were real because radar data proved them to be real, and the government should tell the people the truth. Shame on all of the books I have seen with just Jung's psychological musings and not a word about the later portrayal of what he actually concluded. That is simply misleading scholarship. This occurs on page 100 about two thirds of the way down in *Fire in the Brain*. Ms. Clancy cites Jung several times in her book without pointing out his final views on the subject. Did they even investigate enough to find out what Jung had ultimately said about UFOs, or did they deliberately present a distorted view that they agreed with? I did not bring up Sagan without letting the reader know exactly where he stood on the subject of UFOs as a skeptic with the most elegant sweeping dismissal proclamations ever written or spoken.

Siegel shows very clearly in his own musings that he is open only to a psychological explanation while pretending to be open to other possibilities. The case clearly gave him a chance to present his particular type of psychological explanation that he had thought about for a long time, and present it he does. He lays out the case as: here's what was presented to me; I researched it thoroughly in the best Sherlock Holmes fashion; found a plausible optical effect for the sighting; and then put together the most appropriate psychological explanations and presented my conclusions without disclosing anything that could be used to reconstruct the scene or challenge my explanation.

Along this vein, no date is given in the account, so we do not know the month, day, or year, or the highway or specific location where the event occurred. We do have the knowledge that a 1978 study was cited to the Wilsons in the book and the book was copyrighted in1992. So, we can place an approximate date of 1985 as the halfway point between 1978 and 1992 for the incident, and probably not be too far from the year of occurrence. What we do know is the time of day, as that is critical in the explanation and could not be left out.

The account unfolds as follows: Jack Wilson and his son, Peter, were traveling from Florida to San Diego, California, and had just entered the State of Arizona when they heard a mechanical noise coming from the rear of the car. Wilson stopped the car along the side of the road, which was heading west, and got out to investigate (he stood up too quickly and experienced some dizziness), and was met with a blinding light *overhead* when he turned toward the rear of the car. The account in the book credits Wilson with saying a blinding light ... overhead. From this point on in the book, however, the light becomes the early morning sun at a low angle. In other books retelling the story, it becomes the sun at a low angle not overhead, which is not exactly a bright light *overhead,* as Wilson said. Dr. Siegel points out that Wilson would have been facing to the east when he got out, so the bright light would be the rising sun, and Wilson suffers from mild light sensitivity so it would be worse than for the typical person.

The time of day was two hours after sunrise. I tested this and found that the angle of the sun in the sky would be at about a 20 to 25° angle, depending upon the time of year at 32° N latitude, which is about where I believe the incident happened. About a 25° angle to the horizon is hardly a low angle of 5° or less that would be needed if you looked back at the trunk of the car and had the sun directly in your eyes. For someone like myself who grew up in the country on a farm, the sun is fairly well up by two hours after sunrise. We would not consider it as early in the morning as Dr. Siegel does in the book. If Jack Wilson was looking down at the car, the sun would not be directly in his eyes, but a bit troublesome nonetheless. In a reenactment, I gave a quick hand movement up to my brow and completely blocked the sun without thinking about it. Maybe because he was tired after driving for eleven hours straight, and due to his light sensitivity, the sun overpowered him. With all that said, the bright light probably is the sun, and at two hours after official sunrise it can be described as being overhead, even though this is not literally true. In my SUV driving due east at exactly two hours after sunrise in April, 2012, on a street in Sulphur, Louisiana, with the sun visor up not down, I could not see the sun directly. The sun was at a 25° angle, and in the driver's seat sitting up with my back against the seat, I could only see up to about a 20° angle at most, if that, as discussed earlier in this book. Yes, many people would consider the sun overhead at that point and not low near the horizon.

Wilson turned around to avoid the bright light, whether sun or not, at this point. This is when Wilson saw the little gray man that he described as a humanoid three to four feet tall, wearing a gray seamless suit. The little gray man had what looked to be an energy field around him that made him a little blurry, and he had a halo around his head. Peter had gotten out of the car to see to his dad, who had staggered around when first standing up. He came up from behind to support Mr. Wilson, who was facing west, and at that point Peter also saw the little gray man. The little gray man followed Wilson back to his car, at which point the abduction account begins.

Dr. Siegel describes the little gray man as a real image due to an optical effect known as a "glory," which Siegel incorrectly wrote was an optical effect caused by dew on the ground and someone looking west at the head of their shadow called the anti-solar point. Siegel seemed to confuse the glory effect and the Heiligenschein effect. The Heiligenschein effect results from dew on the ground and gives a halo around the head of a person's shadow. The glory effect results from fog in the air and gives a circular rainbow around a person's shadow. The two effects can never occur together, but Siegel needed both of them to fully explain the image of the little gray man that Wilson described. How the glory dressed itself in a gray suit head to toe, as seen by two witnesses, instead of a black or dark suit, is never discussed. That is one point that Dr. Siegel never addressed. I guess that was a minor point, considering the rest of the psychological explanation.

Now, the weather report said dew or mist on the ground as stated on page 107, and Dr. Siegel eases this into early morning mist in the air, but my point is that it was not quite so early morning, which is thought of by most people as about the first hour of daylight (civil twilight for about 25 minutes where it is light enough to see, and 35 minutes after official sunrise), and the mist was on the ground, not in the air. As the story is told in the book, neither witness mentioned that there was mist or fog in the air. This is a curious omission to me. Wilson never stated that it was foggy and he was having trouble seeing the road, or anything of the sort. Two hours after official sunrise, and about two and one half hours after it had become light enough to see, the desert air should have heated up by about four to five degrees in the winter, and eight to ten degrees in the summer. Even a five-degree rise in temperature should have eliminated any mist of the true early morning air (first hour of visibility). The time of year seems to have been during basketball season, about November to March, but I am only guessing from vague clues in the book. The important detail of the exact date is curiously not given, so no one can check the weather records. On average, the area of Arizona where I believe this case occurred has about one foggy day a month. December averages about two foggy days, while one or two months average no foggy days.

The entire psychological explanation of the case rests on an optical effect called a "glory" being present to produce the glory image around Jack Wilson's shadow that Jack and Peter Wilson both interpret as the image of a little gray man. According to Dr. Siegel, this image then triggers a series of psychological steps that produce the impression of an abduction. This sounds very convoluted. The more convoluted an explanation gets the more doubtful it becomes, since each step is purely speculation, but Dr. Siegel's writing is very smooth and convincing. He can get inside your head in a big way.

Peter Brookesmith, in his 1998 book *Alien Abductions*, swallows Dr. Siegel's convoluted explanation hook, line and sinker, and even makes some slight embellishments of it. [134] This is completely understandable, since Derereux and Brookesmith in their1997 book *UFOs and Ufology: The First Fifty Years*, clearly believe in the scenario of a real sighting of something triggering a psychological abduction event. We have already discussed and shown their earth-light magnetic field hypothesis to be founded upon little more than wishful thinking. As shown earlier, they already believed in a purely psychological explanation triggered by some event or an earth-light and, as such, bought into a completely ridiculous convoluted explanation that pulled in rare psychological aberrations and linked them together.

In the book, Dr. Siegel smoothly covers the optical effects, but he left out the second "i" in Heiligenschien. (Sometimes I get nit picky.) He also seemed to reverse the two effects, as stated earlier. The Heiligenschien effect is the much weaker of the two and results from dew on the ground creating a halo around the shadow of a person's head. I tried to see the Heiligenschien effect two hours after sunrise out on my lawn on two separate days in completely clear conditions with heavy dew, but could not really say I saw much of anything.

A glory results from mist in the air that is below the level of the observer, and is usually photographed with someone's shadow falling on the mist from a fountain or waterfall. A glory is much stronger and more dramatic than the Heiligenschien effect. In the glory effect, typically a circular rainbow is seen around the person's head and upper torso. The two effects seem to be mutually exclusive in the same image, and the shadow never comes dressed in a seamless gray suit. A dark or black suit, yes, a gray suit, no. The presence of a glory and the Heiligenschien effect in the same image is somewhat essential for Dr. Siegel's explanation, and it is also essential that both witnesses instantly interpret the glory image as a little gray man enveloped in an energy field. Now, I would be willing to bet that 100 out of 100 people who had never seen or heard of a glory before would, if presented with such an image, describe it in some terms similar to "that's a cool rainbow around my shadow." I would be willing to bet that none of them would jump to the conclusion of "that's a little gray man surrounded by an energy field," even if they were very tired and sleep deprived.

Furthermore, the conditions do not seem right (mist on the ground not in the air) for a glory effect to be present. Strike one against the psychological explanation unless what was meant in the official weather report was a low, ground-hugging fog. It seems a little late in the day for this after nearly 2 1/2 hours of visibility. Also, both Jack and Peter's shadows should have created the same effect so if the conditions had been right, two separate little gray men surrounded by halos should have been seen at some point. It is extremely curious that both would instantly interpret the optical effect in the same way. Brookesmith slightly embellishes this detail in his book, as it was not in the original account, as he says Peter saw his own shadow and therefore a separate little gray man, as if it is automatic that everyone jumps to the conclusion that they are seeing a little gray man in a seamless suit when presented with their shadow surrounded by a rainbow.

No diagram of where they were standing and looking when they saw the little gray man was provided to the reader. None of their drawings were shown in the book. Not their separate drawings of the little gray man or any of their other drawings. Dr. Siegel did provide a sketch to Jack Wilson of his position in relation to the car and the sun when presenting the psychological explanation, but it is not given in the book. The case is curiously lacking in certain details, such as this and the date and the highway, etc., that are withheld from the reader so that no one can verify any of it. Not exactly what I would expect from a scientific investigation or a presentation of the findings to a technical audience. Any competent UFO investigator would provide such details when the case was so dependent upon them.

Dr. Siegel told Wilson that the little man followed him back to the car as any proper shadow would. Here is where I nearly lost it. No one would be so tired that they could not recognize their own shadow under these circumstances. If the little gray man was just a shadow, it would have been in front of Wilson as he headed back from the rear of the car to the driver's side door. In my reenactment of the incident, I end up about five to eight feet to the rear of the driver's door. Nothing in the story indicates that Jack Wilson should have ended up in front of the driver's door. That the little gray man followed me back to the car seems to be the wrong terminology. The little gray man should have led Wilson back to the driver's door, but should have stayed a constant distance away if he were only a shadow. If Wilson ended up actually stepping forward of the driver's door before seeing the little gray man, then his shadow would have followed him back while staying a constant distance away, but this detail is not exactly clear in the book. The following illustration shows how a man's shadow with a glory effect would appear at that time of day. The shadow is so long that for a four- or five-foot glory image, the legs of the little gray man would not show up. Everything is out of proportion in the glory image. The glory image would be too out of proportion to be a viable explanation for what Wilson saw. This is illustrated in the figure that follows below.

The time of year is critical for the glory explanation to have any viability. Standing on the driver's side of a car facing west two hours after sunrise from about mid-October to early March (comprising several holiday seasons), a person's shadow will fall across the car. From mid-May to mid-July, a person's shadow would fall toward the highway. Those shadow positions are not conducive to seeing a glory effect. A person's shadow will fall parallel to the car for a few months in the spring and a few months in the fall where if there were fog near the front of the car, it might give rise to an optical effect.

Illustration of the "glory effect" Cited by Dr. Siegel in *Fire in the Brain*

Left: Author's Shadow 8:16 AM CDT June 7, 2012, 13' 10" Length

Right: Simulated Glory Effect with a 5' Little Gray Man Drawn Alongside
(Outer Band- Red, Inner Band- Blue)

The Diagonal Joint in the driveway points almost due west. This time of year, a person's shadow would fall toward the highway.

From Top of the Head of the Shadow to the End of the Fingers is about 8'

The Glory Explanation Just Does Not Work in the Author's Opinion

Dr. Siegel's writing is almost hypnotic. The reader becomes suggestible. The reader becomes caught within its spell, agreeing with everything as the narrative smoothly washes over their mind. If anyone goes out to look up the optical glory effect, they will find a dark shadow surrounded by basically a circular rainbow according to various pictures online. The outer circle will appear red and the inner circle will be bluish. A dark, almost black, shadow is what is seen, not a little gray shadow, and nothing appearing to be dressed in a suit of any kind. If Wilson had said that the humanoid figure was dressed in a dark or a black tight-fitting suit, then the description could be said to sort of match that of a glory optical effect. If I had not decided to check out these optical effects myself, I may have missed this point.

Dr. Siegel never gave the reader Peter's description of the little gray man compared to his father's description. A second opinion saying that it was dressed in a seamless suit, or describing his eyes or face, would perhaps draw too much attention and disrupt the hypnotic flow of the book at the point where Peter's recollection of events are presented. From my cell phone photos above, the glory image just does not work. The shadow is very elongated, and a four-foot radius glory effect only comes down to the shadow's elbows. No one could mistake such an image for a little gray man in a seamless suit.

Next, according to Dr. Siegel, the two men talked briefly about seeing the little gray man as being a radioactive creature, or a ghost, or a creature from outer space (Dr. Siegel puts much emphasis upon them discussing a creature from outer space before falling asleep), and each drifted to sleep being super-tired. Brookesmith dutifully repeats that they were both super-tired. Hold it right there. Jack Wilson had just driven for 11 hours, presumably with two or three stops for restrooms and gasoline, but mostly driving at night, while Peter had been able to snooze or close his eyes, stretch his legs a little, etc., etc. Driving requires active concentration and holding your body in a position at attention, and it was done for such a long period of time that in this case there is virtually no possible way Peter could be as tired as Jack Wilson was. There is absolutely no way. Peter would presumably be somewhat rested and be ready to give his father a breather by taking over the task of driving. This is another red flag in the psychological explanation. But continuing on in the psychological explanation, according to Dr. Siegel, since both were super-tired (Jack yes; Peter no.), they each entered a hypnagogic state that can precede sleep characterized by flashes of light, and sometimes with circular illuminated images that seem to come nearer and nearer (known as the Isakower phenomena). [135] Rarely does this just happen to someone out of the blue. Many people that experience this are in or seeking psychiatric help at the time for this or some other disorder. In other words, they usually have more mental disorders going on, but perhaps not always. There is absolutely no way for this series of psychological states to happen to two people at the same time, and Siegel surely knew this and proceeded anyway. In the hypnagogic state between sleep and consciousness, a person is supposed to be as suggestible as if in a hypnotic state, so the illuminated circular images became the looming UFO, according to Dr. Seigel (this is all supposition). Dr. Seigel tells us that the hypnagogic state can produce a feeling of floating followed by a feeling of paralysis, followed by drifting into the dream world. Dr. Siegel says that both men then slept for seven hours. This is from around eight or nine in the morning to three or four in the afternoon in a desert, or nearly desert, area under full sun. What?!?! Again, we are confronted with supposition that borders on the ridiculous. There are a number of red flags raised by Dr. Siegel's explanation.

First Red Flag. There is no possible way to have any idea of whether either of them actually had the sequence of mental states proposed. That the two of them had the same sequence of mental states is total speculation and would be quite improbable. Dr. Siegel presents it as if it were verified fact. He is very smooth and convincing, especially to anyone who wants to believe such an explanation.

Second Red Flag. Sleeping seven hours along what I have to assume was Interstate 10, but could possibly be a parallel highway. If they were inside the car with the windows up, they could have easily died, depending upon the time of year. Yet even in winter with a daytime high of 50 degrees when sitting in full sun in the desert, or near desert, it can get hot inside a car with the windows up so that it becomes uncomfortable after a few hours. Therefore, I have to assume that it was cooler weather with a cold wind blowing for them to stay comfortable inside a car that long and be able to sleep for that long in full sun. If it was the cooler time of year, then their shadows would have most likely fallen across the car and there could not have possibly been any type of glory image.

Perhaps they had the windows down or cracked, or the door open if it was hotter weather, but the cramped discomfort of sleeping inside of a car should have caused them to wake up within a few hours. Again I point out how the basic details of the month, day, and year were not given, nor the highway they were on nor a diagram of the scene, nor an outline of their movement outside the car, nor the daytime high temperature, nor if the windows and/or doors were cracked or not. These details would have been recorded by any thorough investigation of the case, and maybe Dr. Siegel had all of these details and more, but he thought it would bore the reader, or maybe he realized that some of these details contradicted his psychological explanation. The time of year could have easily contradicted his glory explanation and/or the time spent sleeping in the car.

Third Red Flag. They are parked along the side of a major highway (I assume I-10, as they were traveling cross country) on the shoulder for seven hours and no one checks on them, two unconscious figures presumably slumped over in their car that can be seen by all who pass by. Not the state police, no motorist, no one checks on them in those seven hours. Now, maybe that can happen, but usually someone would have noticed two unconscious figures slumped over in a car, especially if windows or doors were open.

The Fourth Red Flag. This one refers to the difference in tiredness between Jack Wilson and his son Peter. Peter was resting while his father drove so that he could take over when his father got sufficiently tired. Under a similar situation as the Wilson case where I had been driving for some time and my wife and daughters had been resting, one of them would have been ready to get back on the road probably in less than an hour, and they would have driven while I dozed off. In a typical situation like this, Peter would perhaps have slept an hour or two, or three at most, and then got behind the wheel (after three additional hours that would have been 14 hours not behind the wheel); with any one of my daughters it would have probably happened within about 30 minutes. Even if Peter was staying awake to keep his father from falling asleep while driving, it is unlikely that Peter would be as tired as Jack Wilson.

Now comes an absolutely unbelievable part. When they awoke, they discovered that they'd had very similar dreams about the little gray man, floating in a long tunnel, and flashbacks to their Florida trip. Mr. Wilson had a few more images of geometric figures or patterns, but otherwise the psychological explanation glossed over the point that they had a very similar or the same dream. Danger! Danger! Danger, Will Robinson! An enormous red flag needs to be raised at this point. I cannot believe Dr. Siegel presented that with a straight face. Two people do not have the same dream at the same time. He, of all people, should know that is very improbable. The entire psycho explanation fails on that point alone. Dr. Siegel presents it so smoothly that it seems believable, and he seems to say that Peter was influenced by his father and the two together reinforced the shared perception. Dr. Siegel could easily have been the best used car salesman ever. A much simpler explanation is that they both had a similar real experience, however improbable that experience may seem, and that they each recalled some aspects of that experience.

Now what if it had been just Jack Wilson by himself traveling alone? There would be very little grounds in this case to dispute the psycho explanation except that the optical glory part just does not work at all. Again, this makes cases with two or more witnesses the ones to focus upon if you are looking for a signal among the noise, because the psycho explanation cannot be as easily inflicted upon two or more people as it can upon one. For those that believe that there can be no signal, then they will not even bother to look. I refer to Ms. Clancy here and all others who reside deep in type 2 error territory. Believing that nothing can be there, they will never look and will never find anything. In addition, Dr. Siegel was under a tight time deadline, which really meant that he would not consider anything other than a purely psychological explanation. Anything else would be too time consuming.

There is more to this case, and it gives a perfect example of how the same evidence can be looked at so differently by the skeptic and by the abduction researcher. Dr. Siegel asked both Jack Wilson and Peter to draw the images of what they saw during their experience. Peter had not seen his father's drawings nor vice versa. The drawings were very similar. This fact is quite interesting to me, but not to Dr. Siegel. It could be supporting evidence of a shared real experience, but Dr. Siegel dismisses this because they had discussed what they saw as they continued on to San Diego, so their drawings should be similar. Maybe, maybe not.

Next comes what Dr. Siegel calls his trump card and in a verbatim quote from *Fire in the Brain* on page 110, he states: "For my finale, I showed Mr. Wilson dozens of drawings done by hypnotized subjects who had no significant knowledge of UFOs but were instructed to simply imagine an abduction. These invented drawings showed little gray men, long corridors, geometric patterns, examinations, and TV screens with memory scenes. (All elements occurring in the Wilsons' drawings) They matched the Wilson drawings perfectly."

It sounds like impressive evidence, does it not? It is not quite as impressive as it sounds. The drawings Dr. Siegel is talking about did come from a study done in 1978, before much of the alien abduction phenomena was mass marketed to the public in bestselling books and movies. The study was done by Alvin Lawson, professor of English at California State University at Long Beach. [136] The study included a number of volunteers to see if the abduction phenomenon was built into the unconscious minds of individuals. They were screened to make sure they knew little or nothing about UFOs. In the book *A Secret Life* on page 292, David Jacobs, historian and abduction researcher, discusses the study. [137]

Jacobs believes one of the participants may have been an abductee, as this participant was the one lone individual that drew a little gray man, but this was probably just a random thing. The participants were told to imagine an examination occurring during the abduction, so the fact that there were drawings of examinations is a moot point.

Jacobs makes the point that few of the imagined abduction account narratives matched the typical alien abduction case. In fact, the narrative generated by each subject was different from each other and from most known alien abduction accounts, according to Jacobs. That may be overstated and can be debated, but the essence of Jacob's statement is probably true that the match is not that good. According to Jacobs, the details within each story was different in all imagined accounts, and the aliens in every single account were different from each other: a lizard, a cone-shaped thing with no head, an asymmetrical head with no eyes, a wise man with a beard, and, of course, one little gray man. Jacobs concludes, "Lawson showed that imaginary abductees were just that—imaginary."

How can the statement by Dr. Siegel (along with Ms. Clancy and most skeptics), and the opposite statement by Jacobs, both be correct about the same study? It is clear that Dr. Siegel was looking for a match with the Wilson drawings, and he got them by using drawings from the imaginations of a number of different people, which in his determination to find a psycho explanation this reasoning seemed completely fair, but of course it is not a fair comparison at all. Dr. Siegel was saying that by picking and choosing from the collection of drawings from the hypnotically directed imaginations of a number of people, he could find a match with the Wilson drawings. I would be curious to see if Dr. Siegel needed the imagined work of as few as three or four individuals to match the Wilson drawings perfectly, or if it took several more.

The telling thing is that only one of the imagined abductions showed a little gray man. That an imagined abduction account produced a gray being has been used by some in the skeptical community as implying that any hypnotized subject will come up with a little gray man (Dr. Siegel implied that and said little gray men, implying that more than one imaginary subject drew a little gray man). Brookesmith showed the actual drawing of the imagined little gray man on page 87 in *Alien Abductions*, but said nothing about only one person in the study drawing a little gray man, thus reinforcing the false idea that anyone put under hypnosis at that time would produce a drawing of a little gray man. Today with gray alien images popping up from time to time quite frequently in movies and on TV or commercials, and in cartoons, the frequency of little gray men showing up in a similar study may be much higher. It would be an interesting study to repeat now some 34 years later. Ms. Clancy discusses this in her book, and it seems that a reasonable abduction story and the image of the gray aliens has spread worldwide, so today apparently cultural contamination rules.

I did some further examination of the Lawson study and I find it questionable in a big way. Lawson believed that all alien abductions were the result of a psychotic episode in which the abductee remembers the trauma of their own birth. Due to the lack of development of the infant brain, it is now thought to be impossible to remember one's own birth. Whether possible or not, for multiple witness cases there is a nearly zero probability that two normal people will have the same psychotic episode at the same time. That amounts to explaining a very improbable event with a very improbable or impossible explanation.

I was surprised to find out how the Lawson study was actuality done. The questionable nature of the Lawson study was in the way the hypnotized subjects were led scene by scene through an abduction. The study took each person step by step through a typical abduction sequence of events, and then concluded that a person who had just imagined an alien abduction could give you a version similar to suspected real abductees. It was absurdly done. It was an example of the fallacy of circular logic. To me, the only suggestion that should have been made was: Imagine that you are being abducted by space aliens. Create your own narrative and tell us everything that happens. They were led through the capture to being on board; asked to imagine an examination, asked to imagine that the aliens gave them a message and how the message was given, imagine their return, and to imagine an aftermath of how it affected them. They were also asked to draw each of these scenes, and then Dr. Siegel seems amazed the people that just imagined abductions under hypnosis drew some similar scenes as the Wilson drawings. They were given the scene and told to draw it step by step. Am I the only one that thinks this is completely leading the witness? To get real data you must treat it like a real case. Nothing should have been mentioned until after it was brought up by the hypnotized volunteer. The researchers did all the thinking and imagining for them as far as laying out a reasonable scenario, sequence of events, and scene selection.

In the end, neither Peter nor Jack Wilson seemed to completely believe the psycho explanation, and the mechanical noise coming from the back of the car was never explained (the car had been taken to a garage to no avail), as given in a poignant exchange between

Dr. Siegel and Jack Wilson at the end of the account. Peter believed there had been some kind of contact, if it were only through the mind or astral projection. Why did Peter believe this? I believe because he had seen the little gray man himself. He was in a much more rested state than his father, and he realized that the image of the little gray man was real. He believed some kind of actual contact occurred.

The chapter absolutely could not have been better written from the point of view of the unusual sighting-becoming a delusion-becoming an imagined/dreamed abduction explanation (one form of a psychological explanation). It is, I believe, an example of hypnotic writing where many readers accept whatever the writer is saying without question. It could persuade almost anyone that Dr. Siegel had nailed it, anyone except a true detective or investigator, who probably could see more potential holes in the explanation than I did.

At first, all I saw was someone pretending to be open minded who was not (a pretend straddler of the type 1 and type 2 error boundary line), and that the case was tied up in an all-too-neat and tidy a package that was entertaining and flowing like a who-done-it detective story. I spent several nights not sleeping very well at all because of this case. I had some sleep revelations of my own. First, I realized that when I am very tired, I either do not dream or cannot remember dreaming the next morning. Maybe, contrary to some people, I have never had a dream about what I was thinking about just before I went to sleep, never that I can recall. My dreams seem to be worries, concerns, or nonsense pulled from somewhere else, but not from my conscious mind. And in all of my life, I cannot remember a single dream that I had when I went to bed super-tired. In fact, I remember only a few dreams in any detail at all. I have never suffered a sleep paralysis episode of being awake but unable to move. I recently learned that such episodes are frequently associated with narcolepsy and vivid dreams. [138] I do not suffer from this type of mental disorder, which is where Dr. Siegel pulled most of his dream explanations from.

Finally, Dr. Siegel did not challenge his own explanation (also known as testing your hypothesis). Challenging your own explanation is what someone who straddles the boundary between type 1 and type 2 errors does routinely, if at all possible. For instance, he did not indicate if he asked the two witnesses themselves separately whether or not the air in front of the car was foggy or hazy or misty, and with two eyewitnesses on the scene this is incredible. It was very important to show that the witnesses agreed with that single point. He did not indicate if he asked the witnesses how close the little gray man approached them. If they said that they first saw him some significantly further distance away (20 feet or more) and later he approached quite a bit closer, then the glory explanation is clearly wrong.

These questions and more needed to be asked, but someone who does not believe that there could be any physical reality to the little gray man will not ask these questions unless they are a trained on-scene investigator, or they understand that once a plausible explanation is found, it must be challenged in any way possible or a further experiment needs to be done to test the hypothesis. One must not look for only confirming evidence and ignore possible non-confirming evidence. Dr. Siegel was too eager to embrace the drawings from the Lawson study as proof of something other than that the directed step-by-step manipulation of the imaginations of a group of people can cover the range of things that could happen in one potentially authentic case.

The Wilson case would have been a wonderful case to get the hypnotized version of what went on during the incident from each witness separately. This would be one possible further experiment to challenge the psychological explanation. It would be imperative to follow the rigid protocols similar to those used by Dr. Simon in the Betty and Barney Hill case. If Jack recounted a "*g d gen -u-wine alien abduction*" and Peter did also, then the psychological explanation would be toast. If Jack recalled an alien abduction, but Peter did not and said that they just basically talked a little while and fell asleep in the car, then it would lend some credibility to Dr. Siegel's explanation. The explanation would have been challenged and would have held up to the challenge. A great opportunity to provide insight in the area between the psychological explanation and the alien abduction explanation was missed in this case in my view.

Brookesmith correctly points out how different the case would have probably turned out if Jack Wilson had gone to an abduction researcher. I agree with this point, but not because I believe the alien abduction accounts are implanted by the hypnotist or that the subject makes up an account to please the researcher as Brookesmith and others imply. The accounts come out of the mouths of the experiencers as the transcripts show, and many times the experiencers are greatly surprised by their accounts. The case would probably have come out differently, because you cannot determine whether or not you may have a "g. d. gen-u-wine alien abduction" unless you properly investigate the case from all perspectives, and that was clearly not done.

Challenging an explanation requires a lot more time and work. However, one simple challenge would be to try to recreate the glory optical effects, which the entire psychological explanation hinges upon, and see how easy or difficult it is. You could also see if the image of a little gray man in a seamless suit instantly jumps into your mind. I just could not stop thinking about the glory effect. Dr. Siegel had made it seem so automatic in the book. A little dew, a little mist in the air, and presto—you get a major optical phenomena.

I tried to recreate the glory effect in my backyard on the grass using a water hose with a spray nozzle and three different spray settings over a period of several days with no luck. I did observe a rainbow effect when I approached the spray from the side. I took some measurements of angle of the sun and so forth. I finally produced a weak Heiligenschien effect at one hour after sunrise, and it was not very impressive with a faint halo around the top of my shadow's head.

I even thought for a day that maybe you couldn't see your own glory effect, but then decided I really needed to research what glories were in the physics literature. I found a physics paper describing rainbows and glories. It was a complex paper about 137 pages long with lots of mathematical equations, as the author was seeking the best physics and mathematical description for these phenomena. The paper was entitled "The Mathematical Physics of Rainbows and Glories" by John A. Adam in *Physics Reports* in 2002. [139] He mentioned the rainbow scattering angle of 138° and I remembered that I had approached the spray mist from the side and saw a rainbow. From the measurements taken earlier, I calculated the angle where I had seen the greatest rainbow effect. Amazingly, my measurements gave a calculated angle of 42° for the maximum rainbow effect, the complementary angle to the 138° incident light scattering angle just as it should be, as together they should add up to 180°. What I had done was confirm the rainbow angle, something that optical physicists probably know by the age of five. I would now be able to pass eighth grade physical science.

The paper by John A. Adam was complex, indeed. I estimate that it would take me a few years of study to completely understand it, but it also gave the scattering angle of the glory effect as nearly 180° meaning that the rainbow colors will be tightly wrapped around your shadow. Furthermore, the rainbow effect can be produced by large and small water droplets such as found in rain or a water hose, but the glory effect can only be produced by a dense cloud or dense fog with water droplets of about 0.025 millimeters in diameter or smaller, because of their light-scattering effects. A mist on the ground or a light mist in the air will not work. The glory effect was most often reported by mountain climbers in the past, but is more often reported by pilots and others in airplanes today where the plane's shadow falls upon dense clouds below. A friend of mine once said he had seen a round, almost completely circular rainbow from an airplane, and I now know he had seen an optical glory. It is seldom reported by people on the ground, because the sun needs to be behind the observer at a very low angle, perhaps even with the horizon and bright with no fog between the observer and the sun, and there needs to be a dense fog within the shadow's length from the observer, a set of conditions that are very seldom met.

By the fact that the weather report said a mist on the ground overnight and not a frost, we can surmise that the overnight low was probably about 40° F, and that would typically place the time of year in the spring or fall. Our earlier point that the official temperature was probably at least four degrees higher two hours after sunrise brings us to 44° F by that time, but I remember something to the effect that official temperatures are taken some four feet off the ground in the shade. My backyard measurements show that an additional 6° F can be added to the air about four feet off the ground in direct sunlight, so that now we have reached 50° at the level of the car on the shoulder of the road. My measurement of bare ground temperatures shows a whopping 16° F greater difference above the official air temperature for a spot on the ground receiving sunlight uninterrupted for two hours after official sunrise. The surface of the shoulder of the road would likely be at 60° F by this time. Therefore, there would not likely be much dew or mist remaining on the shoulder of the highway beside the car. The direct witnesses did not mention any mist, much less a dense fog that would have to be virtually enveloping the car hood to produce the optical glory effect described by Dr. Siegel. Does the reader now see how difficult it would be to produce a glory effect along the side of a highway on the shoulder in an area where the sun was shining down at about a 20-25° angle, and how difficult it would be to have a dense fog where the sun had been shining for two hours on an area probably with sparse vegetation such that the area would have rapidly heated up?

To wrap this up, let us do some specific what-if calculations to show how ridiculous the glory explanation actually was. Jack Wilson was a fairly tall man. His shadow would have been about 13 to 17 feet long depending upon the exact time of day and the exact time of year. Let us place Wilson 12 feet from the front of the car when he first saw the little gray man and let us place an impenetrable fog even with the front of the car. The fog bank can be vertical or horizontal (ground hugging) and extend indefinitely. Let us say that Wilson's shadow was 16 feet long. His shadow would hit the fog bank and make a four foot long glory image surrounded by a tight rainbow. The image would have no legs as we demonstrated earlier and would not appear to be a little gray man, but rather would appear to be a dark torso. As we observe the scene we have setup, what happens as Wilson takes a step or two to get back into the car? He would have to move forward four or five feet to get into the car. His shadow would now be creating an eight or nine foot long glory image. The little gray man would have become the big gray giant. The very fact that the glory image had to move in Dr. Siegel's explanation and had to maintain the same proportions as it moved shows that what Wilson saw could not have been due to an optical glory image. I will wager that no atmospheric optical physicists ever read the UFO section of Dr. Siegel's book or if they did, they remained silent or turned off their minds.

With the possibility of a glory affect completely nil, I see a psychological explanation that has completely unraveled. Dr. Siegel's explanation has come apart at the seams. The logical implication is that Jack and Peter Wilson saw a real little gray man and had a UFO encounter with a number of hours of missing time! Peter clearly did not accept Dr. Siegel's explanation of no contact of any kind with a little gray man, and good for him in that he knew what he had seen and refused to be completely bullied by nothing more than psychological guesswork. My final conclusion is that it is more than plausible that Jack and Peter Wilson had experienced a "g. d. gen-u-wine alien encounter" and possibly abduction, and an investigator mired deep in type 2 error territory failed to recognize it or properly investigate it.

Mainstream psychological professionals see many people with mental and emotional problems, and many of these professionals believe that anything paranormal is all in the mind when, in fact, there is good evidence for some type of reality for a number of unexplained phenomena that are termed paranormal. To deny that puts many psychological professionals into a position of never thoroughly investigating an unusual case because of their own prior beliefs.

Their world is a world of coming up with explanations for their patients' experiences, which they believe are all in the patients' minds. Such professionals dwell deep in type 2 error territory where they may never consider another way of testing their explanations for correctness.

In Dr. Siegel's mindset, this was his UFO case and no one else was going to mess with it. He was going to get his psychological explanation out there to be seen and heard. If Dr. Siegel really wanted to test his explanation or really investigate the case, why not contact a reputable UFO organization from day one? Ask them if there were any UFO reports in the vicinity of the incident site. Further, only two days after the incident, a good UFO investigator would try to locate a local motorist that may have seen the Wilson car parked beside the roadway while memories were still fresh. If a witness or two were found that said they saw two men sleeping in a car along the side of the roadway about noon of that day with windows down and a door or two open, then one part of Siegel's explanation would be plausible. On the other hand, if witnesses could be found that said nobody was in the car, then the psychological explanation would completely go up in smoke. Who knows what might have been turned up by an immediate investigation? Maybe some evidence would have turned up to support Dr. Siegel's explanation, but the point is that the investigation was not attempted due to type 2 error thinking. An investigator of a case must not behave as a true believer or as a debunker, but must behave as both simultaneously, and must objectively investigate in as unbiased a manner as possible. An UFO investigator probably would not have been readily available to jump on the case, and any immediate witness information may not have turned up, but one must try anyway. Dr. Siegel seems not to have considered any other possible explanations or tests of his explanation (testing his hypothesis) because he did not think any other explanation was possible.

In the next chapter, I feel that I must discuss three other cases that scientists, skeptics, and debunkers cannot resist commenting on and cannot resist distorting verifiable facts until they can make the case into something they can explain. These three cases are Fatima, the Battle of Los Angeles, and Roswell.

Chapter 29

Fatima, The Battle of Los Angeles, and Roswell

"Science cannot solve the ultimate mystery of nature. And that is because, in the last analysis, we ourselves are a part of the mystery that we are trying to solve."

—*Max Planck (1858-1947) German Theoretical Physicist -quantum theory, Nobel Prize*

Fatima

The events that occurred near Fatima, Portugal, in 1917, were deemed a miracle by the Catholic Church in 1930, something that does not happen without a great deal of investigation by church authorities. To say that something extraordinary happened is an understatement. The events that happened were interpreted in purely religious terms by most of those involved, but strip away the religious fervor and you still have a remarkable story. Oh, I know that most skeptics want to say it was all overactive imaginations and mass hysteria coupled with some sort of rare atmospheric phenomena such as a sundog (light refraction due to ice crystals in thin clouds also known as a parhelion or "mock sun"), or by the crowd staring at the sun too long. Notice I did not say mass hallucination, because there is no such phenomenon as mass hallucination. There is, however, mass hysteria.

Most skeptics, such as Brian Dunning on the website *www.skeptoid.com*, make a series of logical fallacies, sweeping generalizations, and the appeal to reason or to prior prejudices to dismiss the entire thing.[140] Such skeptics think that unexplained phenomena are not possible. I can dismiss almost anything using the same logical fallacies, assumptions, and generalizations, but every time I dismiss authenticated observations, the probability of being correct diminishes.

Dunning assumes everyone there was a fanatical believer, and therefore susceptible to religious fervor suggestibility. Not so. This is just one of his many fallacies strung together. At almost any large gathering of people there will be those that are completely skeptical of what is supposed to transpire. Newspaper reporters were there that had made fun of and satirized the coming event in articles leading up to October 13.[141] Skeptical newspaper reporters later gave some of the most objective reports of what happened. Dunning does not mention any of this, since it does not fit in with his pet hypothesis. Once a skeptic leaves out a very pertinent detail, you know that what follows is an attempt to mislead. People not even attending the event saw something unusual in the sky from miles away. This is another detail that was not mentioned, because it conflicts with Dunning's later conclusions.

Dunning also suggests that most of the photographs taken show a bright and sunny day with huge crowds when it was known to have been raining at the event. Here he is trying to make something of nothing, as accounts clearly state that the sun did come out later. His entire point here shows twisted logic and tries to get the reader to think that the photographs showing the large crowds in bright daylight are not authentic. He tries to rewrite history to cast doubt, or does not investigate any further when he sees what he might exploit as a conflict in the story. He is simply wrong about the daylight and the large crowds, as both were there. He also questions the objectivity of Father John de Marchi, the primary investigator for the church into the incident, implying that he was only looking for supporting evidence and not doing a balanced investigation. Dunning should know all about forming an opinion from an unbalanced investigation or without taking into account what objective witnesses reported seeing. Other investigators said that accounts varied considerably as to what was seen and that is true, but that is also consistent with the phenomena being localized and not a celestial body. For a localized phenomenon, the location of a witness relative to the phenomenon will determine how much or how little was seen.

Father de Marchi used the testimony of credible witnesses, such as skeptical newspaper reporters, a professor of science, seminary professors, and other credible, responsible people. Dunning stated several times that the crowd was composed of all believers, all looking for a miracle, and later implied that the crowd got together and gave a consensus opinion on what happened. Absolutely not true on all points. Invariably, skeptical blogs seem to be incapable of presenting a balanced view of a case. Details that do not agree with their pet hypotheses are ignored. They pick and choose what to share with the reader and contradictory information is left out.

What did Dunning get right? Dunning correctly points out that the secrets about the future told to the children were not revealed to anyone until most of the events had already happened, in fact, long after most of the events actually happened. Therefore, they cannot be verified to have been known beforehand. He also points out that Lucia's own mother did not believe Lucia to be a true 'seer,' as she claimed, and at least one priest claimed that Lucia, by then a nun, suffered from religious hallucinations. Dunning does not deny that the foreknowledge of the October 13th display was real, but he says that the power of suggestion led to the crowd believing that they saw something on that date. The skeptics heavily believe in the power of suggestion to explain what was seen. Never mind that true believers, skeptics, and the indifferent reported seeing similar things. He does not bring up the fact that some of the best accounts came from skeptical reporters who were at a loss to explain what was observed.

Dunning stated that the only photograph of the sun he could find for the event looked normal to him, so nothing strange could have happened. Do I have to point out Dunning's fallacy here? Or the multiple assumptions pulled out of thin air that he is relying on to make a point? He also states that he thinks the crowd suffered from staring at the sun too long. He reveals that he stared at the sun too long as a kid and saw peculiar after images. I have seen the same thing, but these retinal fatigue images do not match what credible witnesses reported seeing at Fatima.

Here is the bottom line: There was real information passed from some unknown source to Lucia on or before July 13, 1917 that told of a future display that was to happen on the specific date of October 13, 1917 at high noon three months into the future. The children said the display would be a sign to all believers. Well, on the given date at the given time of noon with possibly as many as 100,000 people as witnesses, or as few as 30,000, the "Miracle of the Sun" happened. Some witnesses were as far as 12 miles away and going about their business. I have yet to find a skeptical treatment of Fatima that acknowledges this fact, because it conflicts with their central premise that it was all in the minds of fervent believers at the scene. Skeptics generally evoke hysteria and suggestibility at every turn for any type of unexplained phenomena. Many believers, some non-believers, skeptics, newspaper reporters with a skeptical bent, photographers, and the simply curious were present. As reported by credible witnesses, some of which were newspaper reporters who had belittled the event beforehand, the "sun" appeared more like a metallic silver disc turned broadside, was dim enough to be looked at directly, and danced or zigzagged across the sky as it was giving off beams of light of different colors. After the spectacle of the sun ended, the spectators discovered that the muddy field (due to rain earlier) was nearly dried up by the phenomena, as were their clothes. This point is seldom, if ever, mentioned in the skeptical literature concerning the event. Some were terrified during the spectacle, as it appeared that the sun was zigzagging toward the earth from their vantage point. Of course, it was not the sun. There is a photograph of the object seen that day in the book *The Fatima Prophecy: Days of Darkness Promise of Light*. [142] The photograph was taken after the spherical object had descended to near the horizon. What the photograph shows is a spherical object that is darker than the surrounding sky at that point and is at a low angle to the horizon (10 to 15°). It is clearly an unidentified flying object, a UFO by definition, and clearly not the sun.

Skeptic Joe Nickell wrote an article about Fatima in *Skeptical Inquirer* magazine in 2009. [143] Nickell's writing is quite good. He gives a very nice summary of the socio-political situation in Portugal in 1917, and he is persuasive unless you know certain facts about the case, which he leaves out. Of course, no calm, rational person with any knowledge of physics could think for a second that it was the sun. Dunning referenced Nickell's work, and was giving basically a rehash with some added musings in his Skeptoid blog. Dunning is not alone in doing this.

Mainstream psychologist Stephen Benedict-Mason, PhD. does basically the same thing on the website http://www.psychologytoday.com/blog/look-it-way/200910/the-miracle-fatima. [144] Most mainstream psychologists believe that all unexplained phenomena originate solely in the mind. They also universally disregard the testimony of objective, credible witnesses as discussed earlier regarding Ms. Clancy. Following this line of reasoning, we would never have a conviction in a court of law where the case was solely based upon eyewitness testimony if the jury was made up of all psychological professionals. It is really a troubling situation that psychological professionals dismiss so many witnesses with statements that are themselves sweeping generalization fallacies. It seems that they are saying that there can be no objective witnesses to any unexplained phenomena. My own sweeping generalization fallacy is that many of the psychology professionals seem to be psycho-illogicals of the first magnitude concerning unexplained phenomena.

Nickell suggested the possibility that the witnesses may have seen a sundog or parhelion, also known as a "mock sun" (sunlight reflecting off ice crystals in a cloud). A sundog is a safe, scientifically acceptable explanation. A sundog always appears at the same altitude in the sky as the sun offset in azimuth by 22° to the side of the sun. To get an understanding of what a 22° offset is, go outside and point one hand at the sun and hold the other hand parallel to the first and about two and one half feet away. If a sundog were present, the second hand would be pointing at it. Sundogs always occur at the same apparent elevation in the sky or angle to the horizon as the sun, but at an azimuth off to the side at an angle of 22°. At noon on October 13th at Fatima, Portugal, the position of the sun in the sky would be due south (azimuth heading 180°) at about a 42° angle to the horizon.

Fatima is located at 39.617° north latitude. The sun on October 13th is located approximately at 5.9° south latitude. The approximate zenith calculation is 39.617 – (-5.9) = ~45.5°. The approximate angle to the horizon calculation is 90° - 45.5° = 44.5°. These are first approximations that ignore the curvature of the earth and the actual geometry. A number of websites can be consulted to get the actual value. They all give a slightly different value, but all the values are close to 42°. A sundog location on that date at that place will be at a 42° angle relative to the horizon at 22° to either side of the sun or on both sides simultaneously. The position of a sundog would be south by southwest or south by southeast. Why bother the reader with this detail? It is because one only has to look up slightly to see something at a 42° angle to the horizon. Photographs from Fatima show people looking up sharply enough to see something at a 60 to 70° angle or more. This clearly indicates that they were not looking at the sun, but were looking at something higher in the sky. One credible witness said that the sun appeared at its zenith over the crowd. That literally means something bright appeared straight up or 90° above the crowd. The photographs of the crowd, and the angle to which they have their faces tilted back, indicate that at the instant the photograph was taken the phenomenon was at a 60 -70° angle or more to the horizon relative to the people in the photograph.

Many skeptical commentators stay mostly hands off and dismiss cases from afar without getting too close. It is much safer that way. In doing this, they can use sweeping fallacies and they do not risk the embarrassing consequence of their explanation being in conflict with their own objective findings. They would then avoid having evidence that conflicted with their stated explanations. Joe Nickell's own findings published in 1993 conflict with his suggested sundog explanation in the same book. [145] Joe Nickell, in the 1993 book *Looking for a Miracle: Weeping Icons, Relics, Stigmata, Visions, and Healing Cures*, states that based on witness testimony the position of the phenomenon was at the wrong azimuth and elevation to have been the sun. The wrong elevation is confirmed by actual photographs of the people, as stated above. He suggested at that time the phenomenon may have been a sundog or "mock sun," but his own findings of it, having been at the wrong elevation for the sun, also precludes it from having been a sundog, which occurs at the same elevation as the sun just displaced 22° in azimuth to the side. And there are many more problems with the sundog explanation, since a sundog cannot move around, does not look like a disc or sphere, cannot give off heat, and since the ice crystals of the clouds' relative position to the viewer may be critical, I am not sure that a sundog can be seen at one location and simultaneously at another 12 miles away. Also, most sundogs are seen shortly after dawn and near sundown when the sun is low on the horizon. A sundog seen at high noon is the rarest of them all, but the sun is not exactly high overhead in Portugal at noon on October 13th.

My wife, my daughter, and I saw a sundog for the first time on Thanksgiving Day, November 22, 2012. We were traveling by automobile east of Lake Charles, LA in a westerly direction a little after 4:00 p.m. and there were relatively thin cirrus clouds to the left side of the sun. The sun was near a 15° angle of elevation to the horizon. The sundog appeared to the left of the sun as a small patch of rainbow with blue-green to the left side and red nearest to the sun. The sundog was not round, but was a little elongated. It was literally one small segment of a rainbow. In other words, it did not match at all the description of the Fatima phenomena.

In 2009, Nickell touches lightly on the sundog explanation and focuses on the idea that the crowd stared at the sun too long and looked away, causing it to appear that the sun was moving. Not very convincing, since he showed previously that they were not staring at the sun. Another indication that the phenomenon was not the sun and the people were not staring at the sun was from the testimony of witnesses that were miles away from Fatima. The Reverend Joaquim Lourenco was a boy at the time he saw the sun dance in the sky from Alburitel, some 12 miles east of Fatima. He was facing generally west and slightly south at a time when the actual sun would have been due south of his position. He was looking in completely the wrong direction to see the actual sun off to his left, but saw what he thought was the sun dancing in the sky to the west.

The sighting was not of the sun or a sundog, or due to retinal fatigue, but was of an unidentified spherical object flying through the air. It was a UFO. The fact that it could be seen up to 12 miles away means that the object was big, very big, and very high up. If it were not tens of thousands of feet high in the air, then its angle to the horizon seen from 12 miles away would be quite low. If there were reports from three different locations as to angle in the sky and apparent size, then we could calculate the size of the object and its location by triangulation. This information may exist and someone else may have already done these calculations. It might take me years to sort through and find this witness information on my own to obtain a reasonable determination for size, altitude, and location. From the information that I have, I can estimate that an object that would fulfill the requirements of appearing to be approximately the size of the sun (from one-half the size to double the size) from 12 miles away would have to be at least 300 feet in diameter up to possibly 1,200 feet in diameter. It would also need to be at least at a 25,000 foot altitude to be seen 30° above the horizon from 12 miles away, and finally, it would need to be offset from being directly overhead at Fatima by a mile or a mile and one-half. If it were not offset from the crowds' location some, then they would have had to look up nearly vertically or 90° straight up to see it. Now I could be off some, but my estimates bring some perspective to the question of what could produce the reported sighting.

Skeptics and modern psychological science do not embrace the possibility that a person can receive real information about a future event through a vision. It is one form of telepathy, a message received in the mind of an individual from another intelligent source. Mainstream psychological professionals will say the whole thing occurred solely in the mind of the individual, a fire in the brain, to use Dr. Siegel's terminology. In other words, the individual was having a psychotic episode. Modern science has an explanation for every religious vision or any vision of the future ever reported: fire in the brain, individual psychosis, nut cases having hallucinations ... you get the idea. Most of the time, they are probably correct in their diagnosis. However, Fatima was something altogether different; something that mainstream science cannot deal with because of its deeply bunkered type 2 error thinking on the subject. Even though the sign in the sky occurred just as predicted, the beliefs that such things are impossible preclude mainstream science's acceptance of any part of the phenomena. The best that modern science can come up with is that it was just a coincidence that something appeared in the sky that day and this unlikely coincidence, coupled with a crowd worked up in religious fervor, led to an episode of mass hysteria. The idea that the crowd was simply staring at the sun is completely untenable, and was shown to be false by testimony and photographic evidence. This indicates something real other than the sun had to be there. So science is stuck without an explanation for what it could be.

If it were just a coincidence, how unlikely a coincidence was it? What are the odds of predicting the date, time, and nature of the phenomena (a sign in the sky) for something like this? Some skeptics will say 100%, since they believe the crowd imagined everything. It is clear that the skeptics are doing more imagining than anyone else, but the evidence says that something happened. What are the odds that something happened, but that it was totally by random chance? Well, let us limit the date range to before the end of 1917, starting on July 13. That gives about 171 days or 1/171 odds of predicting the correct day. The chance of predicting the hour is 1/24 and since there can only be two or three large-scale events, such as an earthquake, a sign in the sky, or an unexplainable sound, then the odds are about 1/3 of predicting the type of event. So the overall odds become 1/171 x 1/24 x 1/3 or 1/12,312. In other words, there is about one chance in 12,000 that they just got lucky and a completely random event occurred on the predicted date at the predicted hour. If I had used the criteria of predicting things to the exact minute, which seems to have been the case, then the odds against a random event becomes 60 times greater, or about one chance in 720,000; pretty large odds against it having been a random unrelated event.

How did the Fatima story unfold in the first place? The complete story actually started in 1916 or before, and some of this leads skeptics to say it was all due to overactive imaginations on the part of the children that kept escalating. We can all relate to that argument, but that does not mean it is correct. The three children, Lucia dos Santos and her cousins, Jacinta and Francisco Marto, reportedly had seen a being of light, an angel, in 1916 on three occasions when they were nine, eight and six years old, respectively. They first reported seeing a vision of a woman "brighter than the sun" on May 13, 1917, as they were herding sheep on their family's property near the village of Fatima. According to Lucia, the woman wrapped in light floated above a small oak tree at the site. Word soon spread about the children's visions, and people began coming to the area and following the children to the small tree where these encounters occurred on the 13th of each month.

In August, the children could not keep their appointment on the 13th, as an anticlerical local official who was skeptical had them detained and threatened them with death by boiling oil if they did not tell him the secrets. The children were eventually released and had their August encounter a few days later at a different location.

These meetings with the children cannot be said to be visions, which implies something only seen by the children, but rather were sightings, as people who gathered to witness the meetings with the children saw spheres and globes of light. Vallee, in his book *UFOs in Space*, has a discussion of what was seen in the month prior to the mass sighting in October, and it is quite revealing. On page 163, Vallee's quotes the Reverend General Vicar of Leiria who was present to witness the September 13 meeting with the children. The reverend saw an immense, bright sphere that he called "an aero-plane of light" flying at moderate speed. The crowds did not see the light being, who was by now identified as the Virgin Mary, but some reported seeing a small cloud or mist rising up from behind the tree. The essential point here is that the same light sphere phenomena was seen in the months before the October grand finale, but there were many, many fewer witnesses at these earlier meetings.

Much ballyhoo is made by skeptics of the fact that all the communication came only through Lucia. She seemed to go into a trace for communication with the being of light. The lady reportedly told her a number of things that they were to keep secret until the appropriate time, but the prediction of a sign in the sky on October 13th at high noon was not kept secret. There are said to be three secrets associated with Fatima. There was apparently a lot more incidental information transmitted to the children. The children reportedly were told that World War 1 would soon end, and that Jacinta and Francisco would die in the near future, which they did with the worldwide influenza epidemic that occurred shortly after WWI ended. They were also reportedly told that if people did not change their ways a second world war would break out after a light in the sky was seen by much of the world. In 1938, shortly before Hitler started what was to become WWII, the aural borealis or northern lights were seen further south than in 230 years. The problem with all of this is that the information was not revealed until after the events had happened. Lucia had become a nun and revealed parts of this information in 1927 and 1941. Of course, anyone can predict events after they have happened, and skeptics cry foul over this aspect of the case, and rightly so.

Sister Lucia seemed to have died on February 13, 2005, at the age of 97. However, there is controversy about this all over the internet. Photographs of Sister Lucia at 38 and at 80 something years of age seem to be of two different people of different ethnic heritage. The profile images seem to be the most telling, as Sister Lucia's very upturned nose at 38 becomes a downturned nose at 80 with a differently shaped nasal flare, and her chin at 38 was in line with her forehead, but at 80+ the chin is very protruding. The photographs seem to be of two different people, as there are many other differences that I have not discussed. What it means, I do not know. I have no idea why this apparent ruse would be perpetrated or if I have become victim to an internet hoax myself. The photographs of the older Sister Lucia are definitely not the same person as you can see from pictures of Lucia at 10 and Sister Lucia at 38.

What of the three secrets? The first secret concerned a vision of hell that the children were supposedly shown, the second secret dealt with Russia turning its back upon the church and that this would be reversed someday (that seems to have been accurately predicted beforehand!?!?), and the third secret dealt with chaos within the Catholic Church with many other lesser secrets rolled into it. I will say that the broadest version of the last secret concerning difficulties for the Catholic Church seems to have been predicted correctly beforehand as well, but maybe anyone could make this type of prediction based upon the difficulties of administering such a large and diverse organization. It is also true that the secret was written in such a rambling fashion that it is difficult to understand exactly what it does mean. I will mostly stay out of the debate about the timing of the reveal of the secrets, the religious aspects of Fatima, and the apparent Lucia imposter.

What interests me here is the apparent transfer of information telepathically, the light phenomena, and the silver disk sightings and maneuvers. There is much here in common with UFO sightings and reported telepathic communications, such as those associated with abductions. The parallels are striking. As stated previously in the book on the Phoenix Lights by Dr. Lynne Kitei in the 2010 edition on page 227, Jim Dilletoso, in a transcript of a radio broadcast in 2001, states in referring to the spectral properties of the Phoenix Lights, "We can't get a match on flares. We can't get a match on Hale-Bopp (a comet). The only thing we can get a match with are lights photographed on film on November 13, 1917 in Fatima, Portugal!" He may have mistakenly said November instead of October, but if this spectral evidence is correct, it is hard evidence of a link between two occurrences of unexplained phenomena nearly 80 years apart. This is extraordinary. I cannot believe there has not been more public discussion about the light spectrum findings. Certainly no skeptic will bring it up in discussing the case, as it conflicts with their point of view. Also, what is with the 13th of the month and a year ending in seven? The major incident in the Phoenix Lights series of incidents occurred on March 13, 1997, versus the Fatima series of incidents that led to the grand finale display on October 13, 1917. Is somebody or something playing games with us here or is all this just a bizarre coincidence?

The Battle of Los Angeles

On February 24/25, 1942, the episode known as the "Battle of Los Angeles" occurred just over two and one-half months after Pearl Harbor. This was nearly five and one-half years before the modern UFO craze hit in 1947. The case is also one of the rarest types on record: a radar-visual-photographic case. An unbelievable amount of anti-aircraft weaponry was fired at a group of UFOs passing over Los Angeles and Burbank, California. Something on the order of just under 2,000 rounds of anti-aircraft shells were fired. Nothing was shot down, but as many as six people on the ground were reported killed by the fallout. [146] The visual sightings of unknown aircraft and unknown radar hits led to the overwhelming response, as it was thought that Japanese aircraft were attacking the area. One famous photo by a newspaper photographer shows the hazy image of an oval shaped object in the crossbeams of eight spotlights from the ground. The photograph was of unquestioned authenticity, as it was taken by a newspaper photographer as the event happened.

In 2009 or 2010, the Sy-Fy channel show "Fact or Faked" tried to reproduce the photo by taking every suggestion that had been made over the years, and could not reproduce it without putting an oval object in the center of the crossbeams. [147] Bruce Maccabee has done an extensive analysis of the photo as well, concluding that a solid oval object was present. [148] The size of the object was possibly as large as 75 or 100 feet across and 20 to 30 feet in the vertical dimension.

The "Fact or Faked" TV show also tested how easy it was to shoot down a weather balloon, as the official response claimed the object to be a weather balloon, and some skeptics have claimed that a weather balloon could have been hit and kept on drifting overhead (what??). Well, in a fairly realistic simulation, an untrained cast member shot down a weather balloon on the first or second try and it fell, well, like a lead balloon; so much for that particular bit of nonsense. I like the approach this show took, as this is how all testable UFO reports should be investigated, but you have to have a budget to afford to do so. The show returned with new episodes for later in 2011 and 2012. Note that this one case had ground observers, radar, and completely authentic photographic film evidence, and was covered by news organizations as it happened. This is one of the most rock solid cases ever, and it happened before the UFO craze hit in 1947.

On the 65th anniversary of the incident February 24, 2007, the case was reviewed by a local TV station. The episode had the obligatory skeptic or debunker included at the end. He was James Oberg, an NBC News consultant in space-related matters. He started out saying some true things, but ended up talking a complete line of nonsense. He first said that in wartime you see lots of things in the sky. That part is true, because you had many, many more people watching the sky over Los Angeles for possible Japanese bombers at that time just two months after Pearl Harbor. The same group of objects could have come along in peacetime and maybe not be seen by anyone. He ended up saying the documentation was spotty (What? It was documented by the major newspapers with newspaper photographers' photographs.), and continued on to say that the case was too old with no living witnesses. Again, adult witnesses may be 85 years old and older, but undoubtedly some would still be around. It was complete debunker nonsense. He basically said that we should forget about it because the case was too old. I have seldom heard such ridiculous statements made by anyone. The case is on a lot of lists as one of the top ten best UFO cases of all time. Anyone that can dismiss the Battle of Los Angeles case probably belongs to the psycho-illogical crowd.

Roswell

This famous incident occurred sometime before July 8, 1947. The initial flying saucer craze had just begun a few weeks earlier and was in full swing. The entire country was aware of a number of sightings that were publicized, and there were even more unpublicized. The military did not know what these craft were, but believed they were real and actually suspected they were Soviet in origin. How do I know that the military believed that they were real, you ask? The very issuance of a statement by the base public information officer to the effect that a flying disk had been recovered near Roswell, NM shows that the military believed them to be real. You do not issue a statement saying that you have recovered an imaginary object you do not believe to exist. Later documents recovered by the freedom of information act shows that the Air Force believed the UFOs were real as early as 1947, and desperately wanted to know what they were. [149]

A number of ufologists have worked on the Roswell case over the years, including Stanton Friedman, William Moore, Karl Pflock, Kevin Randle and Donald Schmitt. The case was dormant from shortly after it began in 1947 until 1978. Friedman interviewed Jesse Marcel near New Orleans, LA in 1978, and the rest is history. Books have been published over the years by Charles Berlitz and William L. Moore, [150] Randle and Schmitt, [151] Friedman and Don Berliner [152], among others, and a book throwing cold water on the entire affair by Kal Korff. [153] The controversial MJ-12 documents that Stanton Friedman has spent a great deal of time trying to verify are related to Roswell as well. Friedman reports that he won a number of bets with Klass concerning typewriter fonts and similar details relating to the MJ-12 documents. MJ-12 refers to members of a secret group of scientific and technical advisors to the president of the United States concerning UFOs, and specifically the Roswell crash. The existence of the MJ-12 group is still highly controversial.

In his 2012 book, *Reflections of a UFO Researcher*, Kevin Randle discusses his involvement with the Roswell case and the interviewing of many, many witnesses. One gets a sense of the tedious effort that has gone into investigating the Roswell case over the years and the difficulty of evaluating the witnesses' testimony to come up with a coherent story. After all this time and effort, what can I possibly add to the case? Nothing but my brand of logic. It begins by accepting, for the moment, what the Air Force and the skeptics contend was the crash site of a mogul balloon assembly. In 1947, these assemblies were large, extending for 100 feet or more, and involved linking together eight to twelve regular neoprene weather balloons with radar targets or "kites," and instrument packages for recording various types of data. [154] In short order, the regular neoprene weather balloons were dropped in favor of the enormous polyethylene skyhook balloons, but the Roswell incident involves the earlier assemblage.

There is one question that I have wondered about since the first time I heard the story. I have never seen it addressed, or addressed adequately to my satisfaction. It concerns the press release that was issued saying that a flying disk had been recovered by the Army Air Force in New Mexico. How was the decision to release such a statement to the press made? How many people in the chain of command had to be convinced before such a press release could be issued? How does the Army Air Force logically progress from a Mogul balloon crash site to issuing a statement saying a flying disk has been recovered?

Let us look at a hypothetical situation. Suppose I am a military field investigator. I receive a report or an order to go out into the countryside to investigate some kind of crash site. I would probably find another officer or specialist to accompany me on the investigation. We get directions and head out to the site. After some time, we finally find the place and see a lot of debris. We find a debris field that is circular and 600 feet across. Exactly what one would expect for a large Mogul balloon assembly descending vertically downward to crash into the ground. We see small pieces of debris scattered around. We see box kite type radar targets. We see deflated neoprene balloons. We may not see radiosondes, the normal instrument packages associated with weather balloons, but we do see some perhaps larger instrument packages associated with the Mogul project and so on. We start picking up some of the smaller pieces of debris and it is unlike anything we have ever seen. We find it to be a sort of foil covered in a flexible, clear plastic. This seems to be a strange material with unusual properties. We find balsa wood type material covered in hieroglyphics. I convince myself that the material is so strange that it could have only come from a flying disk (the terms flying saucer and UFO were not in general usage at the time). I start jumping around and yelling, "It's a flying disk crash site."

My companion looks at me strangely and says, "Where's the disk?" But I persist. After a while with great effort, I convince him that it is a flying disk crash site. We grab some of the debris and head back to the base. We were so excited that we forgot to take photographs.

We rush the debris to the next officer in the chain of command and say, "This is debris from a flying disk crash site."

He examines it and agrees that it is weird alright, but before he authorizes an order to send a patrol out to recover the debris, he asks for us to describe the scene to him. He asks, "Any photographs?"

We say, "We forgot to take them in our excitement. You'll have to go out there and see for yourself."

We all go back out to the crash site. We get there and he says, "Where's the flying disk?"

We say, "It all disintegrated. See all the debris?"

He says, "It looks like a large weather balloon assembly crash site. There's nothing here that looks like a flying disk."

We say, "Yeah, but see all the weird pieces of foil covered in clear plastic, and the weird balsa wood type material covered in hieroglyphs. It's got to be from one of those flying disks."

It goes on and on like this until he finally relents and says, "Yeah, it must have been a flying disk."

After a great effort, we had convinced him that it is indeed a flying disk crash site. We take photographs. We hurry back to base and get the photographs developed. We are ready to go to the base commander's office. Once there, we excitedly tell him that we have discovered a flying disk crash site. We say, "Look at this weird plastic covered foil and this strange wood like material covered in hieroglyphics. It's got to be from a flying disk."

He says, "Very unusual looking material. I see you also have photographs. Let me take a look at those." We show him the photographs. He stares intently for a time and says, "Where's the disk?" After a longer pause he says, "It looks to me like you discovered the legendary, long searched for weather balloon graveyard. Gentlemen, you're wasting my time!"

We say, "Oh no, sir."

Around and around, the discussion goes on until finally after hours of discussion, he relents and says, "You've convinced me. Let's send out a press release. We have recovered the remains of one of those strange flying disks."

The above is clearly a fairytale, a fairytale that could not happen, but it is a fairytale that you have to believe in if you believe the Mogul balloon story. When the question is asked,

"Where's the disk?" you had damned well better be able to say, "Right here, sir," or you would be carted off for psychological evaluation. You would have to have substantial intact remains of the saucer or disk to show to the chain of command or no press release happens. They have to be convinced of the veracity of what you are saying. The Mogul balloon fairytale could not happen. The real account is a little different in that a local rancher was said to recover the remains of the disk and a little later brought it to the attention of the Air Force. The essence of the story is the same. At some point, someone has to produce the remains of an actual flying disk or no press release happens. The several levels of command that have to be convinced by the evidence will not accept plastic covered metal foil, strange lightweight beams, or the instrument package from a Mogul balloon assembly as the remains of a flying disk. The probability of several levels of command accepting this as evidence of a flying disk is zero. If anyone can show in a logical, reasonable manner how this fairytale could have happened, and it would thus become a credible story, then I would change my mind on Roswell and conclude that it was probably a Mogul balloon crash. At present, I cannot logically see how you can get from a Mogul balloon crash site to a press release saying that a flying saucer or disk has been recovered. Everyone involved would have to have been completely deranged. Logic says that the Army Air Force, or soon to be Air Force, had an unguarded moment where they told us the truth, but then had to quickly retract it.

In August 2012, a former CIA agent claimed to have come across an evidence box at CIA headquarters some years earlier that indicated to him that the crash of a flying saucer near Roswell really happened. However, he was making the rounds in support of a science fiction book he had written, which makes him an easy target for skeptics to claim that he was just trying to get attention for his book. [155] Others have come forward over the years to say Roswell was a real event. The highest ranking officer to acknowledge that the Roswell flying saucer crash really happened was General Arthur Exon. General Exon, who had been stationed at Roswell at the time, told this to Kevin Randle in 1991, and yet the story put out by the mainstream media in the 1990s was mainly about the very improbable story of a Mogul balloon crash as pushed by the Air Force. Some confusion about witness statements and their accuracy have also come to light over the years. However, no matter how convoluted the story has become about witnesses exaggerating their role in the events and the exact timing, cold logic says the overwhelming probability is that the UFO crash at Roswell really happened.

CHAPTER 30

Tying Up Loose Ends
Government Cover-ups; Michio Kaku; Skeptical Analogy; The TV Media

"Secrecy is the first essential in affairs of the State."

—Duc De Richelieu, (1766 -1822) 2nd and 5th Prime Minister of France

Former VP Dick Cheney was asked about UFOs on a radio talk show on April 11, 2001. [156] Cheney was specifically asked if he had been briefed on UFOs. His response was, "Well, if I had been briefed on that, I am sure it was probably classified and I couldn't talk about it." Cheney did not dismiss the question of UFOs. He basically said that the topic was classified and he could not talk about it. Why would the subject be classified if there was nothing of substance concerning UFOs? Governments of all types throughout history have attempted to keep sensitive matters secret or classified. They do not attempt to keep things secret that do not exist or are trivial.

Stanton Friedman is a long-time UFO researcher and probably the world's most extensive researcher of government documents about UFOs. He has uncovered many instances of government misinformation and withholding information. Richard M. Dolan, in his books about *UFOs and the National Security State*, gives case after case of a government or military entity distorting or giving outright false information to the public. [157] What government is going to tell its citizens that there are unknown objects flying through our skies that appear to be under intelligent control and we cannot do anything about them. Over the years, the government has said to ignore those people claiming to see such things because it's all in their heads while secretly documenting hundreds of unexplained military cases. A friend of mine equates this to being similar to the time J. Edgar Hoover said that organized crime or the mafia did not exist when every FBI agent knew that it did.

Timothy Good's 2007 book *Need to Know: UFOs, the Military and Intelligence*, points out case after case of official cover ups, and is mostly at odds with Alexander's 2011 book on the subject. [158] Alexander, who retired as a colonel in the US Army, has an impressive resume of working on classified projects, and as a consultant to the CIA and other agencies and councils, but he could never find an official group in the government that was responsible for UFOs. The bottom line is that Alexander would have found out little if he were asking people about a subject that had a need-to-know security clearance and he did not have that clearance. He does have contacts in high places, but need-to-know security clearances takes precedence. Alexander maintains that most of what the government knows or knew about UFOs has already been released at this point through FOIA requests. He points out that FOIA requests have an unintended side effect of causing agencies to destroy outdated documents as soon as they are declassified so that FOIA searches are less of a burden. In the world's remaining superpower in the modern error of terrorism, security matters are more sensitive than ever. The very fact that various security clearances exist means that the government is withholding information, and has always been withholding information on many subjects, including UFOs.

Leslie Kean's 2010 book *UFOs: Generals, Pilots, and Government Officials Go on the Record* discussed military cases from all over the world. [159] Prominent physicist Michio Kaku even gave a foreword of high praise for the standard of research done in Kean's book, but that didn't stop James Oberg, NBC space analyst, from weighing in on Kean's book. In early August of 2010, on the website www.msnbc.msn.com [160], Oberg made the assertion that Kean's book was based upon a questionable foundation. Oberg suggests that pilots aren't very good observers, despite the fact that they have to quickly recognize whatever is in the air and decide if action is needed or not. He even uses this in his argument against pilots: They have to make decisions too quickly and this will cloud what they thought they saw. It is standard debunker fare: dismiss the witnesses, use one or two cases from some other source where the pilots were seeing balloons, and imply that was the case here. Oberg does make some valid points, such as an investigator must get to the raw sensory impressions of what was seen and sort that out from the witnesses' interpretations. Absolutely! But then Oberg does not address a single case in Kean's book directly. Dismissal by implication of similarity is just another type of fallacy.

Kean appeared briefly on the Dylan Ratigan show on MSNBC later in August 2010 to promote the book. (Regrettably, Ratigan left MSNBC by his own accord in June, 2012). Kean's book was made into a documentary that aired on the History Channel in 2011 entitled "Secret Access: UFOs on the Record," where the actual people involved gathered together and discussed their cases. It was a very powerful presentation. I have not found Oberg's response. Kean's book has been called one of the best ever written on the subject of UFOs by some mainstream scientists, but most mainstream scientists act as if they want to avoid UFOs at all costs.

One of the exceptions is Professor Michio Kaku, the world's leading authority in the area of physics known as unified field theory. He wrote the textbook on the subject. Kaku is on TV and radio programs regularly explaining scientific findings and concepts. In 2005, Kaku said that it is highly unlikely the earth has been visited by extraterrestrials, but in 2010 on a History Channel program he said something to the effect that some UFO cases may be valid and, if they are, that suggested the possibility of extraterrestrial visitation. My ears perked up when he said that some UFO cases may be valid. He was very courageous, even though there were a lot of qualifiers in his statement. He can afford to say something like that without hurting his career, as he has already made it to the top in his profession.

Kaku has appeared a lot in the media talking about things other than UFOs. When it comes to UFOs, the media today mostly ignores them unless it is impossible to do so. Most TV coverage today is very inept or plays the whole thing for a joke. It is great fun to play the whole thing for a joke, as was done in late 2011 on NBC affiliate KPLC-TV in Lake Charles, Louisiana. On or about December 18, 2011, some unusual lights were captured by cell phone video by a Lake Charles resident. A number of other people had also seen the lights. The brief video shown on TV showed several lights in the nighttime sky. One light seemed to be hovering while one or two other lights were seen moving back and forth in the field of view above and behind that light. It appears that there was not any sound associated with the lights. An official at the local Chennault Industrial Airpark was contacted and interviewed on TV. He said they'd had a lot of unusual aircraft returning from Iraq and that is probably what was seen. The time the lights were seen and whether or not there were aircraft actually landing at the airpark at that time was not mentioned in the report. It was not even clear that the official had seen the video. The simplest possible follow-up to the proposed explanation was not done, checking the time of the sighting and the time of arriving flights. The official went on to give the peace sign and pretending to be a space alien, said, "I come in peace," playing the whole thing for a joke.

When the camera came back to news anchor Charlie Halderman, he said, "I saw those lights last night. I am not a disbeliever. They were very much out of the ordinary!" This implied that he did not buy the aircraft landing explanation at all, but playing it for a joke was good TV. The report was uploaded to youtube at www.youtube.com/watch?v=5TT7aRcyJDo . [161] The logical fallacy of calling in an authority figure to comment without any real investigation having been done is called the "appeal to authority fallacy," and it is done all the time in media reports of UFOs and other reports as well. The reporter had what seemed to be a reasonable explanation from an authority figure that the lights were probably due to aircraft landing at Chennault, but the lights shown could not be an aircraft, because the lights appeared near each other and some were going in opposite directions.

Later online, the on-air broadcast video was unwatchable on youtube; nothing could be seen. At least, I could not see anything. An enhancement showed only one light that a blogger said looked like a Chinese lantern (small hot air balloon) and it did, but that was not what was shown on TV on the night of December19, 2011. Two lights that are close to one another and moving in opposite directions cannot be Chinese lanterns or an aircraft coming in for a landing, no matter who suggests that it is. It would have been fine to give the same report, but include that we checked the timing and there was an aircraft landing at Chennault at the same time the video was shot or that there was not.

Dr. Lynne Kitei gave the local media good marks in the Phoenix Lights case, but not the national media. The phenomena actually started as early as 1995, when she and her husband Frank photographed some similar lights near their mountain home in the Phoenix area. In the 1997 incident, the massive boomerang type object was seen by thousands of witnesses, including Fife Symington, a former fighter pilot, who was the governor of Arizona at the time. He was completely puzzled by what he saw, but did not go public with it. He later staged a practical joke with a person dressed as a space alien at a news conference that made all those who reported seeing the phenomena the butt of the joke. I would have enjoyed the ruse if he had come clean about what he had seen. Serious ufologists and the people who saw the huge airship or configuration of lights were more than a little bit upset by his making fun of the entire episode without telling what he had observed himself. The practical joke livened up the news conference and the TV media loved it.

Chapter 31

The Humble Reasoning of a Single Individual (Dr. Bruce Maccabee Dents Science's UFO Paradigm)

"In questions of science, the authority of a thousand is not worth the humble reasoning of a single individual."

—*Galileo Galilei (1564-1642) Italian physicist, mathematician, astronomer, and philosopher who played a major role in the Scientific Revolution.*

"To know the history of science is to recognize the mortality of any claim to universal truth."

—*Evelyn Fox Keller, (1936 -) Physicist, from Reflections on Gender and Science*

I have come to the end of my nearly two-year odyssey much more informed on the subject of UFOs than when I began. During my career, I mostly avoided the subject of UFOs outside of family discussions, with a few exceptions. At times, I would sometimes reflect upon what I had seen and what my family had seen, and I would wonder what it could have been. What could we have seen? In the course of this research journey, I have read some 35 books and visited many internet sites, both pro and con on the subject. In the past, without the internet, my research would have taken much longer if it would have even been possible within a reasonable period of time. I have been saddened as I came to understand the irrational response of the scientific establishment to the UFO phenomena. I have come to understand the scientific establishment's UFO paradigm. The paradigm has been belief driven. The belief is that there cannot be any unexplained phenomena; therefore, they (mainstream scientists, skeptics, debunkers, etc.) do not even have to consider it. There cannot be anything to this UFO business, so they (science, skeptics, and debunkers) do not even have to bother to investigate anything. They feel that they do not have to listen to the witnesses. All they have to do is dismiss the witnesses as unschooled or mistaken, and replace what the witnesses saw with whatever narrative they feel comfortable with. There is no need to do any hypothesis testing, since there cannot be anything to test in their view. They seem to feel completely justified in omitting details and retelling the witness's story how they see fit. Never have I seen so many fallacies strung together as in the scientific and skeptical responses to UFOs and other unexplained phenomena. It is the *Emperor's New Clothes* in reverse. The scientific authority has certainly thrown its weight around concerning unexplained phenomena, but the authority of a thousand means nothing in science. One gram of logic is more powerful than a thousand grams of fallacy.

To those that dismiss the UFO phenomena:

> Dismissal of the testimony of thousands of credible witnesses is not logical.
> Dismissal of dozens upon dozens of radar-visual cases is not logical.
> Dismissal of nuclear radiation spikes is not logical.
> Dismissal of authenticated film and photographs is not logical.
> Dismissal of cases because they are older than some arbitrary date is not logical.
> Making sweeping generalizations/proclamations that are almost always fallacies is not logical.
>
> Dropping facts or witnesses' testimony that contradict your pet explanation is not logical or ethical.
>
> The human frailty of slanting everything to fit into your prior beliefs is emotional, not logical.
>
> Using teenage misadventures in the attempt to discredit a witness is not logical.
> Believing that unexplained phenomena are not possible, therefore, I can dismiss the case without proper investigation is not logical.
>
> Appealing to authority or experts for explanation without investigation is not logical.
>
> Proposing an explanation without testing it or challenging it when it is possible to do so is not logical. (We are all probably guilty of this one at some time, mainly because we are many times not thinking from a broad enough perspective. We are not thinking outside the box.)

Extraordinary claims do not require extraordinary proof, contrary to what Carl Sagan and others have said, at least not to begin scientifically studying those claims. What is required for any claim to stand up to scrutiny, and for science to begin studying them, is just some supporting evidence, and UFOs have a ton of supporting evidence from any objective viewpoint. But the case for what is responsible is still mostly circumstantial with a slew of suspects. Most of these suspects involve extraordinary explanations from the least extreme to the most extreme. Proof implies knowing beyond any doubt and, as Evelyn Fox Keller indicates, this universal truth of science, this absolute proof, has to be revised from time to time. There are intermediate stages in science where evidence is collected and studied, where hypotheses and theories are proposed, tested, rule the day for a time, and are later modified or discarded. It is a long and winding road to get to the stage that we call proven beyond any doubt.

Most well-accepted concepts reside at the stage science calls a viable theory, which means that there is lots of evidence for it, but reaching the final stage of absolute truth in every detail, absolute proof of every tenet of the theory, is reserved for only the select few. Sagan's famous statement, accepted almost universally and quoted countless times, is a proclamation or sweeping generalization fallacy meant to dismiss doing research into unexplained phenomena. It bypasses the normal scientific process. The unexplained phenomena that are UFOs has never been systematically studied, only investigated after the fact. Sort of like a hit-and-run accident where the perpetrator leaves and there is no way to trace them. The one viable plan to actively study UFOs was purposed by Ruppelt in 1952, but was shot down by the CIA controlled Robertson panel's conclusions and some early technological failures of the equipment in the field.

There are only a few sins greater in the eyes of the scientific establishment than a scientist that embraces UFOs and exposes the fallacies of the scientific establishment on the subject. Most serious UFO investigators have concluded that there is something real and unknown to science behind the UFO phenomena. Philip J. Klass, the world's leading skeptic for nearly 40 years, thought so as well, at least early on. Klass's main hypothesis of the real phenomena behind some UFOs was that they were due to ball lightening/plasmas from power lines, but this explanation was weak. It was expanded by others such as Persinger and Derr, along with Devereux and Brookesmith, to include geologically generated earth-lights and later to earth-light plasmas or atmospheric plasmas produced by meteors disintegrating in the atmosphere. The correlation testing of Project Blue Book Unknowns and other considerations found that atmospheric plasmas were not viable explanations for the Blue Book Unknowns. There is also not a generally accepted scientific explanation for any of these phenomena, ball-lightning/earth-lights/atmospheric plasmas. In my view, Klass really believed that there were some unexplained natural phenomena behind UFOs, but he eventually dropped his atmospheric plasma contentions due to peer pressure from fellow skeptics and the science establishment.

The best explanation I have seen for ball-lightning comes from Tore Wessel-Berg's 2003 paper, although it seems to have been mostly ignored up to this point in time. Wessel-Berg presents another hypothesis of upper atmosphere ball-lightning, whereby under the correct conditions, the phenomena could mimic a maneuvering cigar-shaped UFO. It has a limited utility in explaining some UFO sightings. In my opinion, balls of plasma energy such as ball-lightning and/or earth-lights will display random, erratic, drifting behavior dependent upon local electric and magnetic fields, and will not appear as maneuvering UFOs. Some UFO sighting reports of the nocturnal light variety could possibly be generated by ball-lightning or earth-lights observed by novice observers.

Finally, the claim that earth-lights/atmospheric plasmas' associated magnetic fields can induce hallucinations in humans appears to be complete nonsense, and is even contradicted in the very book making the claim. This hypothesis has been soundly refuted by millions of MRI tests performed every year with exposure to much stronger magnetic fields without generating narrative adventure hallucinations in patients.

Most of the central Mississippi UFOs exhibited behavior that excluded them from being earth-lights, and they were certainly not ball-lightning. If they were not ball-lightning/earth-lights or any other prosaic explanation, then what were they?

First, several of my family's sightings were very closely related to the sightings at Flora, Taylorsville, Petal, and Charles Hickson's 1974 Mother's Day sighting. In three of those cases, a classical flying saucer was seen in full detail, and in one case in outline. My family's sightings included one flying saucer in full detail, two in outline, and one with lights around the middle similar to Charles Hickson's Mother's Day 1974 sighting. Most of these sightings involved an object within 100 to 600 feet of the observer. All of these close encounter sightings were in the 1970s in the 1972–1978 timeframe. I can only conclude that my family's sightings, and those in the historical record for central and southeast Mississippi during the 1970s, were generated by the same phenomena, and it seems to be phenomena under intelligent control.

When I began this book in August of 2011, I had heard of the Pascagoula abduction and the 1957 RB-47 case, but I had no knowledge of any of the other documented Mississippi sightings of the 1960s and 70s. I made the national forest connection in the book because all of my family's sightings and the historical Mississippi sightings were near or not far from heavily wooded areas and national forests. Maybe this is a coincidence, but it is a coincidence shared with several other UFO hotspots around the United States where UFOs were often observed hovering and maneuvering close to heavily wooded areas. The amount of forest cover of a given state also correlated moderately at a statistically significant level with the number of Project Blue Book Unknowns for the individual states. One implication is that these UFOs were flying craft of some type with intelligent beings or robots onboard, or were remotely controlled flying craft of some type. If this were the case, being near a heavily wooded area allowed the UFO to quickly dart out of sight whenever the need arose. Of course, our multiple sightings and those of the Uintah basin in Utah, those in Elmwood, Wisconsin, and those repeated sightings described by Mr. Frascogna as still occurring near Flora, MS, could never have made the Project Blue Book list and would have been frowned upon by ufologists throughout much of the modern UFO era because of the repeat witness nature of the sightings. Seeing a UFO was supposed to be a completely random, once-in-a-lifetime thing. I only noted one repeat witness case among the 564 Project Blue Book Unknowns. UFOs were not supposed to be flying by the same witnesses or flying the same routes repeatedly. There are probably hundreds of cases similar to my family's case throughout the world, and many UFO investigators have yet to fully embrace the reality of the situation.

There have been many explanations put forward for UFOs, and specifically flying saucers and their occupants. From this point, I refer only to cases that have been thoroughly investigated and remained unexplained, and show no evidence of being a hoax (proclaiming a case a hoax, because there is not a way to explain it, is not evidence).

I reject the purely psychological explanations or hallucinations because they are completely at odds with the facts of multiple witness cases involving credible witnesses, radar-visual cases, and authenticated photographs. The psycho explanation has such an appeal to many skeptics and psychological professionals that they, themselves, have become victims of a psychosis, the psycho explanation psychosis. It is a natural result of filling one's mind with type 2 error thinking (being completely closed minded on the subject of unexplained phenomena) and the underlying belief that all people are irrational underneath a thin veneer of logical objectivity. All I will say is that the veneer of logical objectivity is thicker for some people than for others.

I reject ball-lightning/earth-lights/atmospheric plasmas as the primary explanation of UFOs because of their own inherent limitations, but I do accept them as an explanation for some nocturnal light sightings. They cannot appear to be a structured object with a tapered, rounded bottom, a center section that appears to be two saucers placed lip to lip with a ridge around the middle where they join, and then appear to have a transparent copula on the top. They cannot appear to be black triangles with bottom lights at each corner and in the middle. One would have to have incredibly strong magnetic field gradients to possibly create such effects, and that is not going to happen outside of a laboratory, if it is even possible to do inside a laboratory. And so with all of the psycho-illogical explanations out of the way, the skyhook balloon explanations limited to about one percent of the unexplained sightings, and ball-lightning/earth-lights/atmospheric plasmas given their just due in explaining some nocturnal lights, what are we left with? Quite a lot actually.

Some say what we are seeing is time travelers. I don't want to say that time travel as we see in the movies is impossible, but it sure seems to be. The classic paradox called the grandfather paradox was discussed on the Discovery Channel's website *www.dsc.discovery.com/tv-shows/curiousity/topics/parallel-universe-theory-quiz.htm#mkcpgn=otbn1* . [162] If you travelled back in time and killed your grandfather as a baby, then you could not have been born, but if you were not born, how could you have killed him. There are many time travel paradoxes.

I do not think that the past can be changed. I think it may be possible to view the past somehow, but interaction would be impossible. Can you imagine the boon to law enforcement? Imagine a police detective requesting a court order to go back in time (time travel viewing would be strictly regulated and restricted to historians and law enforcement) to view the crime scene in order to identify the perpetrator. I do not think that UFO phenomena are due to time travelers.

Some say all Unexplained Phenomena are related. UFOs would then be just one aspect of the paranormal, such as ghosts, poltergeists, etc. Some say we are seeing visitors from another dimension, but if there is a spirit world wouldn't it be from another dimension? Modern physics and Professor Michio Kaku seem to say that there can be parallel universes and other concepts that are pretty much beyond my comprehension. The Discovery Channel website mentioned earlier indicates that "dark energy," or what is now called "dark flow," is supporting evidence for the "string theory" model of the universe. The significance of this is that "string theory" allows parallel universes to interact at some level with one another. Could advanced beings from a parallel universe find a way to hop into our universe and hop out again at will?

My favorite idea out of those proposed is that UFOs come from societies that developed right here on earth and are millions of years older and more advanced than humans. No, it's not logical. Where do they live? What do they eat? How have they solved the problems of disease and over population? Where are their technical and manufacturing centers? No it's not logical, but it might be the only explanation that does not invoke the paranormal, or super-physics.

The idea that UFOs are the near earth vehicles of space travelers seems almost mundane compared to some of the others. However, space travel to another planet outside one's own solar system would be a most difficult undertaking. Mankind has no current technology (or any feasible technology) that could allow us to travel in space as they do on *Star Trek* and similar TV programs or in the movies. We appear to be many breakthroughs away from developing a true space traveling technology, and might require thousands or tens of thousands of years of technical development to get to that stage. At present, we do not know that true interstellar space travel is even possible. Space travel just beyond the edge of our solar system is just barely possible at present, and then by using space probes and robots. The human body is not designed for a long-term zero gravity environment or for the radiation levels in space. The distances involved in travelling just 10 light years away from your home planet are just so enormous as to be incomprehensible. With our current best technology, it would take over 100,000 years to get there. Stanton Friedman and Carl Sagan both discussed nuclear powered spacecraft that could possibly travel at $1/10^{th}$ the speed of light or more to cut the trip down to perhaps 100 years or less. Traversing the Milky Way galaxy, which is about 100,000 light years across and about 1,000 light years thick, would take about a million years of continuous travel with a $1/10^{th}$ light speed vessel or about a billion years with current technology.

Some say we could traverse the Milky Way galaxy with our current technology in a billion years and establish colonies on suitable planets along the way. If it were technically possible, it means that a thousand generations would have to have been born and died on a trip from one suitable planet to another. This requires a focus with one goal in mind across thousands of generations, and the solving of innumerable technical problems such as artificial gravity. Communications with the mother planet would become more and more troublesome as the ships went deeper and deeper into the galaxy. Eventually, after a few tens of light years, each ship would be completely on their own for long periods of time. The overall logistics of coordinating this quickly becomes impossible.

These are enormous technical problems that have to be solved just to get to the nearest habitable planet outside one's own star system. These considerations are why many scientists say that UFOs cannot be the vehicles of space travelers. The distances are so vast that star system to star system space travel seems impossible. Not only that, but suppose you achieve what seems to be impossible and develop a spacecraft capable of travelling just enough under the speed of light to avoid a serious increase in mass or weight (Einstein's theory of relativity-mass increases to infinity at light speed), then you have some other problems as well. One is this: If you are traveling through space near the speed of light and happen to hit a grain of sand weighing just one milligram, then my calculations indicate it will rip through your ship like a high-powered rifle bullet through thin sheet metal killing anyone or destroying anything in its path. If you do develop a spacecraft that can travel that fast, you have to have an infallible method of clearing your flight path. Although the probability of hitting that grain of sand or micrometeorite may be extremely low in the vastness of space, the consequences that would result if it did happen would be catastrophic.

Another problem is that even light speed is not fast enough. Yes, light speed is not fast enough to really explore the Milky Way Galaxy and come back to report on it, but it is fast enough to interact with your nearest neighboring planets out to perhaps maybe 50 or100 light years or so. While the astronauts travelling at near light speed may not age that much, the folks back home will age at the normal rate. When our hypothetical astronauts come back from a 50-light year trek out and back, there will be no adult alive that saw them off unless the average lifespan has been increased to 125 years by that time. Oh, and another thing. You do not attain light speed instantly as they do in the movies and science fiction shows. A six-month acceleration at about 1 G or 22 mph/sec^2 is needed to get up to light speed, and a six-month deceleration is needed to slow down to normal speeds. This little detail adds another two years to any trip out and back. As an astronaut, you could have no other life outside of the space program. Similar aspects and much, much more were explored by Paul R. Hill in his 1995 book *Unconventional Flying Objects: a scientific analysis*.

So the whole idea of travelling from one star system to another seems so fraught with problems as to be impossible. However, just because we have not figured out how to accomplish space travel to another star system does not mean that it is impossible. If space travel is somehow possible, someone somewhere will have figured out how to do it. This is based upon the estimate that there are probably millions, if not billions, of planets in the Milky Way galaxy capable of sustaining life. The famous Drake equation was developed by Frank Drake in 1961 to address this very thing and get to the possible number of planets in the Milky Way

Galaxy with which we might be able to communicate using radio astronomy. [163] It can be used to also estimate the number of intelligent civilizations more advanced than we are on the earth, but most inputs into the equation are just guesses. Here is an easy to understand narrative that is just as valid. If there are millions of planets in the Milky Way Galaxy capable of sustaining life, then there would be tens of thousands of planets with intelligent life. If that is so, there will be hundreds of civilizations more advanced than Planet Earth, dozens that are millions of years more advanced, and a few that are hundreds of millions or billions of years more advanced. If they figured out how to travel in space, whether it is by brute force, wormholes, time-warps, or whatever, then eventually they would visit a very inviting planet called Earth. Who am I to say that this has not happened? Certainly the testimony of numerous credible witnesses, radar, and other events and evidence suggests this as one possibility. Remember one of George Carlin's best quotes:

"If it's true that our species is alone in the universe, then I'd have to say that the universe aimed rather low and settled for very little."

—*George Carlin (1937-2008), Comedian, Social Commentator*

So what are we left with in our hypothetical explanation of UFOs? Well, there is the advanced human aircraft explanation favored by many people. Certainly a lot of UFO reports are just misidentified aircraft of all types, but we are talking about the cases that remain unexplained after careful investigation. Some argue that top secret or "Black Program" aircraft are responsible for many of the unexplained UFO reports. A favorite idea of this group is that the black triangle sightings are of a top-secret black triangular craft belonging to the US military. Top-secret triangular craft that the US military is said to have include the SR 91 Aurora airplane, the suborbital TR-3 Black Manta, and the orbital Blackstar spacecraft. These craft are supposed to have suppressed or reverse engineered, anti-gravity technology (reverse engineered from flying saucers) as discussed on the website *www.wikipedia.org/w/index.php?title=black_triangle_(UFO)* . [164] The B-2 Spirit Bomber is another aircraft with supposed reverse engineering from alien technology. Another not-so-secret craft that some say was responsible for some UFO sightings was the F-117 Nighthawk or the so-called "Stealth Fighter" (actually a ground attack aircraft), which was revealed to the public in 1988 some five years after it was operational. It was retired about 20 years later in 2008. [165]

A black triangle UFO was videotaped as it out-maneuvered several jet interceptors sometime in the mid-2000s. The videotape was shown to an aviation expert on the "Giant Triangles" episode of *UFO Hunters* that originally aired on March 18, 2009, on the History Channel. [166] The tape appeared to be authentic. The resident skeptic on the show, Kevin Cook, commented that maybe the jet interceptors were escorting the black triangle, implying that it was our own top-secret aircraft. It was a peculiar escort operation, if it was one, as the jet interceptors had the afterburners going full blast. In other words, they were flying as fast as they could go. The aviation expert looked carefully and said that the black triangle craft was not flying aerodynamically. This is something that is only possible for a conventional manmade craft equipped with jets pointing up, down, and all around. It is sort of like some spacecraft that we have launched with tiny rockets to fine tune their trajectory, but with rockets providing vertical lift, forward motion, and directional control. The black triangle exhibited none of these required rocket or jet exhausts. The black triangle made a flat 90° turn at high speed, which is impossible at any speed for an aircraft flying aerodynamically. The jets had to bank hard to try to maintain contact with the black triangle, as the turning maneuver left them far behind. They were all just about out of camera range at that point. This flat, non-aerodynamic turn maneuver has been reported for UFOs since the 1940s and for the black triangles since they first appeared en masse in the late 1980s over Belgium. This maneuver creates a sizable G-force problem for any human pilots that might be on board, the documented acceleration from other cases of black triangles taking off from essentially a hovering position creates a huge G-force problem as well, and the black triangles have no recognizable means of propulsion, as they fly silently without visible rocket or jet exhausts. This all argues that the black triangles are not manmade or flown by humans; unless we believe that the military has obtained anti-gravity technology somehow. I am not ready to believe that just yet. In any event, it makes no sense for us to have this technology and not be currently using it on the battlefield or elsewhere, as stated by Alexander in his 2011 book. One other point: Atmospheric plasmas cannot flee from jet interceptors nor can they be escorted by them, so the UK Ministry of Defense's plasma explanation for black triangles is, to put it politely, rubbish.

One other possibility that we touched on earlier in the book was the idea that UFOs may be creatures/energy beings or some sort of life forms or things that live in the air. It is not that unreasonable to speculate that some unexplained UFO reports could be due to an unknown life form. This hypothesis is credited to Trevor J. Constable, who called the unknown life forms "Critters" or "Heat Critters." [167] It might be possible if there are no physical remains to fall to the earth when they die, or possibly they completely disintegrate in the air before reaching the earth. Otherwise, some of them would likely fall to the ground and be discovered, but the probability of recovery would be low if the population were quite low. If they were purely energy beings, there may not be a carcass to recover. The idea has some appeal, but testing it or studying it might require a lifetime of fieldwork without any results.

There are other hypotheses out there, such as the "Cosmic Trickster" and "Ultraterrestrial" hypotheses that were/are promoted by John Keel, Jacques Vallee, Robert Anton Wilson, and Terence McKenna. [168] The basic tenet of the hypothesis is that UFOs have an objective reality, but humans cannot fully comprehend or understand it. This hypothesis says basically that the same thing behind UFOs today was what was behind angels, demons, fairies, and other supernatural beings reported in the past. They're right; I don't understand it. If beings from hidden dimensions and parallel universes cannot explain angels, demons, and supernatural beings, what is there that can? To accept this we have to say that there is a supernatural realm to our world that is un-described as yet by physics, or that cannot be described by physics.

Here's where I stand on possible explanations of truly unexplained UFO sightings with multiple credible witnesses and other supporting evidence:

Possible Explanations for Unexplained UFO Sightings
(Sightings that cannot be explained after thorough investigation)

Explanation	Comment
Psychological, Mistaken Observers, Hoaxes	Already Weeded Out by Investigation
Atmospheric Life Forms/Energy Beings	Do Not Know if Possible or Not
Ball-lightning/earth-lights/atmospheric plasmas	Proven Responsible for Some Nocturnal Lights
Supernatural Beings	Disputed Evidence
Advanced Earth Civilizations pre-dating Humans	Possible, very improbable
Beings/Craft from another dimension/parallel Universe	Possible, String & M-brane theory
Extraterrestrial Space Travelers, ETH	Possible, Diffusion theory of (The ET Hypothesis) Galaxy Exploration

Notice that I included advanced Earth civilizations predating the current human civilization as a possible explanation. Most people in the UFO arena do not include this extremely improbable explanation, but we are talking about something that most people would designate as extremely improbable. I harbor an irrational fondness for this particular explanation. If we are listing possibilities, why not include it? What if UFOs are just the comings and goings of very advanced beings that arose right here on this earth that usually do not want to have open direct contact with humanity.

Notice that I included mainstream science's explanations for UFOs and listed them as already weeded out by thorough investigation. Science, skeptics and debunkers tend to want to embrace the psychological explanation for most UFO sightings whenever possible. Mainstream science tried to paint everyone who saw or reported a UFO as "nut cases," even though highly credible people have reported seeing UFOs, including two individuals, Carter and Reagan, who went on to become president of the United States. In those cases, mainstream science says that they were simply mistaken. If all else fails, cry hoax. There you have the scientific establishment's response to UFOs: "nut cases," mistaken witnesses, and hoaxes.

However, it was mainly the editors of scientific journals that refused to publish anything about UFOs and vehemently attacked UFOs in editorials over the years. Very few people are as deeply bunkered in type 2 error thinking as are scientific journal editors and for good reason, because they and their reviewers are responsible for maintaining the high scientific standards of their journals. They cannot allow themselves to commit a type 1 error of accepting something bogus into the literature (accepting a false positive). But something incredible happened on December 31, 1978 near New Zealand, and this was an incredible UFO sighting from a freighter aircraft with eight witnesses in all in the air and on the ground, including a TV reporter and a cameraman with 16 mm film and audio recordings onboard the aircraft. The case included radar-visual conformation, and a color film of the unknown light. And in 1979, something perhaps even more incredible happened. Dr. Bruce Maccabee, USN Optical Physicist and UFO investigator on his own time, was able to publish an article in the scientific journal *Applied Optics* about the optical power output of the New Zealand UFO. [169] Maccabee describes the entire publication process in an article in *Journal of Scientific Exploration* in 2005. [170] The title of this article is called "Challenging the Paradigm," and Maccabee refers to what he did as denting the paradigm or mindset against UFOs, but that this anti-UFO mindset is very much still intact.

Maccabee describes how the chief debunkers for the New Zealand UFO proposed that it was due to a swarm of electrically charged insects flying between clouds in the atmosphere, or that it was due to light reflection or refraction from the powerful lights of Japanese squid fishing boats. All of these proposals were easily shown to be incapable of duplicating the lights behavior, but the debunkers were given much more leeway, and Maccabee almost did not get to publish a rebuttal to his critics. An article in another journal claimed that the New Zealand light was due to atmospheric refraction based upon nothing but newspaper articles stating that as a possibility. Even though Maccabee had thoroughly investigated and shown that the lights could not be atmospheric refraction, he was not allowed a rebuttal. An author that was anti-UFO was allowed to publish an article that amounted to explanation without investigation and Maccabee, who had thoroughly investigated the incident, was rejected.

And so it goes, except that Maccabee and others struck again in 2005, and in a pretty incredible fashion. Deaedorff, Haisch, Maccabee, and Puthoff published an article in the *Journal of the British Interplanetary Society* with the title "Inflation-Theory Implications for Extraterrestrial Visitation." [171] Inflation-Theory is now the accepted version of the Big-

Bang-Theory of the universe. Basically modern cosmology says that the universe and the Milky Way Galaxy should be teeming with planets that support life. Also, when combined with string theory and M-brane theory, there should be parallel universes that we are literally sharing the same space with and that advanced occupants of these parallel worlds may be able to slip into our world and back out again. Physics has also advanced to the point of talking about traversable wormholes and other mechanisms of travelling about the galaxy or universe. The conclusion we could reach is that Earth should be being visited by many different advanced civilizations from outer and inner space. The fact that no advanced civilization has made formal contact is referred to as Fermi's paradox from a statement by Enrico Fermi in 1950 while at lunch with colleagues at Los Alamos National Labs. [172] Fermi asked, "Where is everybody?" He was basically asking, "Where are all the space aliens?"? Maccabee, et al, proposed that evidence of extraterrestrial visitation may be found in certain high-quality UFO reports, and these may satisfy the predictions of modern cosmology. Certainly the cast of alleged space visitors from UFO abduction reports would fit the bill rather nicely.

If you accept that some sort of extraterrestrial is here, the question always arises about why there has been no formal public contact. The answer usually comes back as if from a scene of *Wayne's World*. [173] "We're not worthy" or "We're not ready." I cannot argue that humanity would appear worthy or ready for formal contact from the viewpoint of a civilization perhaps a million years more technologically advanced than our current technology. We have made great progress in some ways, but any objective evaluation will find that we have not reached the destination we strive for. The problems of getting along with others in all of its aspects, and the elimination of world hunger and poverty, still have a long way to go. It seems that about one-half of the world's population is struggling to survive. In the United States of America, some 40-50% of the population struggles financially to make ends meet and cannot earn a true living wage, but many still have a standard of living on paper that is better than most of the people who have ever lived on Planet Earth, although some 17% of US residents go to bed hungry on any given night. This is a paradox of modern life. Life can be very good or it can be very bad in the most technologically advanced society on Earth.

In any grand overview, we have shown ourselves to be selfish and reckless in our stewardship of Planet Earth. What self-respecting advanced civilization would openly condone what they have witnessed over the last 100 years? We display what must appear from afar as endless warfare and endless attempts at suppression of the weak by the powerful in both a military sense and an economic sense. No outside neutral observer would think we were worthy of much of anything except to keep a leery eye upon, and would be sure to never engage us in a situation where we had the upper hand. Humanity's history has been to take what we want by force or by unethical maneuvering, whether legal or illegal.

Given that they have the technological advantage, why wouldn't technologically advanced space visitors squish us like bugs and take whatever resources they wanted as so many Hollywood movies portray? Why would they not display the same bad behavior that humanity has displayed over the centuries? After all, this is what more technologically advanced groups of humans did to less technologically advanced humans up until very recent times. Some are still trying to do this. Perhaps advanced visitors would not have to deal with us at all to take whatever they need. Perhaps advanced space travelers can mine asteroids and uninhabited planets to obtain what they need easier than dealing with us. Perhaps they see us as primitive curiosities to be studied, and at some distant time in the future they will openly contact our civilization when we have advanced to a state of being worthy. Or maybe it has nothing to do with any of this. Maybe the verified UFOs are from a source that we cannot fully understand or comprehend in our present condition, as some have suggested.

Perhaps the most troubling cases to many people are alien abduction cases (troubling whether real or imagined), which some say have a very ominous tone. Project Blue Book generally trash canned alien abduction reports in its day without consideration, but did list five UFO occupant cases in the list of unexplained cases. One of the occupant cases would be listed today as a Nordic entity case and bordered upon a contactee case if not an abduction. The Nordic entity was reported to speak to the witness in an unintelligible language, pat the witness on the back, jump into an egg-shaped UFO, and take off into the wild blue yonder.

The most typical type of alien reported in association with alien abduction cases in the US are the gray aliens with the huge all-black eyes. In the realm of true believers, some charge that the gray aliens are evil and are using human genetic material to create gray alien-human hybrid beings to save their own race. It is also believed by some that this hybrid race will come to dominate over humans and the earth.

There are also those that say the Grays are good and are watching over us, and their reported genetic experiments are meant to help mankind. Modern space-age religions have sprung up where the Grays serve the role of angels and the goal of everyone is to become worthy enough to stand in the presence of the being of eternal light and energy (God). In this way, the space alien religions come to resemble traditional religions. Some groups claim that the supreme religion is the space alien religion, and that the traditional human religions are just subsets of this supreme religion. [174]

I am not in the business of evaluating philosophy and religion, nor do I know how to evaluate such things. I do not know where to begin to test the various hypotheses that are involved in the space alien religions. All that I am certain of is that from my and my family's direct observations, some UFOs do appear to be real flying craft of unknown origin with capabilities beyond humanity's current level of technology. Furthermore, the UFO phenomena seen by my family was clearly related to the Mississippi cases reported in the1970s, and there appears to be a Mississippi UFO corridor where unexplained UFO phenomena occur regularly, even to the present day. The corridor should probably be enlarged a bit to include the Santa Rosa County/Gulf Breeze, Florida area. My hypothesis that UFO activity is usually greatest near heavily forested areas is borne out by the data from the entire United States for UFO hotspots and the Project Blue Book correlations. If someone else can show that they proposed the idea first, then I will acknowledge that in the future.

I invite anyone to purpose reasonable explanations for what was seen by my family. I reached out early on (1977) for the scientific and skeptical explanations, but none of the explanations that the skeptical community has produced to the present time even come close to explaining what was seen. The "silent "or lower noise helicopter explanation was proposed by an engineer aware of such craft, not a professional skeptic. This hypothesis was evaluated and shown not to be able to explain the sightings (when far enough away to be truly silent to an observer, it could not accelerate fast enough to move from a position north of the observer to a position due east in 10 seconds or less and when close enough to an observer to travel from north to east in 10 seconds, it would be close enough to make sufficient noise for the observer to hear it quite easily), but it is the best conventional explanation to be put forward so far for any of our observations. The "silent" helicopter explanation does not work at all for the cases where an object hovered without any sound 50 to 100 feet directly over the observers.

Contrary to what the skeptics may believe at this point, I can be convinced if they can put forth a logical, cohesive argument that explains the observations. Being told we were having psychotic episodes, or were mistaken, or that we are perpetuating a hoax are not viable explanations. The craft/lights we saw were flying noiselessly and maneuvering purposefully, apparently under intelligent control. That has to be the starting point for any acceptable explanation.

After a logical review of the Project Blue Book Unknowns and the earthquake lights data of Persinger and Derr, how can there be any skeptics left out there about the existence of some true unknowns flying around in our skies? Some are apparently flying craft and others balls of energy. There can still be skeptics, because each and every human being creates their own reality narrative, picking and choosing what to include as they go along. That reality narrative does not necessarily match absolute objective reality at all. Some may chose to ignore the correlation results concerning the Project Blue Book data that clearly indicate something real and unknown was flying around. It is their prerogative, but the deeper one goes into type 2 error territory (rejecting everything unusual regardless of the evidence) the further away from reality they may get. We should live on the boundary of the two opposing mindsets and be constantly reassessing where that boundary is in order to be anywhere near actual reality. The vast majority of us never make the effort to assess the world this way. We just find a comfortable spot until something jolts us to move one way or the other, and for most of us that jolt never comes. It is anybody's guess as to when mainstream science will come out of its type 2 error mentality and at least admit that UFOs are worthy of scientific investigation. How many more witnesses have to be told that they did not see what they saw? How many more honest people have to be belittled and ignored by the scientific and skeptical establishments? How many more stones have to fall from the sky?

Epilogue

The above sentence was the original ending of this book, but something incredible happened in the summer of 2013 while the book was at the editors. Clark called and said, "Things are starting to happen again." He went on to say, "The lights are back. Chase and I lived out there (Tiffany's house) for ten years and never saw anything even though we looked for the lights on many nights and saw nothing, but they're back now." Clark went on to tell what had transpired over the course of about eight days from June 14th to June 21st 2013 involving three different witnesses. The last sighting was just two days prior to his call and occurred on Friday June 21st at about 10:50 at night.

On the night of June 13 and 14, Tiffany's husband Mark was having trouble sleeping so about 1 AM he got up and went into the kitchen for water or a snack. His attention was drawn to lights flashing outside that were coming through the kitchen window. He looked out the window at the woods behind the house and saw a white light beam coming from something on the ground through the trees perhaps three or four hundred yards away. The house site is on a hill some forty feet in elevation higher than the woods that do not begin until a hundred yards or more down the hill. A normal person's eyes when standing in front of the kitchen window are nearly 50 feet higher than the ground elevation where the woods are located. It is similar to being in a 50 foot tower looking out over a level country side. He described the light as crazy in behavior. It was flashing in a sporadic manner from the ground upward. The pattern made no sense and was coming from a fixed point on the ground or near the ground. He went outside and started walking toward the light source. At some point, he had second thoughts about entering the woods not knowing what he might be getting into. Coyotes and the occasional Florida panther (a slightly smaller version of a mountain lion) have been seen on the property in the past and a black bear had been spotted in 2012 just a few miles away, not to mention snakes and every type of wildlife in between. He returned home and tried to wake up Tiffany who was nine months pregnant, but she could not rouse up to see the lights, which had been sporadically flashing for at least an hour at that point.

The baby was born on Monday June 17th and Mark's grandmother had come out the next day to help with the baby sitting chores. About mid-night on the 18th, she saw the random flickering light show as she looked out the kitchen window, only this time it moved as she watched from the ground to up in the air at about tree top level. She observed the strange light show for about 15 minutes.

These two sightings were unlike anything we had seen back in the 1970s and 80s. Clark had at first thought maybe they had seen something that our cousin, who owns the neighboring property, may have setup near his deer hunting stand down that way. Clark had come out on Friday the 21st to help some with babysitting and to check out the area behind Tiffany's house later that night. He checked the area out on his four wheeler during the day and found nothing had been added. There were no light stands and there was nothing to strongly reflect light either. The area had not been changed and there was not any sign of someone camping out in the woods. About ten o'clock that night, Clark went out to check on things fully expecting that he would find a mundane explanation for the flashing light show. He took a flashlight and a rifle with a scope for protection or possibly for dispensing the armadillo that had been tearing up the yard at night, but mainly for viewing the lights through the scope if they chose to appear. He set up in a deep shadow so as not to be seen, but he had worn a white tee shirt that could almost be seen even in the shadow because the moon was so bright. The night was clear and still and a very bright moon was out (the super bright moon of 2013 occurred the next night or so). Clark was soon reminded that he had forgotten to apply any mosquito repellent, and after about an hour, was ready to call it a night. The only thing he had seen was an airplane that had flown over about ten minutes before in a southwesterly direction, possibly toward New Orleans. It was about 10:50 PM when he saw, at an estimated distance of 300 to 400 yards away, a white light rise up a little above the tree tops and start flying slowly in his general direction. As it flew from the northwest toward the southeast as we had seen a number of times in the 1970s, Clark thought to himself in all of the times he had seen the light that he had never tried to photograph it. He grabbed his cell phone from his pocket, opened it, and set it on video. There was a flash of light from the cell phone lighting up his white tee shirt as he did this. He was glancing back and forth at the flying light as he was doing this and saw that it seemed to react to his cell phone light as it instantly began zigzagging and changed course to a more southerly course away from his position. There was no sound at any time. As he tried to frame the light in the viewfinder of the cell phone, the light slipped away or went out. When he looked for it with the naked eye, he could not locate it again. He did not get anything on video.

Skeptics say why don't we have more pictures and videos (actually there are some nearly every week) of UFOs since nearly everyone has a cell phone today. The answer is that the cell phone is a very poor choice for recording a light at night or something at a distance. Try to take a picture of the moon and see what you get. It is usually just a glob of light with no detail without the lighter and darker areas, and it looks much smaller and further away than what the eye sees. The typical cell phone is a very poor optical recording device for UFOs, but they may get better as their optics keep improving.

Due to the very bright moon, which was about an hour away from its maximum position in the sky that night (moonrise June 21st at that location was 6:22 PM CDT and moonset was 4:03 AM June 22nd), Clark was able to see a structure behind the light. He did not see much detail, but did get an estimate of the size. His impression was that the light itself was somewhat smaller than what we saw in the past. He estimated the size of the structure behind the light as about half the size of a refrigerator. Typically witnesses of phenomena of this type in controlled tests usually range in estimating something from one-half to double its' actual size. So we can say that the object was from about two feet across to something approaching ten feet across, but most likely was considerably less than ten feet across.

After Clark related these sightings to me, I checked the moonrise and moonset for the 21st and the wind speed and direction at the Jackson airport about 30 miles away. The wind speed at about 11 PM on the 21st was 4 mph from the east-southeast meaning that the object Clark saw was flying into the wind. This eliminates lighted weather balloons, kites, and anything wind driven. It was an unidentified object flying under its own power. Indeed, the lights appear to be back with a bang.

All three witnesses had seen the light phenomena originate at approximately the same location in the woods that they described as 300 to 400 yards away. Out of curiosity, I looked on Mapquest Satellite View to view the woods at that distance in a direct line of sight from the kitchen window. I was immediately surprised to see a large clearing in the woods some 520 yards away in a direct line of sight with the kitchen window. The clearing is some 500 feet long with widths that vary from 60 to 100 feet. I had been unaware of this clearing before. It is surrounded by thick woods in every direction although I later learned that there is a tractor path to it. I learned that the clearing is a rye grass field planted to attract deer and there is a deer stand nearby. The field is usually planted in October and hunters are routinely present during hunting season from November through December. The nearest house is about a quarter mile away and 400 of those yards are thick woods. The woods extend in every direction around the clearing up to about a half mile in one direction.

Something caught my eye on Mapquest as I zoomed in. A white structure came into view against the green color of the rest of the image. I later measured this structure using the distance indicator to have a 40 foot long base and a 40 foot high or higher by 8 to 12 feet wide "tower", for lack of a better word, rising from it. With the image zoomed to the max on my computer monitor, I looked through a magnifying glass to enlarge the image even more. The image looked quite strange. The base actually appears to have a pedestal sitting on it and above this pedestal is a sort of ball and socket joint that looks as if it could fit into the top of the pedestal. The rest of the image is completely bizarre. I could not make the image out to be any type of known building or heavy equipment that I am familiar with. The clearing and the white structure image is shown below along with another satellite image of the exact spot taken four months later showing the structure, or whatever it was, was gone by that time.

It took some effort to find the date of the Mapquest image and the later image. First I contacted Mapquest and they directed me to an imaging company called i-cubed. The i-cubed representative was most helpful. He quickly found out that the Mapquest image was taken in April 2007 by DigitalGlobe and that it was aerial photography instead of satellite imagery. I had inquired about getting prior images and later images and he directed me to Google earth Historical Images. The images for the clearing in the woods ranged from 1996

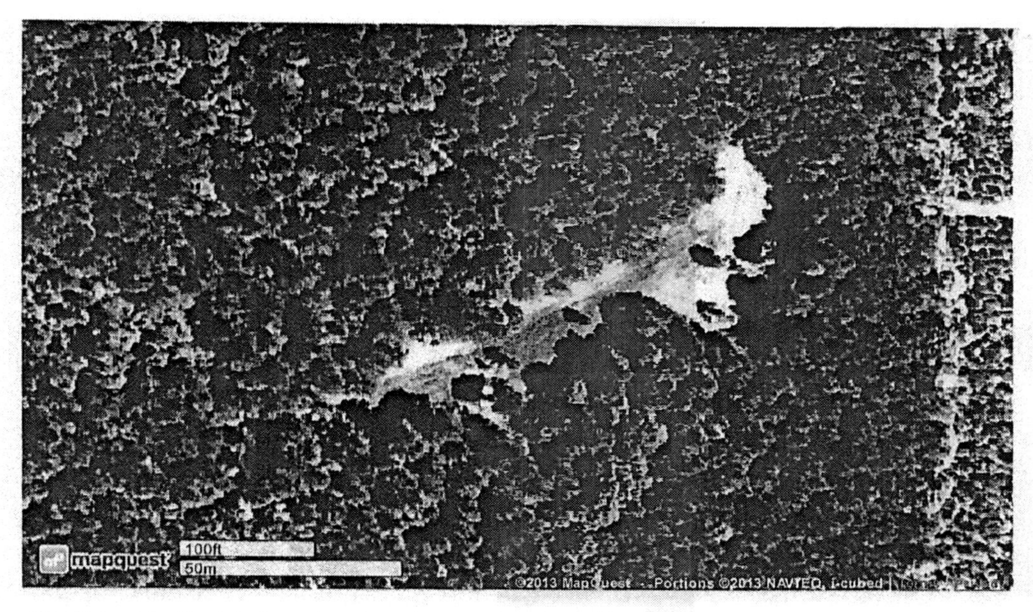

Mapquest image ©2013 Mapquest - Portions ©2013 NAVTEQ, i-cubed
All Images Used under Terms of Use Agreement

Google earth

Left: Mapquest April 2007 Right: Google Earth August 2007

to 2012. The Google earth site did not have the April 2007 image, but had images from June 2006 and from August 2007 just four months after the Mapquest image. The historical images showed that the white structure was not there in June 2006 or in August 2007. I have no explanation for what can give rise to such an image. It looks quite alien to me.

The August 2007 image of the clearing had its own unusual image, one that I almost missed. Looking at an overview I noticed that one of the trees located inside the clearing looked to be more than double in size in the August 2007 image than in the images before and after that date. Zooming in closer, I was able to see that the larger image was not a tree. It was almost white in color and it had tail fins in a double crescent configuration. There on the screen was an image of an unidentified flying craft above where a tree was in the other images. The craft was casting a longer shadow than the tree would have cast. Zooming in until the Google earth view started to transition from satellite to ground view, I stopped it halfway between. The view was now from over the shoulder of the craft showing it to be above the trees surrounding the clearing maybe 80 to 100 feet up, with its' image overlaying the shadow cast by some tall trees surrounding the clearing. The image looks similar to blended body and wing craft such as the recently retired Stealth Fighter (F-117A aircraft) and others. Clark pointed out the similarity of the image to that of the new Navy stealth drone aircraft, the X-47B, just making its debut in the summer of 2013. Another craft of this type is currently seen in some Boeing TV ads running in the summer of 2013, the Boeing blended body drone known as the X-48B. All these aircraft are drawn to scale in the overhead view in the second illustration below. The unknown does not match any of these. It is much smaller than the stealth fighter and the Navy X-47B drone, and is larger than the Boeing X-48B experimental aircraft. If this is the image of a known human manufactured craft, someone will recognize it. If it is a known craft, why would it be over this clearing in the woods, just above the tree tops, where it is so low in altitude? A flying craft that is significantly above the ground will show a separation between its image and its' shadow on the ground.

Also the summer of 2013 saw the publication of Don Donderi's book *UFOs, ETs and Alien Abductions: A Scientist Looks at the Evidence.*[175] (The title sounds somewhat similar to this book's title, which was chosen months ago.) Donderi pointed out on page 48 the similarity between the famous McMinnville, Oregon photograph taken by Paul Trent in 1950 and a photograph taken from a fighter jet of a UFO over France in 1957. The two photographs are virtually identical showing that these are authentic photographs. Klass spent a lot of time and energy trying to show the Trent photographs were faked. What a waste of time and effort, and he never mentioned the fighter jet photo. Authentic photographs exist and yet mainstream science and skeptics keep rejecting the evidence.

In our case we have an unknown flying craft's image on Google earth collected by DigitalGlobe for the USDA Forest Service Agency. I consider this to be an unimpeachable source. This is only the second image of an unknown to be made public from an aerial photographic

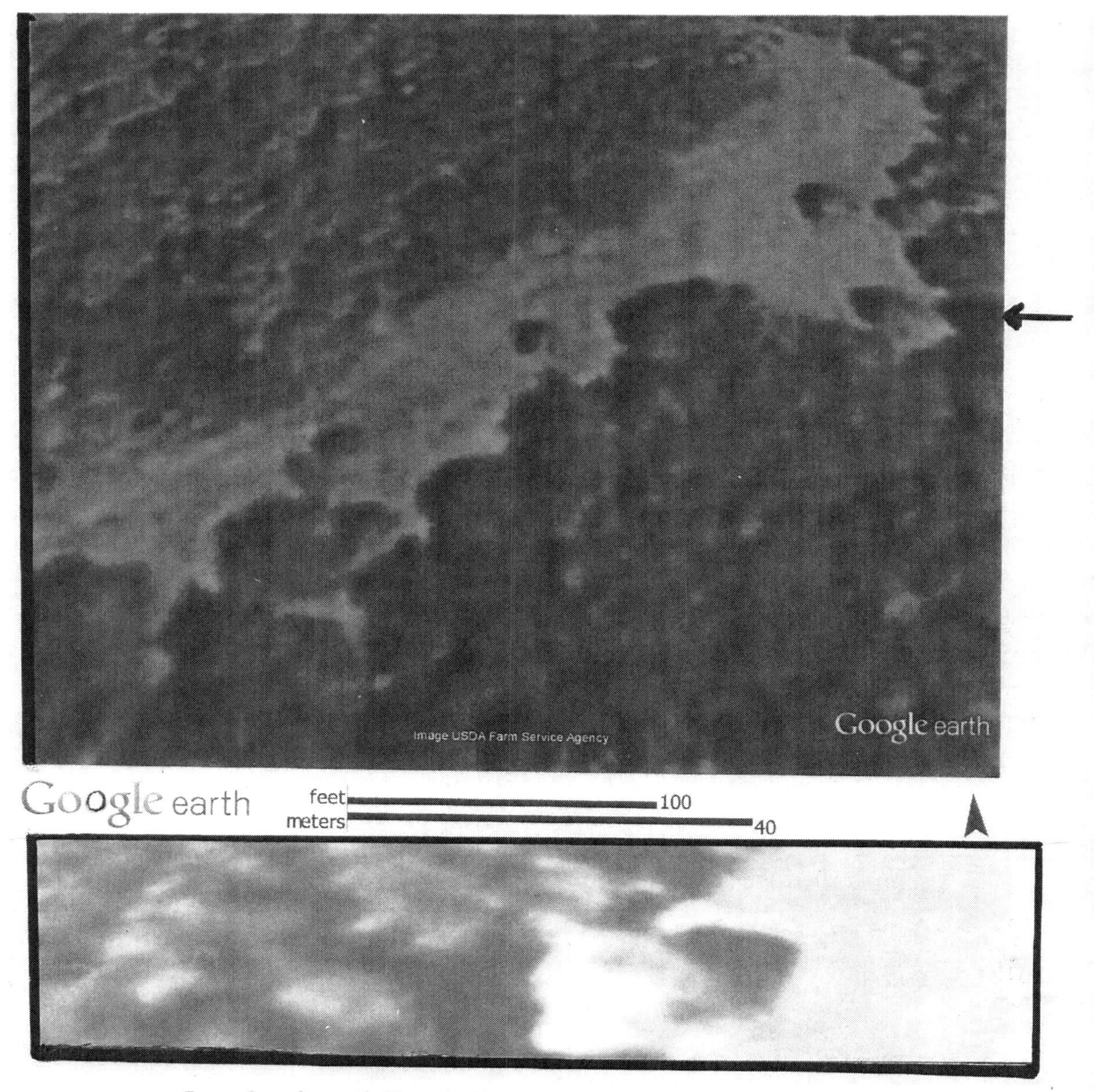

From Google earth Historical Images.
Acquired for the USDA Farm Service Agency by DigitalGlobe 8/7/2007.

Bottom View: Zoom in on the Unknown Craft from the Correct Viewing Angle According to the Author

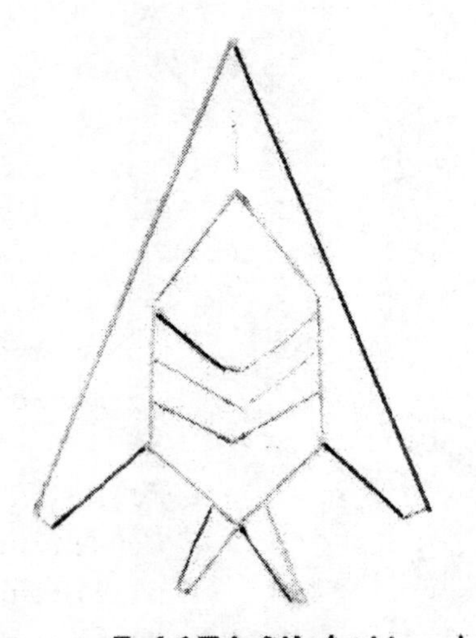

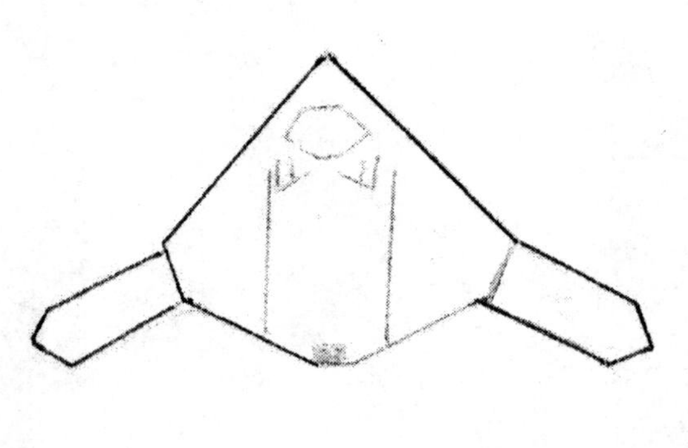

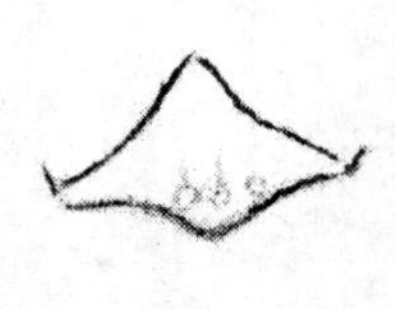

F-117A NightHawk
(Stealth Fighter retired in 2008)

Navy X-47B
(New Drone 2013)

Boeing X-48B
(Blended Body)

Unknown Craft over the Clearing in 2007

Blended Body and Wing Aircraft

(Overhead View, All to the Same Scale)

survey that I am aware of. The first was taken in 1971 in Costa Rica by a government plane using a high resolution camera on an official mapping mission. [176] The aerial photograph showed a very large circular UFO that was either entering or leaving Lago de Cote (Lake Cote). I suspect that there are probably a number of images of UFOs out there in satellite views released to the public that haven't been found as yet. It suddenly dawned on me that our various government security agencies probably have hundreds of these images on file. I would not be surprised if they had the capability to almost instantaneously direct a satellite to take a picture of something unusual showing up on radar. Photographs of UFOs have probably been taken in this way so many times by now that it is common place, but none of it will be released to the public.

I had to look at the image from all directions to determine what I believe to be the correct view of the object from the east, which is shown in a zoomed-in view above. I used the shadow to help determine the correct view. The illustration below shows the top view of the object and the shadow showing a conical section on the top of the object. From these, I determined a probable side view and a probable silhouette for the object. I measured the object to be about 30 feet long by 30 feet wide and possibly 20 feet thick. The depth is just a guess, but the other dimensions are reasonably firm.

From these views, I believe that an object similar to this could be what my parents saw backlit by the moon that night some 35 years ago. They probably would not have noticed the tailfins for an object that was approaching them and in view for about eight seconds total time, but the overall image matches up pretty well. Even the size of about 30 feet by 30 feet by 20 feet looks pretty large when viewed from about 60 feet directly underneath it.

The view of the craft that is in the satellite image on Google earth looks similar in some ways to the very first one drawn by Kenneth Arnold back in 1947. This is shown in the illustration below. Donderi includes the letter Arnold wrote to the Air Force showing a drawing of the object. Arnold later said that one of the nine objects he saw had the distinct double crescent shaped rear section (the same as the craft over the tree in the clearing). Over time all nine of Arnold's craft came to be depicted similar to the bottom more flashy image with the double crescent rear section. The 2007 craft is considerably smaller at a maximum of 30 feet in length compared to Arnold's estimate in the hundreds of feet range. The Project Blue Book explanation for Arnold's sighting was that he was seeing a mirage, which is highly unlikely due to the duration of the sighting and because the objects changed position relative to each other and to Arnold's plane.

So there you have it, unimpeachable photographic evidence of an object over the clearing in the woods in August 2007, but I kept looking, and I found more, much more. I found what looked to be some of the things in the April 2007 Mapquest image, only this time they were flying around. All were solid white in color and appeared to be flying just above the tree tops. One looked similar to an airplane with ridiculously small wings and an exaggerated tail

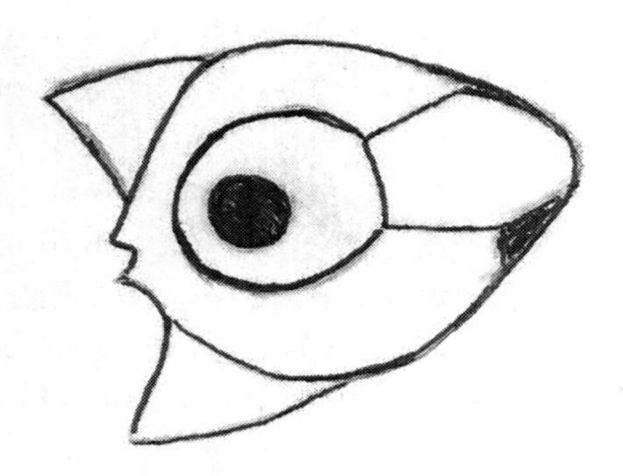

Author's Drawing of 2007 Object Seen From Above

Shadow Showing Conical Top and Notch

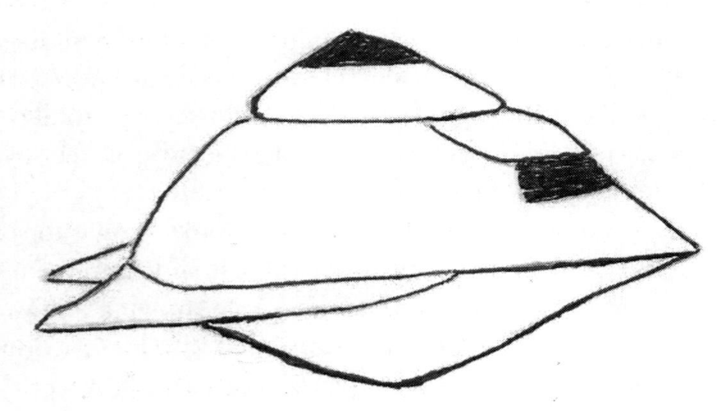

A Possible Side View Based On Top View And The Shadow

A Possible Silhouette For The 2007 Object

with tiny rear stabilizer wings. Another looked to be a platform with a pedestal on it, just as in the clearing, and two others appeared to be classic flying saucers with a copula on top. I was shocked one day to zoom in on Google earth Historical Images to an area I had not searched and see what appeared to be a gray alien standing next to the coolest looking bat-mobile type of vehicle that I have ever seen. My daughter pointed out that the alien image was huge, and it was at 15 feet in height. A human image located in the clearing in the woods on the same Google earth image measured 30 feet in height and did not cast a shadow. Everything else in the image, cars, houses, roadways, measured correctly. What was going on? I do not know. I am considering what to do next. I will probably contact a UFO organization at some point and disclose everything that I have found. I will gladly meet with skeptics, and let them use their own computers to access the images, so they will know everything is on the up and up. One puzzling thing has happened since I first viewed the August 7, 2007 satellite view on or about June 25, 2013. Google earth updated the image sometime on or before July 30, 2013. The brightness of the image has been lessened and the resolution has been blurred somewhat to the point that some of the unusual images cannot be seen clearly. I do not know why this change was made.

And there are more strange images that I have not disclosed here, and I am sure skeptics will say that it is easy to find strange images because we are viewing things from miles to many, many miles above the earth depending upon whether the image was taken by aerial photography or by satellite. However, I panned out and looked as much as 20 miles away in several directions on a number of the images. I searched for quite a long time and found nothing strange.

One final note, an October 2012 higher resolution image of the clearing looks deserted at first glance, but zooming in closer one can see a bright metallic object sitting on the ground. Zooming in to the max and using the ruler on the toolbar on Google earth, one can see a three foot long and three foot wide object tapered toward one end with a double crescent structure on the other. It appears that it could be a one-tenth scale miniature of the object seen over the clearing in the August 2007 image. It is the correct size to be the object seen by Clark on the night of June 21, 2013 flying in his general direction. Skeptics will probably say it is a discarded tractor fender, and maybe I am stretching the resolution a wee bit, but I can see it clearly enough and I believe most people will be able to as well.

Clark believes the clearing in the woods was there in the 1970s. It is now easy to see why we saw "The Light" so often. It probably was not that interested in us. It appears that its base of operations was in the clearing, which was less than one-half mile away. We could not help but see some of its comings and goings.

All the unusual satellite images taken together have forced me to change my perspective. I had assumed the objects responsible for "The Light" flew mostly at night, but the scene where four or five of these things were merrily flying all over in broad daylight has changed

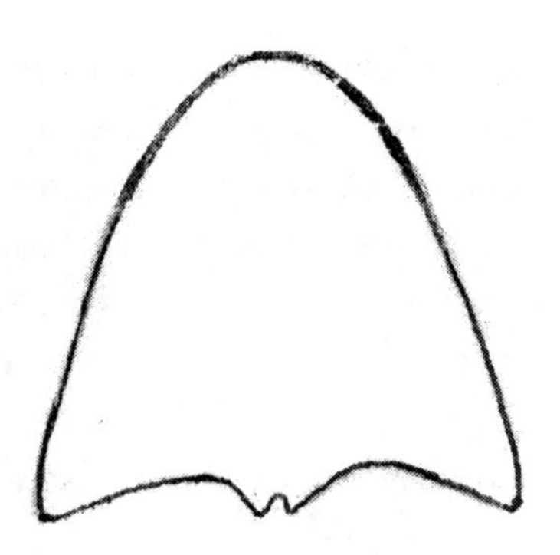

Author's Drawing of August 2007 Object
Viewed From Above

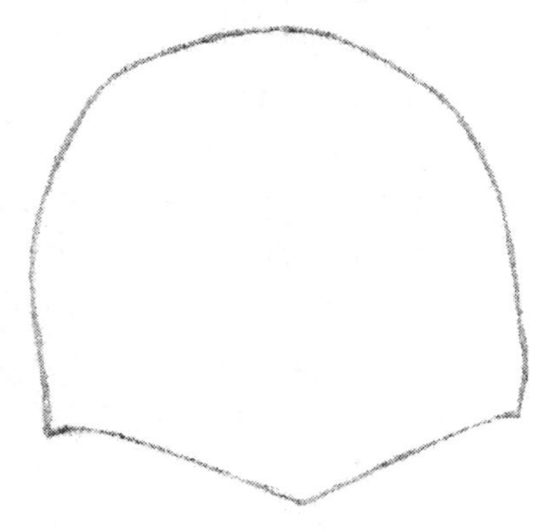

Kenneth Arnold's First Drawing in 1947

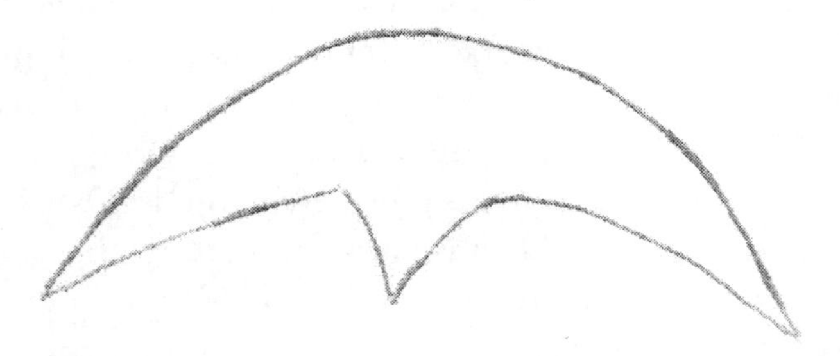

Highly Stylized Later Image of Arnold's Objects

my thinking. It dawned on me that a light at night draws attention. Our eyes are drawn to it, and it can be seen from a long distance away. The same craft, flying silently in daylight just above the treetops and crossing lightly travelled roads in heavily wooded areas, is probably less likely to be noticed than a bright light at night. All of these images, when taken together, look decidedly stranger than just humanoids from another planet exploring our world. Maybe Vallee and others are on to something in the cosmic trickster explanation, or, perhaps, Sagan was correct in saying that alien beings will be virtually incomprehensible. An uneasy feeling has come over me. I cannot shake the feeling that I am "outing" whatever it is that is sometimes in the forest clearing, but my father (1921-2005) can rest in peace, because someone will definitely believe us now.

References

1. Philip J. Klass, *UFOs Explained*, New York: Random House: 1974.

2. NOVA Online Interview with Philip J. Klass from the website www.pbs.org/wgbh/nova/aliens/philipklass.html. Accessed 08/11/2011.

3. Discussion of the Psychosocial Hypothesis for UFOs, PSH, and the Psycho-cultural Hypothesis, PCH, that became quite popular with some UFO researchers in the UK and France such as David Clarke, Hillary Evans, and Michel Monnerie. www.en.wikipedia.org/wiki/Psychosocial_hypothesis Accessed 11/30/2012.

4. John G. Fuller, *Incident at Exeter*, New York: G.P. Putnam's Sons; Reprint. Edition: 1966 and www.ufos.about.com/od/bestufocasefiles/p/exeter.htm Accessed 11/30/2012.

5. J. Allen Hynek, (1910-1986) Astronomer, Professor, Debunker, and Ufologist, www.en.wikipedia.org/wiki/J._Allen_Hynek Accessed 10/20/2011.

6. Donald H. Menzel, (1901-1976), Astronomy, Astrophysics, UFO Debunker www.en.wikipeda.org/wiki/Donald_Howard_Menzel Accessed 10/11/2011.

7. Problems With Steuart Campbell's "Amateur Science Solves the UFO Mystery" Another attempt to explain all unexplained UFO sightings as mirages. From Jerry Cohen's website www.cohenufo.org/amtr_hypoth.html#origarticle Accessed 7/20/2012.

8. Edward U. Condon (1902-1974), Nuclear Physicist, Radar Researcher, Quantum Mechanics Pioneer, Principal Author of the Condon Report, www.en.wikipedia.org/wikw/Edward_Condon Accessed 4/23/2012.

9. Robert J. Low's "Trick Memo" from 8/9/1966 can be viewed at www.nicap.org/dosc/660809lowmemo.htm Accessed 11/8/2012.

10. Donald E. Keyhoe (1897-1988), Marine Aviator, Aviation Author, Leading Ufologist (initially a skeptic) 1950s and 1960s, published the "Trick Memo" showing mainstream science's emphasis upon debunking and calling UFOs all in the mind before the project had started www.en.wikipedia.org/wiki/Donald_Keyhoe Accessed 10/11/2012.

11. Air Force wanted out of Project Blue Book www.ufocasebook.com/projectbluebook.htm Accessed 7/20/2012.

12. Philip Morrison, "Books: Other conceptions of the search for extraterrestrial intelligence," *Scientific American*, May 1975, pp. 117-118.

13. Philip J. Klass, *UFOs—Identified*. New York: Random House: 1968.

14. Paul Devereux and Peter Brookesmith, *UFOs and Ufology: The First Fifty Years*, London, UK: Wellington House: 1997.

15. Jerome Clark, "Phil Klass vs. The 'UFO Promoters'" *Fate*, February 1981, from the website: www.ufoevidence.org/documents/doc402.htm Accessed 11/08/2012.

16 Travis Walton, *Fire in the Sky: The Walton Experience*, New York: Marlowe & Co.: 1997.

17. Philip J. Klass, FBI file, *www.cufon.org* Accessed 07/25/2012.

18. UFO Researcher lists *www.noufors.com/who's_who_in_ufology.htm* and www.ufoevidence.org/researchers/page-1.htm Accessed 12/3/2012.

19. Ted Bloecher, *Report on the UFO Wave of 1947*, originally published in 1967, reproduced on the website in 2005, *www.noufors.com/Documents/ReportOnWaveOf1947.pdf* Accessed 12/3/2012.

20. Twining Memo, September 23, 1947 from Lieutenant General N. F. Twining to Brig. General George Schulgen, "The Reported Phenomena Are Real" from the website www.nicap.org/twining_letter.htm

21. Billy Booth, About.com Project Sign: UFO Study Group from the website www.ufos.about.com/od/projectsign/a/projectsign.htm Accessed 12/4/2012.

22. A. 1946: "Ghost Rockets" Over Scandinavia, The UFO Briefing Document Case histories from the website www.bibliotecapleyades.net/ciencia/ufo_briefingdocument/1946.htm Accessed 12/4/2012

    B. Liljegren, Anders & Svahn, Clas, "Ghost Rockets and Phantom Aircraft," paper in the anthology *Phenomenon - Forty Years of Flying Saucers*, New York: Avon Books: 1989.

    C. M. J Philippus, *Ghost rockets, foo-fighters and flying saucers*, Fort Lauderdale, Florida: PPI: 1992 (out of print).

23. Michael D. Swords, Project Sign and the Estimate of the Situation, from the website www.bibliotecapleyades.net/sociopolitica/sign/sign.htm Accessed 12/4/2012.

24. Edward J. Ruppelt, *Report on Unidentified Flying Objects* London: Victor Gollancz: 1956. 2nd ed, expanded, New York: Ballantine: 1960. First edition is on the website (www.nicap.org/rufo/contents.htm).

25. Karl T. Pflock, "The Best Hoax In UFO History?" 1997 from the website www.nicap.org/bhoax.htm Accessed 7/20/2012.

26. Flying Saucers CIA Declassified Document: Sept. 7, 1952, (security implications inherent in the flying saucer problem) from the website www.wanttoknow.info/ufos/19520907_flying_saucers_cia_document_declassified Accessed 12/4/2012.

27. Milton Brener, "Revisiting Edward J. Ruppelt," from the website www.ufodigest.com/article/revisiting-edward-j-ruppelt Accessed 12/4/2012. (Author of *Our Interplanetary Future: A UFO Primer For Skeptics*, BookSurge Publishing an Amazon.com Company: 2009).

28. The Robertson Panel from the website www.en.wikipedia.org/wiki/Robertson_Panel Accessed 7/23/2012.

29. *Project Blue Book Special Report Number 14*. New York: Battelle Memorial Institute for Project Blue Book (UFORI), 1955.

30. The Green Fireballs, from the website www.theufochronicles.com/2012/07/return-of-incredible-green-fireballs.html Accessed 12/4/2012.

31. Radiation Spikes associated with UFOs over National Laboratories:

A. Lawrence Fawcett and Barry J. Greenwood, *UFO Cover-up: What the Government Won't Say*, Austin, TX: Touchstone: 1990.

B. David Rudiak, "Radar Inspired National Alert December 6, 1950" from the website www.nicap.org/reports/rina3.htm Accessed 7/25/2012.

C. David Rudiak, "Rocket Science 101" 3/28/1996 page 4 on the website www.nacomm.com/news/rs101.htm Accessed 7/25/2012.

D. David A. Johnson, "Do Nuclear Facilities Attract UFOs?" on the website www.cufon.org/contributors/DJ/Do Nuclear Facilities Attract UFOs.htm Accessed 7/25/2012.

32. Stanton T. Friedman and Kathleen Marden, *Science Was Wrong: Startling Truths About Cures, Theories, and Inventions "They" Declared Impossible*, Pompton Plains, NJ: Career Press New Page Books, Paperback: 2010.

33. "The Impossible Rocks that Fell from the Sky" An excellent history of the scientific acceptance of meteors on the website www.unmuseum.org/rocksky.htm Accessed 7/28/2012.

34. Brian Marsden, Eugene M. Shoemaker (1928-1997) bio on the website www.2.jpl.nasa.gov/sl9/news81.html Accessed 12/4/2012.

35. Ted Bunn, Black Holes FAQ (Frequently Asked Questions) on the website www.cosmology.berkeley.edu/Education/BHfaq.html Accessed 12/4/2012.

36. Joint Army-Navy-Air Force Publication (JANAP) 146 issued in 1953 forbid military or civilian pilots from discussing UFO sightings with the public, from the website http://uforesearch.wikia.com/wiki/Joint_Army-Navy-Air_Force_Publication_146 Accessed 12/4./2012.

37. Air Force Regulation 200-2 (or AFR 200-2) version August 1954, supersedes AFR 200-2, 26 August 1953, Including Change 200-2A, 2 November 1953 (Actual Document) on the website http://en.wikisource.org/wiki/Air_Force_Regulation_200-2 Accessed 12/4/2012.

38. Stanton Friedman, *Flying Saucers and Science: A Scientist Investigates the Mysteries of UFOs*, Franklin Lakes, NJ: New Page Books Div., The Career Press, Inc.: 2008.

39. Kevin Randle, *Reflections of a UFO Investigator*. New York & San Antonio, TX: Anomalist Books: 2012.

40. "We Have Contact" All about the so called UFO contactees of the 1950s and later, on the website https://webspace.utexas.edu/cokerwr/www/index.html/sbrothers.shtml Accessed on 11/14/2012.

41. Carl G. Jung, *Flying Saucers: A Modern Myth of Things Seen in the Skies*, Princeton, NJ: Princeton University Press: 1979 (First published in USA in 1959).

42. Gordon Creighton," Dr. Carl Jung and the UFOs: The Real Story," *Flying Saucer Review*, Vol. 46/4, Winter 2001, pp. 7-11. From the website www.ufoevidence.org/documents/doc734.htm Accessed 10/18/2012.

43. Jacques Vallee, *Anatomy of a Phenomenon: UFOs in Space*, (Paperback) New York: Ballantine Books, Random House: 1974, p121.

44. Jon Voisey, "Was the 'First Photographed UFO' a Comet?" - actual photographs from 1883 by Jose Bonilla on the website www.universetoday.com/89911/was-the -first-photographed-ufo-a-comet/ Accessed on 7/15/2012.

45. Hynek's Astronomers Survey, website www.en.wikipedia.org/wiki/Unidentified_flying_object page 13, Accessed on 10/3/2011.

46. Peter Sturrock, Astronomers survey in Friedman, *Flying Saucers and Science*, p. 210.

47. Gert Helb and J. Allen Hynek, Amateur astronomer poll results reprinted in International UFO Reporter (CUFOS), May 2006, pp. 14-16.

48. Carl Sagan, "Direct Contact Among Galactic Civilizations by Relativistic Interstellar Spaceflight," *Journal Planetary and Space Science*, 1963, Vol. 11. pp. 485 to 498.

49. Seth Shostak (1943-), Senior Astronomer at the SETI Institute in Mountain View, CA from the website www.en.wikipedia.org/wiki/Seth_Shostak Accessed 10/11/2011.

50. Paul R. Hill, *Unconventional Flying Objects*, Charlottesville, VA: Hampton Roads Publishing Co.: 1995.

51. Richard F. Haines,"Photoanalysis of Digital Images of Anomalous Aerial Object Taken Sept. 17, 2010 Above Santiago Chile (CEFAA-Banderas)," International Air Safety Report: NARCAP IR-3, 2011 from the website www.narcap.org/ (National Aviation Reporting Center on Anomalous Phenomena) Accessed 11/22/2011.

52. Richard M. Dolan and Bryce Zabel, *A. D.: After Disclosure, When the Government Finally Reveals the Truth About Alien Contact*, Pompton Plains, NJ: Career Press, New Page Books: 2012.

53. John B. Alexander, *UFOs: myths, conspiracies, and realities*, New York: St. Martin's Press: 2011.

54. David S. McKay, Everett K. Gibson Jr., Kathie L. Thomas-Keprta, Hojatollah Vali, Christopher S. Romanek, Simon J. Clemett, Xavier D. F. Chillier, Claude R. Maechling, and Richard N. Zare, "Search for Past Life on Mars: Possible Relic Biogenic Activity in Martian Meteorite ALH84001" *Science*, 16 August 1996: Vol. 273 no. 5277 pp. 849-0.

55. Howard Blum, *Out There: The Government's Secret Quest for Extraterrestrials*, New York: Simon and Schuster: 1990.

56. Peter A. Sturrock, "Physical Evidence Related to UFO Reports: The Proceedings of a Workshop Held at the Pocantico Conference Center, Tarrytown, New York, Sept. 29-Oct. 4 , 1997" from the website www.jse.com/ufo_reports/Sturrock/toc.html Accessed 07/03/2000.

57. Chicago O'Hare Airport UFO Incident of Nov. 2006 from the website www.en.wikipedia.org/wiki/2006_O'Hare_International_Airport_UFO_sighting Accessed 12/5/2012.

58. A. Christian Wolff, Radar Tutorial & History of Radar on the website www.radartutorial.eu Accessed 10/23/2012.

B. Target Recognition Process, 12.4.3 Unknown Radar Targets, from the website www.scribd.com/91716918/Introduction-to-Radar-Target-Recognition Accessed 10/23/2012.

59. James E. McDonald, "Meteorological Factors in Unidentified Radar Returns," 14th Radar Meteorological Conference, American Meteorological Society, November 17-20, 1970 [held at the University of Arizona ,Tucson, Arizona].

60. Dr. Lynne D. Kitei, *The Phoenix Lights: A Skeptic's Discovery That We Are Not Alone*, Charlottesville, VA: Hampton Roads Publishing Company: 2000, 2004, 2nd ed. 2010.

61. Kevin D. Randle, "Carl Hart and the Lubbock Lights," from the website www.ufoscience.org/lubbock.html Accessed 7/26/2012.

62. Smithsonian Migratory Bird Center Fact Sheet- Neotropical Migratory Bird Basics, from the Website: www.nationalzoo.si.edu/scbi/migratorybirds/fact_sheets Accessed 3/25/2013.

63. Michael Swords and Robert Powell (Principal Authors), Clas Svahn, Vicente-Jaun Ballester Olmos, Bill Chalker, Barry Greenwood, Richard Thieme, Jan Aldrich, Steve Purcell (Contributors), *UFOs And Government: A Historical Inquiry*, San Antonio, TX, Charlottesville, VA: Anomalist Books: 2012.

64. Ground Saucer Watch validation of the Lubbock Lights, from the Lubbock Avalanche Journal website www.lubbockonline.com/stories/030109/fea_399395090.shtml Accessed 7/29/2012.

65. Stephenville, TX case, complete transcript of CNN *Larry King Live*, UFO Hunters Investigate Sightings, Aired July 11, 2008 - 21:00 ET, from the website www.ufo-

blog.com/pdf/CNN_Larry_King_Live_11th_July_2008_Stephenville_investigation.pdf Accessed 12/5/2012.

66. Stephenville, TX Radar Study from the website www.stephenvillelights.com/ Accessed 10/5/2011.

67. UFO Flap of the 1970's, Columbia, Mississippi Civil Defense Radar tracking of unknowns, from the website www.burlingtonnews.net/ufoflap-70s.html Accessed 8/30/2011.

68. Carl Sagan, *Cosmos*, New York: Random House: 1980, p. 303.

69. History Channel, *UFO Files*, "Russian Roswell," Weller/Grossman Productions, First Aired October 31, 2005.

70. History Channel, *UFO Files*, "Deep Sea UFOs," Weller/Grossman Productions, First Aired January 23, 2006. Also, History Channel, *UFO Hunters*, "USOs," Motion Picture Production, with William J. Birnes, Pat Uskert, and Ted Ackworth, First aired February 6, 2008.

71. Edward Bullard in Jerome Clark's, *The Emergence of a Phenomenon: UFOs from the Beginning Through 1959 (The UFO Encyclopedia)*, Detroit, Michigan: Omnigraphics: 1992.

72. "A Century of UFOs," *UFO Roundup*, Vol. 4, No. 36, December 30, 1999, from the website www.zetatalk.com/theword/tworx134.htm Accessed 3/11/2010.

73. Number of UFO reports over last 65 years, 150,000 from the website www. hyper.net/ufo/simmary.html page 3, Accessed 10/3/2011.

74. Jenny Randles in C. D. B. Bryan's, *Close Encounters of The Fourth Kind: A Reporter's Notebook on Alien Abduction, UFOs, and The Conference at M. I. T.*, New York: Penguin Books (Paperback): 1995, page 68.

75. The Autokinetic Effect (a small point of light against a dark featureless background appears to move), on the website www.en.wikipedia.org/wiki/Autokinetic_effect Accessed 3/29/2012.

76. A. Helicopter Speed Record 249 mph, Westland Lynx piloted by John Trevor Eggington (UK) 8/11/1986, on the website www.enwikipedia.org/wiki/Helicopter Accessed 10/28/2011.

B. A tiltrotor craft combination helicopter/airplane developed for the military by Bell Helicopter starting in 1981 and first flown in 1989 is capable of over 315 mph. Military model is called V-22 Osprey and civilian model is the Bell-Agusta 609. From the website www.helis.com/database/modelorg/842/ Accessed 12/5/2012.

C. Stealth Helicopter, www.globalsecurity.org Accessed 1/2/2013.

D. Robert K. Anoll and Edwin D. McConkey, "Rotorcraft Acceleration and Climb Performance," DOT/FAA/RD-90/6, 1991. From the website www.dtic.mil/cgi-bin/GetTRDoc?/AD=ADA243737 Accessed 1/6/2013.

E. Michael Falco and Roger Smith, "Influence of Maneuverability on Helicopter Combat Effectiveness," 1982. From the website www.ntrs.nasa.gov/archive/nasa/casi/19820015338_1982015338.pdf Accessed 1/6/2013.

F. Damping of Sound Level with Distance, www.sengpielaudio.com/caculator.htm Accessed 1/14/2013.

77. Julius Henry "Groucho" Marx, (1890-1977), Comedian, Master of quick witted replies, TV, movies.

78. Leah Berkman, "What is the Limit of Resolution of the Human Eye?" on the website www.ehow.com/about_6603780_limit-resolution-human-eye_.html Accessed 7/31/2012.

79. Length of Chinook Helicopter (CH-47D), www.helicopterpage.com/html Accessed 2/27/2012.

80. Black Helicopter Conspiracies are probably DEA Helicopters, from the website www.urbandictionary.com/define.php?term=black+helicopter+theory Accessed 12/5/2012.

81. UAV (Predator Drones) Developed in the 1990s, from the website www.fas.org/irp/program/collect/uav.htm Accessed 12/5/2012.

82. A. Edmund T. Rolls And Martin J. Tovee, "Processing speed in the cerebral cortex and neurophysiology of visual masking," *Proceedings of the Royal Society of London*, B, 1994, pp. 9-15, from the website www.jstor.org Accessed 6/4/2012.

B. Martin J. Tovee, "How Fast is the speed of thought?" *Current Biology*, 1994, Vol. 4 No. 12, pp. 1125-1127.

C. R. VanRullan and S. J. Thorpe, "The time course of visual processing: from early perception to decision-making" (Abstract), *J. Cogn. Neurosci.*, 2001 May 15; 13(4): 454-61 from the website www.ncbi.nim.nih.gov/pubmed/11388919 Accessed 6/4/2012.

83. A. William J. Birnes, *The UFO Magazine: UFO Encyclopedia*, New York: Pocket Books: 2004, pp. 33-36.

B. Also see Isaac Koi's UFO website(encyclopedic listing of top UFO cases and references) www.isaackoi.com/ufo/19801227-rendlesham-forest-incident.html Accessed 3/14/2012.

84. A. The Malmstrom AFB UFO/Missile Incident in central Montana in 1967 from the website www.cufon.org/cufon/malmstrom/malm1.htm Accessed 11/8/2012.

B. The alleged Warren AFB UFO/Missile incident in Wyoming in 2010 from the website www.dailymail.co.uk/news/.../UFO-cause-power-failure-Wyoming-nuclear-missile-base-say-technicians.html Accessed 11/8/2012.

C. Retired Military Officers: UFOs Eyed U.S. Nuclear Facilities, Missile Silos www.news.blogs.cnn.com/2010/09/27...el-say/?hpt=T2 Accessed 11/8/2012.

85. Mississippi UFO Conference April 2011, from the website www.msufo.com/?page_id=106 Accessed 9/11/2011.

86. A. Project Blue Book "Unknowns" from the website: www.nicap.org/bluebook/bluelist.htm Accessed 1/15/2013.

B. Brad Sparks, A Comprehensive Catalog of 1500 Project Blue Book UFO Unknowns has been or is being compiled. Work in Progress Version 1.7. Dec 31, 2003.

C. Don Berliner, "UFO Reports Marked as Unidentified" Compiled for the Fund for UFO Research (with brief synopsis of each case) from the website: www.oocities.orgsecret-zonegr/projectbluebook.htm Accessed 3/7/2013.

87. Charles Hickson and William Mendez, *UFO Contact at Pascagoula*, Gautier, MS: Charles Hickson: 1983.

88. Joe Eszterhas, "The Claw Men from the Outer Space," *Rolling Stone*, January 17, 1974, pp. 27, 38-47.

89. Gary and Ruth McDowell, *Mississippi Secrets*, Bloomington, IN : IUniverse, Inc.: 2007, 2011.

90. Flora, Mississippi Incident, "UFO Spotting Focuses on Deputy Creel," Madison County Herald (Canton, MS), February 17, 1977. From the website www.ufoevidence.org/cases/case425.htm Accessed 9/18/2011.

91. Billy Booth, "1959 - The Papua, New Guinea UFOs" April 5, 1959 (Father Gill Case) from the website www.ufos.about.com/od/bestufocasefiles/p/papua.htm Accessed 12/6/2012.

92. Taylorsville, MS Incident, "Officer Calls UFO 'Work of the Lord'," *Jackson Daily News*, May 25, 1977, from the website www.msufo.com/?page_id=106 Accessed 9/11/2011.

93. Petal, MS Incident, October 7, 1973 from Hickson and Mendez, *UFO Contact at Pascagoula*, pp. 152-55. Also see the website www.theblackvault.com/encyclopedia/documents/MUFON/Journals/1974/February_1974.pdf page 16, right column accessed 12/6/2012.

94. *Fire in the Sky* Motion Picture, distributed by Paramount Pictures, 1993.

95. John G. Fuller, *The Interrupted Journey*, (*Incident at Exeter* two volume set) New York: MJF Books: 1966 (original copyright).

96. Wisconsin Department of Natural Resources *www.dnr.wi.gov/topic/landscapes/index.asp?mode=detail&landscape=11* Accessed 12/6/2012.

97. Val Johnson UFO case from the website www.ufos.about.com/od/physicalproofcases/p/valjohnson.htm Accessed 12/6/2012.

98. Gulf Breeze, Florida UFO Sightings from the *website www.ufocasebook.com/gulfbreeze.html* Accessed 12/6/2012.

99. Frank B. Salisbury, *The Utah UFO Display*, Springville, Utah: Bonneville Books/Cedar Fort Inc.: 2010 (First published by Devin-Adair Publishing Co., 1974).

100. Colm A. Kelleher and George Knapp, *Hunt For The Skinwalker: Science Confronts the Unexplained at a Remote Ranch in Utah*, New York: Paraview Pocket Books: 2005.

101. Donald H. Menzel and Ernest H. Taves, *The UFO Enigma-the Definitive Explanation of the UFO Phenomenon*, New York: Doubleday: 1977. Also see the website *www.isaackoi.com/ufo-books/menzel-donald-h-and-taves-ernest-the-ufo-enigma.html* Accessed 12/6/2012.

102. Budd Hopkins, *Missing Time: A Documented Study of UFO Abductions*, New York: Richard Marek Publishers: 1981.

103. Michael A. Persinger and John S. Derr, "Luminous Shapes with Unusual Motions as Potential Predictors of Earthquakes: A Historical Summary of the Validity and Application of the Tectonic Strain Theory," *International Journal of Geosciences*, 2013, 4, 387-396.

104. Michael A. Persinger (1945-) neuroscience researcher and university professor, Persinger's God Helmet, on the website www.en.wikipedia.org/wiki/Michael_Persinger Accessed 3/29/2012.

105. M. Larsson, D. Larhammarb, M. Fredrikson, and P. Granqvist, "Reply to M.A. Persinger and S. A. Koren's response to Granqvist et al. 'Sensed presence and mystical experiences are predicted by suggestibility, not by the application of transcranial weak magnetic fields'," *Neuroscience Letters* 380 (3): 348–350,2005.

106. Tore Wessel-Berg, "Ball Lightning and Atmospheric Light Phenomena: A Common Origin," *J. Scientific Exploration*, Vol. 18, No. 3, pp. 439-481, 2004.

107. Tore Wessel-Berg, "A Proposed Theory of the Phenomena of Ball Lightning," *Elsevier, Physica D, volume 182, pp. 223-253.* 2003.

108. Ball Lightning discussed by Wikipedia, At least 10 different hypotheses are discussed, but not Wessel-Berg's, www.en.wikipedia.org/wiki/Ball_lightning Accessed 6/1/2012.

109. Peter A. Sturrock, Editor, *The UFO Enigma: A New Review of the Physical Evidence*, New York: Warner Books: 2000.

110. Ministry of Defense, MOD, UK, (Secret Document Now Unclassified 2006) "Unidentified Aerial Phenomena in the UK Air Defence Region: Executive Summary," Scientific and Technical Memorandum-No. 55/2/00 (Claims that Black Triangle UFOs are structured atmospheric plasmas caused by meteors breaking up in the atmosphere).

111. Magnetic field Strength varies by 1/distance cubed for a magnetic dipole, from the website *www.ccmc.gsfc.nasa.gov/RoR_WWW/presentations/Dipole.pdf* Accessed 12/6/2012.

112. Wilson, S. C. & Barber, T. X. (1983). "The fantasy-prone personality: Implications for understanding imagery, hypnosis, and parapsychological phenomen*a.*" In, A. A. Sheikh (editor), *Imagery: Current Theory, Research and Application* (pp. 340-390). New York: Wiley: 1983.

113. Magnet Field Strengths, of the earth, of a refrigerator magnet, of a sunspot, of an MRI, on the website www.en.wikipedia.org/windex.php?title=Orders_of_magnitude_(magnetic_field)&printable=yes Accessed 10/13/2012.

114. Benjamin Crowell, "Lectures on Physics: Hallucinations during an MRI Scan" from the website www.vias.org/physics/example_4_7_o1.tml Accessed 10/13/2012.

115. "Intelligence of Crows Rivals Chimps?" The Nature of Hiking website www.thenatureofhiking.com/intelligence-of-crows.html#.UMIbIYPAdfx Accessed 12/7/2012.

116. A. Trevor James Constable, *Sky Creatures-Living UFO's*, New York : Pocket Book Library: 1978.

B. Trevor J. Constable's "Heat Critters" www.fimufon.wordpress.com/ufology/ Accessed 12/7/2012.

117. Marfa Lights Research website www.nightorbs.net/ Accessed 12/7/2012.

118. Anthony J. Crone and Russell L. Wheeler, Open-File Report 00-260, "Data for Quaternary Faults, liquefaction features, and possible tectonic features in the Central and Eastern United States, east of the Rocky Mountain front," U. S. Department of the Interior, U. S. Geological Survey, 2000.

119. Class B Fault definition from Crone and Wheeler, 2000, P. 5. Geologic evidence demonstrates the existence of a fault or suggests Quaternary deformation, but either (1) the fault might not extend deeply enough to be a potential source of significant earthquakes, or (2) the currently available geologic evidence is too strong to confidently assign the feature to Class C , but not strong enough to assign it to Class A.

120. Busiest UFO Period in History for the Deep South was Oct. 10-17, 1973, from the website www.larryhatch.net/LAMSAR.html Accessed on 12/7/2012.

121. GeoNet: Gravity and Magnetic Dataset Repository, Pan American Center for Earth and Environmental Sciences (PACES), USGS, website www.irpsrvgis00.utep.edu/repository-website/Default.aspx Accessed 5/22/2012.

122. The Yakima UFO Sightings, Discussion of Yakima Indian Reservation UFOs, www.abovetopsecret.com/forum/thread663961 Accessed 05/11/2013.

123. Phil Patton(Principal Reporter) with David Coburn, Erin McCarthy, Joe Pappalardo, and Erik Sofge, "U. S. Map of the Top UFO Hotspots and How to Report a Sighting," *Popular Mechanics* website, December 18, 2009. www.popularmechanics.com/technology/aviation/ufo/4304768 Accessed 05/11/2013.

124. Greg Long, *Examining The Earthlight Theory: The Yakima UFO Microcosm*, The J. Allen Hynek Center for UFO Studies: Chicago, IL: 1990.

125. Noe Torres and Ruben Uriarte, *Aliens in the Forest: The Cisco Grove UFO Encounter*, Roswell Books: Edinberg, TX: 2011.

126. B. D. Gildenberg, "The Cold War's Classified Skyhook Program: A Participant's Revelations." *Skeptical Inquirer*, Vol. 28.3, May/June 2004. From the website www.csicop.org/si/show/cold_warrsquos_classified_skyhook_program/ Accessed 8/1/2012.

127. National Investigation Committee on Aerial Phenomena, "United States Air Force Projects Grudge and Bluebook Reports 1-12," Washington, D.C., 1968. From the website: www.nicap.org/docs/pbb/nicap_pbr1-12_srch.pdf Accessed 3/7/2013.

128. J. Allen Hynek, Philip J. Imbrogno, and Bob Pratt, *Night Siege: The Hudson Valley UFO Sightings*, St. Paul, MN: Llewellyn Publications, 2nd Ed. Expanded & Revised: 1998.

129. C. D. B. Bryan, *Close Encounters of The Fourth Kind: A Reporter's Notebook on Alien Abduction, UFOs, and the Conference at M. I. T.*, New York: Penguin Books (Paperback): 1995.

130. Susan A. Clancy, *Abducted: How People Come to Believe They Were Kidnapped by Aliens*, Cambridge, Mass.: Harvard University Press, 2005.

131. Joe Nickell, "Investigative Files: Catching Ghosts," *Skeptical Inquirer*, Volume 18.2, June 2008.

132. The Belgium UFO Wave of 1989 on the website www.ufocasebook.com/Belgium.html Accessed 12/10/2012.

133. Ronald K. Siegel, *Fire in the Brain: Clinical Tales of Hallucination*, New York: Plume (Div. Penguin Books): 1993.

134. Peter Brookesmith, *Alien Abductions*, New York: Barnes & Noble (First published in UK by Brown Packaging Books, Ltd.): 1998.

135. Isakower Phenomenon from Dictionary of Hallucinations www.hallucinations.enacademic.com/1011/Isakower_phenomenon Accessed 12/10/2012.

136. Alvin H. Lawson, "Hypnosis of Imaginary UFO 'Abductees'"," in Curtis Fuller (Ed.), *Proceedings of the First International UFO Congress*, New York: Warner Books: 1980.

137. David Jacobs, *Secret Life: Firsthand Documented Accounts of UFO Abductions*, New York: Simon & Schuster: 1992.

138. "What is Narcolepsy?" National Heart and Lung Institute, Narcolepsy, Hallucinations (Vivid Dreams), and Sleep Paralysis from the website: www.nhlbi.nih.gov/health/health-topics/nar/ Accessed 4/30/2012.

139. John A. Adam, "The Mathematical Physics of Rainbows and Glories," Physics Reports, Vol. 356, 2002, pp. 229-365.

140. Brian Dunning, "Illuminating the Fatima 'Miracle of the Sun'" from the website www.skeptoid.com Accessed 9/21/2012.

141. "Miracle of the Sun," Fatima, Portugal, 10/13/1917, from the website www.en.wikipedia.org/wiki/Miracle_of_the_Sun Accessed 9/21/2012.

142. Ray Stanford, *Fatima Prophecy: Days of Darkness Promise of Light*, Austin, TX: Assoc. Understanding Man: 1974.

143. Joe Nickell, "The Real Secrets of Fatima," *Skeptical Inquirer*, Vol. 33.6, November/December 2009. From the website www.csicop.org/si/show/real_secrets_of_fatima Accessed 9/21/2012.

144. Stephen Benedict-Mason, PhD., Fatima Skeptic website: www.psychologytoday.com/blog/look-it-way/200910/the-miracle-fatima Accessed 9/21/2012.

145. Joe Nickel, *Looking for a Miracle: Weeping Icons, Relics, Stigmata, Visions, and Healing Cures*, Amherst: Prometheus Books: 1993, pp. 176-181.

146. The Battle of Los Angles Feb. 23, 1942, "Best Documented Cases for UFO and Alien Encounters" www.churchofcriticalthinking.org/alien_visitation.html Accessed 10/28/2011.

147. Sy-Fy Channel, *Fact or Faked: Paranormal File*, Base Productions, Episode: "The Real Battle of LA," first broadcast 3/23/2011.

148. Bruce Maccabee, "The Battle of Los Angeles: Photo Analysis,"www.brumac.8k.com/BATTLEOFLA/BOLA1.HTML Accessed 1/26/2012.

149. Colonel H. M. McCoy, Chief of Intelligence USAF Air Material Command, Wright-Patterson AFB, Briefing to the Air Force Scientific Advisory Board on 3/17/1948, "…unidentified flying objects or discs …can't be laughed off. We have over 300 reports which have not been publicized in the papers from very competent personnel… I cannot tell you how much we would give to have one of those crash in an area so that we could recover whatever they are." Published by *Skeptics UFO Newsletter*, No. 39, May 1996 in the attempt to

discredit the Roswell Incident. From Devereux and Brookesmith, *UFOs and Ufology*, p. 135. (It proves UFOs were considered to be real by USAF in 1948. Need to know security clearance would be needed to discuss Roswell, if it had happened, and most of the audience did not have that.)

150. Charles Berlitz and William L. Moore, *The Roswell Incident*: New York: Grosset & Dunlap (First Edition): 1980.

151. A. Kevin D. Randle and Donald R. Schmitt, *UFO Crash at Roswell*, New York: Avon: 1991.

B. Kevin D. Randle and Donald R. Schmitt, *The Truth about the UFO Crash at Roswell*, New York: Avon: 1994.

C. For the controversy about Donald R. Schmitt see: www.roswelfiles.com/storytellers/RandleSchmitt.htm Accessed 12/13/2012.

152. Stanton T. Friedman and Don Berliner, *Crash at Corona: the U.S. military retrieval and cover-up of a UFO*, Saint Paul, MN: Paragon House: 1992.

153. Kal K. Korff, *The Roswell UFO Crash: What They Don't Want You to Know*, Prometheus Books: 1997.

154. A. Robert A. Galganski, "An Engineer Looks at the Project Mogul Hypothesis," on the website www.cufos.org/ros4.html Accessed 8/1/2012.

B. Dave Thomas, "The Roswell Incident and Project Mogul," Skeptical Inquirer, Vol. 19.4 July/August 1995 on the website www.csicop.org/si/show/roswell_incident_and_project_mogul/ Accessed 8/1/2012.

155. Chase Brandon Ex-CIA Agent, "Roswell UFO Was Not of This Earth and There Were ET Cadavers: Ex-CIA Agent Says," Lee Speigel Blog on the website www.main.aol.com/2012/07/08/roswell-ufo-cia-agent-chase-brandon_n_1660095.htm Accessed 7/8/2012.

156. Dick Cheney's response on a radio program to being asked a question about ever being briefed on the subject of UFOs (near the end of the TV program), History Channel TV Series, *UFO Files*, "UFOs and the Whitehouse" first aired 7/25/2005.

157. Richard M. Dolan, *UFOs and the National Security State: Chronology of a Cover-Up 1941-1973*, Revised Edition, Charlottesville, VA: Hampton Roads Publishing Co., Inc.: 2002.

158. Timothy Good, *Need To Know: UFOs, The Military and Intelligence*, New York: Pegasus Books LLC, 2007.

159. Leslie Kean, *UFOs: Generals, Pilots, and Government Officials Go on the Record*, New York: Random House, Inc.: 2010.

160. James Oberg, NBC Space Analyst comments about Leslie Kean's *UFOs: Generals, Pilots, and Government Officials Go On The Record*, www.msnbc.msn.com Accessed 10/04/2012.

161. KPLC TV, Report on Strange Lights Seen in the Nighttime Sky, Lake Charles, LA 12/18/2011 www.youtube.com/watch?v=5TT7aRcyJDo. Accessed 09/28/2012.

162. Paradox of Time Travel on Discovery Channel website: www.dsc.discovery.com/tv-shows/curiousity/topics/parallel-universe-theory-quiz.htm#mkcpgn=otbn1 Accessed 9/1/2012.

163. Frank Drake, 1961, developed the Drake equation to estimate the number of intelligent, communicating civilizations in the Milky Way Galaxy, Forerunner project to SETI, from the website www.activemind.com/Mysteries/Topics/seti/drake_equation.html Accessed 10/7/2011.

164. Black Triangle UFO Explanations, on the website www.en.wikipedia.org/w/index.php?title=Black_triangle_(UFO) and www.reinep.files.wordpress.com/2010/05/11/top-secret-black-triangle-sighted-in-almost-all-parts-of-the-world/ Accessed 8/7/2012.

165. Lockheed F-117 Nighthawk (Stealth Fighter) revealed to the world in 1988 and retired in 2008, from the website www.en.wikipedia.org/wiki/Lockheed_F-117_Nighthawk Accessed 8/7/2012.

166. History Channel, "Giant Triangles" episode of *UFO Hunters* that originally aired on 3/18/2009.

167. Trevor J. Constable, *The Cosmic Pulse of Life: The Revolutionary Biological Power Behind UFOs*, London: Merlin Press: 1989.

168. Jacques Vallee, *Passport to Magonia: from folklore to flying saucers*, Washington, D. C.: H. Regnery Co: 1969.

169. B. Maccabee, "Photometric properties of an unidentified bright object seen off the coast of New Zealand," *Applied Optics*, 18, 1979, 2527-2528.

170. Bruce Maccabee, "Challenging the Paradigm," *Journal of Scientific Exploration*, Vol. 19, No. 2, 2005, PP. 185-193.

171. J. Deardorff, B, Haisch, B. Maccabee, and H. E. Puthoff, "Inflation-Theory Implications for Extraterrestrial Visitation," *J. British Interplanetary Society*, Vol. 58, pp. 43-50, 2005.

172. George Dvorsky, "The Fermi Paradox: Back with a vengeance," August 4, 2007 on the website www.sentientdevelopments.com/2007/08/fermi-paradox-back-with-vengeance.html Accessed 12/13/2012.

173. *Wayne's World* Movie (created by Mike Myers), Paramount Pictures, 1992.

174. Dolores Cannon, *The Custodians: "Beyond Abduction,"* Huntsville, Arkansas: Ozark Mountain Publishers: (First printing 1999, Sixth Printing) 2010.

175. Don Donderi, PhD, *UFOs, ETs, and Alien Abductions: A Scientist Looks at the Evidence*, Charlottesville, VA: Hampton Roads: 2013.

176. B. J. Booth, *UFOs Caught on Film: Amazing Evidence of Alien Visitors to Earth*: David & Charles Book, F&W Media International: 2012.

Appendix I

Microsoft Excel Spreadsheet Data

Project Blue Book Correlation Data, 1947-1954

1947-1954 Data Correl	PBB			Percent	Square Miles	Air	Highlands	
State	Unknowns	1950 Pop	Area Sq M	Forest Cover	Forest area	Bases	% Mounta	Sq Miles Mtn
Alaska	3	138000	663267	33	218878.1	10	75	497450
Alabama	2	3060000	52419	71	37217.5	13	25	13105
Arknasas	2	1906000	53178	55	29247.9	10	45	23930
Arizona	12	756000	113998	27	30779.5	20	50	56999
California	32	10586223	163695	40	65478.0	75	60	98217
Colorado	3	1337000	104093	20	20818.6	9	70	72865
Connecticut	0	2007280	5543	55	3048.7	1	40	2217
Washington DC	4	814000	58	50	29.0	4	0	0
Delaware	1	321000	2489	30	746.7	2	0	0
Florida	7	2821000	65755	39	25644.5	40	0	0
Georgia	7	3451000	59425	64	38032.0	16	30	17828
Hawaii	1	491000	10931	40	4372.4	5	50	5466
Idaho	2	592000	83570	35	29249.5	4	70	58499
Illinois	5	8712176	57914	12	6949.7	4	0	0
Indiana	8	3952000	36418	20	7283.6	9	0	0
Iowa	0	2621000	56271	5	2813.6	1	0	0
Kansas	5	1915000	82277	10	8227.7	20	0	0
Kentucky	2	2957000	40409	49	19800.4	3	10	4041
Louisiana	4	2701000	51840	50	25920.0	6	0	0
Massachucetts	11	4690000	10554	53	5593.6	12	20	2111
Maryland	4	2376000	12407	40	4962.8	5	20	2481
Maine	5	911000	35385	85	30077.3	8	50	17693
Michigan	9	6421000	96716	51	49325.2	14	0	0
Minnesota	4	2995000	86939	30	26081.7	7	25	21735
Missouri	7	3946000	69704	31	21608.2	13	25	17426
Mississippi	5	2169000	48430	62	30026.6	6	5	2422
Montana	4	598000	147042	22	32349.2	5	40	58817
North Carolina	3	4060000	53818	60	32290.8	8	40	21527
North Dakota	1	616000	70700	2	1414.0	2	30	21210
Nebraska	1	1324000	77354	3	2320.6	4	0	0
New Hampshire	2	531000	9350	83	7760.5	2	50	4675
New Jersey	7	4860000	8721	42	3662.8	5	0	0
New Mexico	23	687000	121589	21	25533.7	8	70	85112
Nevada	3	162000	110561	15	16584.2	5	70	77393
New York	5	14830192	54,566	55	30011.3	14	30	16370
Ohio	8	7946627	44825	30	13447.5	8	10	4483
Oklahoma	2	2193000	69898	20	13979.6	6	10	6990
Oregon	5	1532000	98380	45	44271.0	15	75	73785
Pennsylvania	11	10498012	46055	57	26251.4	4	30	13817
Rhode Island	0	779000	1545	55	849.8	2	0	0
South carolina	2	2119000	32020	64	20492.8	8	20	6404
South Dakota	3	652000	77116	3	2313.5	5	70	53981
Tennessee	8	3304000	42143	53	22335.8	7	25	10536
Texas	34	7748000	268820	38	102151.6	48	10	26882
Utah	0	696000	84899	33	28016.7	6	60	50939
Virginia	6	3262000	42774	61	26092.1	8	45	19248
Vermont	1	377000	9614	76	7306.6	1	75	7211
Washington	11	2386000	71300	49	34937.0	20	60	42780
Wiconsin	7	3449000	65498	46	30129.1	2	30	19649
West Virginia	1	2006000	24230	77	18657.1	0	45	10904
Wyoming	0	292000	97813	16	15650.1	3	50	48907
sum	293							
	Correl co	0.486	0.654	-0.020	0.700	0.783	0.051	0.376
Average	5.8	3028330.20	62620.98	41	21642.86	10	30.80	21973.02

1947-1954 Data	PBB	Geo Fault		Miles C	percent	percent	1960	1960
State	Unknowns	Activity	Pop Der	O & GL	Rural	Urban	Rural Pop	Urban pop
Alaska	3	7	0.208	6000	62.1	37.9	85698	52302
Alabama	2	3	58.38	30	45.2	54.8	1383120	1676880
Arknasas	2	6	35.84	0	57.2	42.8	1090232	815768
Arizona	12	5	6.632	0	25.5	74.5	192780	563220
California	32	7	64.67	840	13.6	86.4	1439726	9146497
Colorado	3	4	12.84	0	26	74	347620	989380
Connecticut	0	1	362.1	96	21.7	78.3	435580	1571700
Washington DC	4	2	14034	0	0	100	0	814000
Delaware	1	1	129	28	34.4	65.6	110424	210576
Florida	7	1	42.9	1350	26.1	73.9	736281	2084719
Georgia	7	3	58.07	100	44.7	55.3	1542597	1908403
Hawaii	1	7	44.92	750	23.5	76.5	115385	375615
Idaho	2	5	7.084	0	52.5	47.5	310800	281200
Illinois	5	6	150.4	63	19.3	80.7	1681450	7030726
Indiana	8	5	108.5	45	37.6	62.4	1485952	2466048
Iowa	0	1	46.58	0	47	53	1231870	1389130
Kansas	5	4	23.28	0	39	61	746850	1168150
Kentucky	2	6	73.18	0	55.5	44.5	1641135	1315865
Louisiana	4	4	52.1	300	36.7	63.3	991267	1709733
Massachucetts	11	1	444.4	192	16.4	83.6	769160	3920840
Maryland	4	3	191.5	31	27.3	72.7	648648	1727352
Maine	5	1	25.75	228	48.7	51.3	443657	467343
Michigan	9	3	66.39	800	26.6	73.4	1707986	4713014
Minnesota	4	1	34.45	75	37.8	62.2	1132110	1862890
Missouri	7	6	56.61	0	33.4	66.6	1317964	2628036
Mississippi	5	4	44.79	80	62.3	37.7	1351287	817713
Montana	4	4	4.067	0	49.8	50.2	297804	300196
North Carolina	3	3	75.44	301	60.5	39.5	2456300	1603700
North Dakota	1	2	8.713	0	64.8	35.2	399168	216832
Nebraska	1	2	17.12	0	45.7	54.3	605068	718932
New Hampshire	2	1	56.79	18	41.7	58.3	221427	309573
New Jersey	7	2	557.3	130	11.4	88.6	554040	4305960
New Mexico	23	5	5.65	0	34.1	65.9	234267	452733
Nevada	3	5	1.465	0	29.6	70.4	47952	114048
New York	5	2	271.8	127	14.6	85.4	2165208	12664984
Ohio	8	4	177.3	150	26.6	73.4	2113803	5832824
Oklahoma	2	4	31.37	0	37.1	62.9	813603	1379397
Oregon	5	6	15.57	296	37.8	62.2	579096	952904
Pennsylvania	11	2	227.9	10	28.4	71.6	2981435	7516577
Rhode Island	0	1	504.2	40	13.6	86.4	105944	673056
South carolina	2	4	66.18	187	58.8	41.2	1245972	873028
South Dakota	3	3	8.455	0	60.7	39.3	395764	256236
Tennessee	8	6	78.4	0	47.7	52.3	1576008	1727992
Texas	34	4	28.82	367	25	75	1937000	5811000
Utah	0	5	8.198	0	25.1	74.9	174696	521304
Virginia	6	3	76.26	112	44.4	55.6	1448328	1813672
Vermont	1	1	39.21	0	61.5	38.5	231855	145145
Washington	11	6	33.46	157	31.9	68.1	761134	1624866
Wiconsin	7	1	52.66	50	36.2	63.8	1248538	2200462
West Virginia	1	3	82.79	0	61.8	38.2	1239708	766292
Wyoming	0	4	2.985	0	43.2	56.8	126144	165856
sum	293							
	Correl co	0.278	-0.039	0.349	-0.339	0.339	0.30631	0.496933
Average	5.8	3.46	372.1	139.1	37	63	936283	2092047

1947-1954 Dat STATES	PBB Unknowns Actual		Balloon Launch Radial Model Skyhook
Iowa	0		3
Connecticut	0		3
Delaware	1		3
Rhode Island	0		3
North Dakota	1		10
Vermont	1		3
New Hampshire	2		3
West Virginia	1		3
Washington DC	4		3
Nebraska	1		5
Illinois	5		5
South Dakota	3		5
New Jersey	7		3
Hawaii	1		3
Maryland	4		3
Wyoming	0		10
Kentucky	2		5
Wisconsin	7		5
Nevada	3		5
Oklahoma	2		5
Pennsylvania	11		5
Idaho	2		5
Indiana	8		5
Ohio	8		3
Louisiana	4		5
Tennessee	8		5
Montana	4		10
Utah	0		5
Mississippi	5		3
South carolina	2		5
Minnesota	4		10
Massachucetts	11		3
New Mexico	23		10
Colorado	3		10
Virginia	6		3
Maine	5		3
North Carolina	3		3
Arknasas	2		5
Missouri	7		10
New York	5		3
Alabama	2		5
Kansas	5		5
Georgia	7		10
Oregon	5		5
Michigan	9		3
Arizona	12		10
Washington	11		3
Florida	7		5
Alaska	3		3
Texas	34		5
California	32		10
		Correl Co	0.274450982
		Correl Sq	0.075323341

Cross Correlations 1947-1954	Correl. Co. R	
(Excludes Alaska)		
Pop & Geo Fault	0.056	No Correlation
Pop & Mtn Area	-0.103	No Correlation
Pop & L Area	0.186	No Correlation
Pop & Miles of Coast	0.265	
Pop & F Area	0.390	
Pop & A Bases	0.435	
L Area & Miles of Coast	0.178	No Correlation
L Area & Geo Fault	0.377	
L Area & Air Bases	0.606	
L area & Sq Miles Mtn	0.632	
L Area & Forest Area	0.762	
Forest Area & Geo Fault	0.317	
Forest Area & Miles of Coast	0.371	
Forest Area & Mtn Area	0.408	
Forest Area & Air Bases	0.690	
Air Bases & Geo Faults	0.288	
Air Bases & Mtn Area	0.350	
Air Bases & Miles of Coast	0.632	

Appendix II

Microsoft Excel Spreadsheet Data

Project Blue Book Correlation Data - 1955-1969

	Project Blue Book Unexplained Cases 1955 - 1969				Percent	Square Mile	Air	Highlai	
Skyhook	State	Unknown	1960 Pop	Area Sq M	Forest C	Forest area	Bases	% Moi	Sq Miles Mtn
3	Alaska	3	228000	663267	33	218878.1	10	75	497450
5	Alabama	1	3273000	52419	71	37217.5	13	25	13105
5	Arknasas	0	1788000	53178	55	29247.9	10	45	23930
10	Arizona	1	1318000	113998	27	30779.5	20	50	56999
10	California	7	15850000	163695	40	65478.0	75	60	98217
10	Colorado	5	1758000	104093	20	20818.6	9	70	72865
3	Connecticut	0	2548000	5543	55	3048.7	1	40	2217
3	Washington DC	2	762000	58	50	29.0	4	0	0
3	Delaware	1	449000	2489	30	746.7	2	0	0
5	Florida	7	4951000	65755	39	25644.5	40	0	0
10	Georgia	2	3949000	59425	64	38032.0	16	30	17828
3	Hawaii	0	642000	10931	40	4372.4	5	50	5466
5	Idaho	1	671000	83570	35	29249.5	4	70	58499
5	Illinois	9	10113000	57914	12	6949.7	4	0	0
3	Indiana	4	4677000	36418	20	7283.6	9	0	0
5	Iowa	5	2761000	56271	5	2813.6	1	0	0
5	Kansas	3	2178000	82277	10	8227.7	20	0	0
5	Kentucky	1	3047000	40409	49	19800.4	3	10	4041
5	Louisiana	3	3270000	51840	50	25920.0	6	0	0
3	Massachucetts	2	5167000	10554	53	5593.6	12	20	2111
3	Maryland	5	3116000	12407	40	4962.8	5	20	2481
3	Maine	1	974000	35385	85	30077.3	8	50	17693
3	Michigan	8	7848000	96716	51	49325.2	14	0	0
10	Minnesota	2	3426000	86939	30	26081.7	7	25	21735
10	Missouri	4	4331000	69704	31	21608.2	13	25	17426
3	Mississippi	5	2180000	48430	62	30026.6	6	5	2422
10	Montana	2	678000	147042	22	32349.2	5	40	58817
3	North Carolina	7	4563000	53818	60	32290.8	8	40	21527
10	North Dakota	3	634000	70700	2	1414.0	2	30	21210
5	Nebraska	2	1414000	77354	3	2320.6	4	0	0
3	New Hampshire	1	609000	9350	83	7760.5	2	50	4675
3	New Jersey	4	6099000	8721	42	3662.8	5	0	0
10	New Mexico	5	958000	121589	21	25533.7	8	70	85112
5	Nevada	0	288000	110561	15	16584.2	5	70	77393
3	New York	14	16827000	54,566	55	30011.3	14	30	16370
5	Ohio	13	9739000	44825	30	13447.5	8	10	4483
5	Oklahoma	2	2333000	69898	20	13979.6	6	10	6990
5	Oregon	4	1773000	98380	45	44271.0	15	75	73785
3	Pennsylvania	6	11343000	46055	57	26251.4	4	30	13817
3	Rhode Island	0	857000	1545	55	849.8	2	0	0
5	South carolina	1	2392000	32020	64	20492.8	8	20	6404
5	South Dakota	0	682000	77116	3	2313.5	5	70	53981
5	Tennessee	3	3573000	42143	53	22335.8	7	25	10536
10	Texas	13	9617000	268820	38	102151.6	48	10	26882
5	Utah	1	896000	84899	33	28016.7	6	60	50939
3	Virginia	7	3978000	42774	61	26092.1	8	45	19248
3	Vermont	0	391000	9614	76	7306.6	1	75	7211
3	Washington	2	2860000	71300	49	34937.0	20	60	42780
5	Wiconsin	8	3964000	65498	46	30129.1	2	30	19649
3	West Virginia	0	1857000	24230	77	18657.1	0	45	10904
5	Wyoming	1	332000	97813	16	15650.1	3	50	48907
0.11	sum	181							
0.01		Correl co	0.786	0.358	-0.082	0.438	0.391	-0.281	-0.049
		Average	3594080.00	62620.98	41	21642.86	10	30.80	21973.02

Project Blue Book Unexplained Cases 1955 - 1969		Geo Fault		Miles Coast	percent	percent	1960	1960
State	Unknowns	Activity	Pop Density	O & GL	Rural	Urban	Rural Pop	Urban pop
Alaska	3	7	0.3	6000	62.1	37.9	141588	86412
Alabama	1	3	62.4	30	45.2	54.8	1479396	1793604
Arknasas	0	6	33.6	0	57.2	42.8	1022736	765264
Arizona	1	5	11.6	0	25.5	74.5	336090	981910
California	7	7	96.8	840	13.6	86.4	2155600	13694400
Colorado	5	4	16.9	0	26	74	457080	1300920
Connecticut	0	1	459.7	96	21.7	78.3	552916	1995084
Washington DC	2	2	13137.9	0	0	100	0	762000
Delaware	1	1	180.4	28	34.4	65.6	154456	294544
Florida	7	1	75.3	1350	26.1	73.9	1292211	3658789
Georgia	2	3	66.5	100	44.7	55.3	1765203	2183797
Hawaii	0	7	58.7	750	23.5	76.5	150870	491130
Idaho	1	5	8.0	0	52.5	47.5	352275	318725
Illinois	9	6	174.6	63	19.3	80.7	1951809	8161191
Indiana	4	5	128.4	45	37.6	62.4	1758552	2918448
Iowa	5	1	49.1	0	47	53	1297670	1463330
Kansas	3	4	26.5	0	39	61	849420	1328580
Kentucky	1	6	75.4	0	55.5	44.5	1691085	1355915
Louisiana	3	4	63.1	300	36.7	63.3	1200090	2069910
Massachucetts	2	3	489.6	192	16.4	83.6	847388	4319612
Maryland	5	3	251.1	31	27.3	72.7	850668	2265332
Maine	1	1	27.5	228	48.7	51.3	474338	499662
Michigan	8	3	81.1	800	26.6	73.4	2087568	5760432
Minnesota	2	1	39.4	75	37.8	62.2	1295028	2130972
Missouri	4	6	62.1	0	33.4	66.6	1446554	2884446
Mississippi	5	4	45.0	80	62.3	37.7	1358140	821860
Montana	2	4	4.6	0	49.8	50.2	337644	340356
North Carolina	7	3	84.8	301	60.5	39.5	2760615	1802385
North Dakota	3	2	9.0	0	64.8	35.2	410832	223168
Nebraska	2	2	18.3	0	45.7	54.3	646198	767802
New Hampshire	1	2	65.1	18	41.7	58.3	253953	355047
New Jersey	4	2	699.3	130	11.4	88.6	695286	5403714
New Mexico	5	5	7.9	0	34.1	65.9	326678	631322
Nevada	0	5	2.6	0	29.6	70.4	85248	202752
New York	14	2	308.4	127	14.6	85.4	2456742	14370258
Ohio	13	4	217.3	150	26.6	73.4	2590574	7148426
Oklahoma	2	4	33.4	0	37.1	62.9	865543	1467457
Oregon	4	6	18.0	296	37.8	62.2	670194	1102806
Pennsylvania	6	2	246.3	10	28.4	71.6	3221412	8121588
Rhode Island	0	1	554.7	40	13.6	86.4	116552	740448
South carolina	1	4	74.7	187	58.8	41.2	1406496	985504
South Dakota	0	3	8.8	0	60.7	39.3	413974	268026
Tennessee	3	6	84.8	0	47.7	52.3	1704321	1868679
Texas	13	4	35.8	367	25	75	2404250	7212750
Utah	1	5	10.6	0	25.1	74.9	224896	671104
Virginia	7	3	93.0	112	44.4	55.6	1766232	2211768
Vermont	0	1	40.7	0	61.5	38.5	240465	150535
Washington	2	6	40.1	157	31.9	68.1	912340	1947660
Wiconsin	8	1	60.5	50	36.2	63.8	1434968	2529032
West Virginia	0	3	76.6	0	61.8	38.2	1147626	709374
Wyoming	1	4	3.4	0	43.2	56.8	143424	188576
sum	181							
	Correl co	-0.011	-0.055	0.313	-0.321	0.321	0.717	0.754
	Average	3.52	370.39	139.06	37	63.00	1081272	2512807.88

Cross Correlations	1955-1969	
(Excludes Alaska)	Correl Co. r	
Pop & Mountains	-0.038	No Correlation
Pop & Skyhook Ballns	0.044	No Correlation
Pop & Geo Faults	0.086	No Correlation
Pop & Land Area	0.240	Weak Correlation
Pop & Miles Coast	0.360	
Pop & Forest Area	0.430	
Pop& Air Bases	0.542	
Land Area & Forest Area	0.762	
Air Bases & Forest Area	0.690	

Index

CPSIA information can be obtained at www.ICGtesting.com
Printed in the USA
LVOW11s1826210814

400288LV00009B/397/P